Methods in Molecular Biology

Series Editor
John M. Walker
School of Life and Medical Sciences
University of Hertfordshire
Hatfield, Hertfordshire, UK

For further volumes:
http://www.springer.com/series/7651

For over 35 years, biological scientists have come to rely on the research protocols and methodologies in the critically acclaimed *Methods in Molecular Biology* series. The series was the first to introduce the step-by-step protocols approach that has become the standard in all biomedical protocol publishing. Each protocol is provided in readily-reproducible step-by-step fashion, opening with an introductory overview, a list of the materials and reagents needed to complete the experiment, and followed by a detailed procedure that is supported with a helpful notes section offering tips and tricks of the trade as well as troubleshooting advice. These hallmark features were introduced by series editor Dr. John Walker and constitute the key ingredient in each and every volume of the *Methods in Molecular Biology* series. Tested and trusted, comprehensive and reliable, all protocols from the series are indexed in PubMed.

Protein Evolution

Methods and Protocols

Edited by

Shahid M. Khan

Molecular Biology Consortium, Lawrence Berkeley National Laboratory, Berkeley, CA, USA

Florencio Pazos

Computational Systems Biology Group, National Centre for Biotechnology (CNB-CSIC), Madrid, Spain

Humana Press

Editors
Shahid M. Khan
Molecular Biology Consortium
Lawrence Berkeley National Laboratory
Berkeley, CA, USA

Florencio Pazos
Computational Systems Biology Group
National Centre for Biotechnology (CNB-CSIC)
Madrid, Spain

ISSN 1064-3745 ISSN 1940-6029 (electronic)
Methods in Molecular Biology
ISBN 978-1-0716-4827-8 ISBN 978-1-0716-4828-5 (eBook)
https://doi.org/10.1007/978-1-0716-4828-5

This Humana imprint is published by the registered company Springer Science+Business Media, LLC, part of Springer Nature.
The registered company address is: 1 New York Plaza, New York, NY 10004, U.S.A.

Preface

Proteins are involved in all cellular processes, responsible for most functions encoded in the genome. The computational study of proteins has a storied history. It is 60 years since Margaret Dayhoff published her 1965 atlas of the 65 known protein sequences. The Protein Data Bank (PDB) of 3D structures was subsequently established in 1971. Today, protein sequence databases such as UniProt have over 3 million unique entries, while the PDB has over 100,000 entries.

Proteins are the engines that drive cellular metabolism in living organisms. The 20 amino acid building blocks that form these polypeptides allow for a combinatorial explosion of possible architectures, for both single chains and their assemblies. The protein databases have been leveraged over the years to address two fundamental issues based on the realization that proteins can be categorized into a finite number of families with distinct 3D-folds. First, can the physicochemical basis for protein conformational landscapes be understood in sufficient detail to design novel proteins? Second, what can be learned from natural evolution on how environmental factors drive the protein family diversity? The knowledge obtained has enormous consequences for human health and the ability to harness energy in all its forms for the benefit of our planet.

The past decade has seen impressive advances in in silico methodologies for the study and prediction of different functional and structural aspects of proteins, but much remains to be done. The chapters in this volume on computational "protein evolution" proliferate the recent strategies used in protein science by the broader computational biology community to advance knowledge in new areas. They are organized into four groups: (1) the analysis of 1D protein sequences for coevolutionary fold prediction and protein-nucleic acid interactions, the discovery of distant homologs and functional motifs, and the characterization of fold energetics, disordered proteins and mutagenesis-driven modulation of conformational landscapes; (2) the analysis of 3D protein domains for their structural or dynamic networks, the design of novel enzymes and fold-switching mechanisms; (3) the prediction of 3D protein-protein interactions, inhibitor docking, and the role of water in the architectural dynamics of membrane proteins; (4) hybrid methods that mine both 1D and 3D genomic and protein databases for gene editing and alternative splicing.

This volume could not have been assembled without the efforts of all authors who wrote the chapters. We greatly appreciate their contribution. We hope this book will be a valuable guide for research in protein bioinformatics, as well as a general resource for the large, interdisciplinary community interested in this area.

Berkeley, CA, USA — *Shahid M. Khan*
Madrid, Spain — *Florencio Pazos*

Contents

Preface .... *v*
Contributors .... *ix*

PART I 1D SEQUENCE ANALYSIS: FOLD PREDICTION AND DYNAMICS

1 Distant Homology BLASTP Searches 3
Donard S. Dwyer

2 Predicting Protein Folding Dynamics Using Sequence Information 17
Ezequiel A. Galpern, Federico Caamaño, and Diego U. Ferreiro

3 Predictive Model for Single-Site Mutations That Change Intrinsically Disordered Protein Binding Energetics 33
Soumyanetra Chandra and Raghavan Varadarajan

4 Detection of Specificity-Determining Positions in Protein Families 49
Borja Pitarch, David Juan, and Florencio Pazos

5 Co-Var, a Measure to Determine Intra-protein, Inter-protein, Protein–DNA, or Protein–RNA Coevolution 65
Ishita Mukherjee and Saikat Chakrabarti

6 `adabmDCA 2.0`—A Flexible but Easy-to-Use Package for Direct Coupling Analysis 83
Lorenzo Rosset, Roberto Netti, Anna Paola Muntoni, Martin Weigt, and Francesco Zamponi

7 Unlocking Protein Evolution Insights: Efficient and Interpretable Mutational Effect Predictions with GEMME 105
A. Carbone, E. Laine, and G. Lombardi

PART II 3D STRUCTURE ANALYSIS: DESIGN AND DYNAMICS

8 Using Total Network Dissimilarity to Analyze Protein Structures Across Homologs 125
Vasam Manjveekar Prabantu, Vasundhara Gadiyaram, and Saraswathi Vishveshwara

9 Allosteric Network Analysis Toolkit for Single-Domain Phosphoproteins 139
Maham Hamid, Safee Ullah Chaudhary, Alessandro Pandini, and Shahid M. Khan

10 Computational Identification of Potential Shape-Shifting Proteins from Structures 157
Francisco M. Pérez-Canales, Enrique Alanís-Domínguez, and Ana M. Rojas

11 Transition State-Based Computational Enzyme Design 165
Thomas Gaillard and Thomas Simonson

Part III 3D Structure Analysis: Molecular Recognition

12 Quantification of Conformational Changes of Kinase Regulators, Kinases, and Regulator–Kinase Complexes Using the TSR Algorithm 189
Tarikul I. Milon, Poorya Khajouie, Feng Chen, Tyler A. Borel, Kathleen D. Knierim, Vijay Raghavan, and Wu Xu

13 Identifying Functional Water Encapsulation Sites Within the Transmembrane Domain of Sensor Kinases 213
Rahmi Yusuf, Robert J. Lawrence, Fadhael Alrahman Hasan, Peltine A. Ndifon, Joy Jun Yan 雋恩 Yau 邱, Monabel May Birao, Maetrese Beley, George Baker-Valla, and Roger R. Draheim

14 Protein–Protein Interaction Site Prediction via EquiPPIS and Its Application in Studying Viral Replication 225
Rahmatullah Roche, Sumit Tarafder, Parkesh Suseendran, Xiaofeng Wang, and Debswapna Bhattacharya

15 Modeling Protein–Protein Complexes by Combining pyDock and AlphaFold 237
Luis Ángel Rodríguez-Lumbreras, Víctor Monteagudo, and Juan Fernández-Recio

16 Binding Affinity Prediction of Membrane Protein–Protein Complexes Using MPA-Pred 267
Fathima Ridha and M. Michael Gromiha

Part IV Integrated 1D-Sequence/3D-Structure Pipelines

17 Computational Methods to Engineer Cas Proteins for Efficient Genome Editing 279
Muhammad Qaiser Fatmi, Aneeqa Nadeem, Mehraj Abbasov, and Muhammad Sajjad

18 Exploring the Functional Consequences of Protein-Coding Alternative Splice Variants at a Genome Scale: A Guide to Utilizing EXPANSION 301
Chakit Arora, Natalia De Oliveira Rosa, Marin Matic, Mariastella Cascone, Pasquale Miglionico, and Francesco Raimondi

Index *315*

Contributors

MEHRAJ ABBASOV • *Genetic Resources Institute of the Ministry of Science and Education, Baku, Azerbaijan*

ENRIQUE ALANÍS-DOMÍNGUEZ • *CABD-CSIC, Computational Biology and Bioinformatics Group, Seville, Spain*

CHAKIT ARORA • *Laboratorio di Biologia Bio@SNS, Scuola Normale Superiore, Pisa, Italy*

GEORGE BAKER-VALLA • *School of Medicine, Pharmacy and Biomedical Science, University of Portsmouth, Portsmouth, UK*

MAETRESE BELEY • *School of Medicine, Pharmacy and Biomedical Science, University of Portsmouth, Portsmouth, UK*

DEBSWAPNA BHATTACHARYA • *Department of Computer Science, Virginia Tech, Blacksburg, VA, USA*

MONABEL MAY BIRAO • *School of Medicine, Pharmacy and Biomedical Science, University of Portsmouth, Portsmouth, UK*

TYLER A. BOREL • *Department of Chemistry, University of Louisiana at Lafayette, Lafayette, LA, USA*

FEDERICO CAAMAÑO • *Laboratorio de Fisiología de Proteínas, Facultad de Ciencias Exactas y Naturales, Universidad de Buenos Aires and Consejo Nacional de Investigaciones Científicas y Técnicas, Instituto de Química Biológica de la Facultad de Ciencias Exactas y Naturales (IQUIBICEN-CONICET), Buenos Aires, Argentina*

A. CARBONE • *Sorbonne Université, CNRS, IBPS, Computational, Quantitative and Synthetic Biology Department - UMR 7238, Paris, France; Institut Universitaire de France, Paris, France*

MARIASTELLA CASCONE • *Laboratorio di Biologia Bio@SNS, Scuola Normale Superiore, Pisa, Italy*

SAIKAT CHAKRABARTI • *Structural Biology and Bioinformatics Division, Council for Scientific and Industrial Research (CSIR)–Indian Institute of Chemical Biology (IICB), Kolkata, West Bengal, India*

SOUMYANETRA CHANDRA • *Molecular Biophysics Unit, Indian Institute of Science, Bangalore, Karnataka, India*

SAFEE ULLAH CHAUDHARY • *Biomedical Informatics and Engineering Research Laboratory (BIRL), Department of Life Sciences, Syed Babar Ali School of Science and Engineering, Lahore University of Management Sciences (LUMS), Lahore, Pakistan*

FENG CHEN • *High Performance Computing, Frey Computing Services Center, Louisiana State University, Baton Rouge, LA, USA*

NATALIA DE OLIVEIRA ROSA • *Laboratorio di Biologia Bio@SNS, Scuola Normale Superiore, Pisa, Italy*

ROGER R. DRAHEIM • *School of Medicine, Pharmacy and Biomedical Science, University of Portsmouth, Portsmouth, UK*

DONARD S. DWYER • *Departments of Psychiatry and Behavioral Medicine and Pharmacology, Toxicology and Neuroscience, LSU Health Shreveport, Shreveport, LA, USA*

MUHAMMAD QAISER FATMI • *Department of Biosciences, COMSATS University Islamabad, Islamabad, Pakistan*

Juan Fernández-Recio • *Instituto de Ciencias de la Vid y del Vino (ICVV), Consejo Superior de Investigaciones Científicas (CSIC)—Universidad de La Rioja—Gobierno de La Rioja, Logroño, Spain*

Diego U. Ferreiro • *Laboratorio de Fisiología de Proteínas, Facultad de Ciencias Exactas y Naturales, Universidad de Buenos Aires and Consejo Nacional de Investigaciones Científicas y Técnicas, Instituto de Química Biológica de la Facultad de Ciencias Exactas y Naturales (IQUIBICEN-CONICET), Buenos Aires, Argentina*

Vasundhara Gadiyaram • *Molecular Biophysics Unit, Indian Institute of Science, Bangalore, Karnataka, India*

Thomas Gaillard • *Laboratoire de Biologie Structurale de la Cellule (CNRS UMR7654), Department of Biology, Ecole Polytechnique, Palaiseau, France*

Ezequiel A. Galpern • *Laboratorio de Fisiología de Proteínas, Facultad de Ciencias Exactas y Naturales, Universidad de Buenos Aires and Consejo Nacional de Investigaciones Científicas y Técnicas, Instituto de Química Biológica de la Facultad de Ciencias Exactas y Naturales (IQUIBICEN-CONICET), Buenos Aires, Argentina*

M. Michael Gromiha • *Department of Biotechnology, Bhupat and Jyoti Mehta School of Biosciences, Indian Institute of Technology Madras, Chennai, India*

Maham Hamid • *Biomedical Informatics and Engineering Research Laboratory (BIRL), Department of Life Sciences, Syed Babar Ali School of Science and Engineering, Lahore University of Management Sciences (LUMS), Lahore, Pakistan*

Fadhael Alrahman Hasan • *Al Ahqaqee Medical Center, Kuwait City, Kuwait*

David Juan • *Functional and Comparative Multiomics Group, National Centre for Biotechnology (CNB-CSIC), Madrid, Spain*

Poorya Khajouie • *Department of Chemistry, University of Louisiana at Lafayette, Lafayette, LA, USA; The Center for Advanced Computer Studies, University of Louisiana at Lafayette, Lafayette, LA, USA*

Shahid M. Khan • *Syed Babar Ali School of Science & Engineering, Lahore University of Management Sciences (LUMS), Lahore, Pakistan; Molecular Biology Consortium, Lawrence Berkeley National Laboratory, Berkeley, CA, USA*

Kathleen D. Knierim • *Department of Chemistry, University of Louisiana at Lafayette, Lafayette, LA, USA*

E. Laine • *Sorbonne Université, CNRS, IBPS, Computational, Quantitative and Synthetic Biology Department - UMR 7238, Paris, France; Institut Universitaire de France, Paris, France*

Robert J. Lawrence • *School of Medicine, Pharmacy and Biomedical Science, University of Portsmouth, Portsmouth, UK*

G. Lombardi • *Sorbonne Université, CNRS, IBPS, Computational, Quantitative and Synthetic Biology Department - UMR 7238, Paris, France*

Marin Matic • *Laboratorio di Biologia Bio@SNS, Scuola Normale Superiore, Pisa, Italy*

Pasquale Miglionico • *Laboratorio di Biologia Bio@SNS, Scuola Normale Superiore, Pisa, Italy*

Tarikul I. Milon • *Department of Chemistry, University of Louisiana at Lafayette, Lafayette, LA, USA*

Víctor Monteagudo • *Instituto de Ciencias de la Vid y del Vino (ICVV), Consejo Superior de Investigaciones Científicas (CSIC)—Universidad de La Rioja—Gobierno de La Rioja, Logroño, Spain*

ISHITA MUKHERJEE • *Structural Biology and Bioinformatics Division, Council for Scientific and Industrial Research (CSIR)–Indian Institute of Chemical Biology (IICB), Kolkata, West Bengal, India; Institute for Systems Biology, Seattle, WA, USA*
ANNA PAOLA MUNTONI • *DISAT, Politecnico di Torino, Torino, Italy*
ANEEQA NADEEM • *Department of Biosciences, COMSATS University Islamabad, Islamabad, Pakistan*
PELTINE A. NDIFON • *School of Applied Science, University of Huddersfield, Huddersfield, UK*
ROBERTO NETTI • *Department of Computational, Quantitative and Synthetic Biology, Sorbonne Université, CNRS, Paris, France*
ALESSANDRO PANDINI • *Department of Computer Science, Brunel University of London, Uxbridge, UK*
FLORENCIO PAZOS • *Computational Systems Biology Group, National Centre for Biotechnology (CNB- CSIC), Madrid, Spain*
FRANCISCO M. PÉREZ-CANALES • *CABD-CSIC, Computational Biology and Bioinformatics Group, Seville, Spain*
BORJA PITARCH • *Computational Systems Biology Group, National Centre for Biotechnology (CNB- CSIC), Madrid, Spain*
VASAM MANJVEEKAR PRABANTU • *Molecular Biophysics Unit, Indian Institute of Science, Bangalore, Karnataka, India*
VIJAY RAGHAVAN • *The Center for Advanced Computer Studies, University of Louisiana at Lafayette, Lafayette, LA, USA*
FRANCESCO RAIMONDI • *Laboratorio di Biologia Bio@SNS, Scuola Normale Superiore, Pisa, Italy*
FATHIMA RIDHA • *Department of Biotechnology, Bhupat and Jyoti Mehta School of Biosciences, Indian Institute of Technology Madras, Chennai, India*
RAHMATULLAH ROCHE • *TSYS School of Computer Science, Columbus State University, Columbus, GA, USA*
LUIS ÁNGEL RODRÍGUEZ-LUMBRERAS • *Instituto de Ciencias de la Vid y del Vino (ICVV), Consejo Superior de Investigaciones Científicas (CSIC)—Universidad de La Rioja—Gobierno de La Rioja, Logroño, Spain*
ANA M. ROJAS • *CABD-CSIC, Computational Biology and Bioinformatics Group, Seville, Spain*
LORENZO ROSSET • *Department of Computational, Quantitative and Synthetic Biology, Sorbonne Université, CNRS, Paris, France; Laboratoire de Physique Théorique, École Normale Supérieure, Paris, France*
MUHAMMAD SAJJAD • *Department of Biosciences, COMSATS University Islamabad, Islamabad, Pakistan*
THOMAS SIMONSON • *Laboratoire de Biologie Structurale de la Cellule (CNRS UMR7654), Department of Biology, Ecole Polytechnique, Palaiseau, France*
PARKESH SUSEENDRAN • *School of Plant and Environmental Sciences, Virginia Tech, Blacksburg, VA, USA*
SUMIT TARAFDER • *Department of Computer Science, Virginia Tech, Blacksburg, VA, USA*
RAGHAVAN VARADARAJAN • *Molecular Biophysics Unit, Indian Institute of Science, Bangalore, Karnataka, India*
SARASWATHI VISHVESHWARA • *Molecular Biophysics Unit, Indian Institute of Science, Bangalore, Karnataka, India*

XIAOFENG WANG • *School of Plant and Environmental Sciences, Virginia Tech, Blacksburg, VA, USA*
MARTIN WEIGT • *Department of Computational, Quantitative and Synthetic Biology, Sorbonne Université, CNRS, Paris, France*
WU XU • *Department of Chemistry, University of Louisiana at Lafayette, Lafayette, LA, USA*
JOY JUN YAN 雋恩 YAU 邱 • *School of Medicine, Pharmacy and Biomedical Science, University of Portsmouth, Portsmouth, UK*
RAHMI YUSUF • *Leicester School of Pharmacy, De Montfort University, Leicester, UK*
FRANCESCO ZAMPONI • *Dipartimento di Fisica, Sapienza Università di Roma, Rome, Italy*

Part I

1D Sequence Analysis: Fold Prediction and Dynamics

Chapter 1

Distant Homology BLASTP Searches

Donard S. Dwyer

Abstract

Many neuropeptides have been widely conserved across species during evolution. Valuable insights may be gained by studying their biological activity in different species and identifying common functions. Standard basic local alignment search tool for proteins (BLASTP) searches often fail to detect homologous counterparts in distantly related species such as *Drosophila* and *Caenorhabditis elegans*, despite evidence for functional equivalents. This can be remedied by employing distant homology BLASTP with less stringent search parameters. Here, I describe the use of BLASTP searches with relaxed settings to identify the functional counterparts of kisspeptin and amylin or islet amyloid polypeptide (IAPP) in *C. elegans*. By increasing the E-value and relaxing other parameters, the list of potential homologs was expanded. Follow-up analysis to validate attractive hits included searches with sequences of neuropeptides from other species and convergence of the search results on recurring hits. Strong candidates for functional counterparts of kisspeptin and IAPP were found with this approach. Moreover, the hybrid structure (part RFamide and part calcitonin-like hormone) of one hit—FLP-12—suggested that a shared distant ancestor may have been duplicated and the two copies then diverged into separate RFamide and peptide hormone families. Low homology BLASTP searching is a powerful method for identifying distant relatives of neuropeptides and for characterizing their evolutionary relationship to ancient common ancestors.

Key words Basic local alignment search tool (BLAST), Evolutionary conservation, Neuropeptides, RFamide, Sequence homology

1 Introduction

Peptide hormones/neurotransmitters and their protein receptors form an ancient and essential signaling system involved in growth, development, and physiologic responses and brain function [1]. Conservation of protein sequence and functional activity from *Caenorhabditis elegans* to man has been well-documented for insulin, neuropeptide Y (NPY), transforming growth factor (TGF) superfamily members, and others [1–3]. New insight can be obtained by studying neuropeptide responses in model organisms with less complex nervous systems and physiology providing that the organism possesses functionally equivalent molecules. Although progress has been made in identifying orthologs of

Shahid M. Khan and Florencio Pazos (eds.), *Protein Evolution: Methods and Protocols*, Methods in Molecular Biology, vol. 2979, https://doi.org/10.1007/978-1-0716-4828-5_1,

neuropeptides in model organisms, functional counterparts still remain to be established for many peptides and the evolutionary history is often lacking [4]. Identification of homologous counterparts in evolutionarily distant species requires adjustment of automated search programs such as basic local alignment search tool for proteins (BLASTP) [5].

BLASTP searches are most effective when there is substantial homology (>40%) between the query and hit proteins. However, it can also be used to identify more distant relationships between proteins when search parameters are relaxed and convergent confirmation is applied [6, 7]. Shorter protein sequences, such as those of neuropeptides, pose a challenge to homology search efforts owing to the fact that matches over limited segments are underweighted in the absence of an extended alignment context [5, 8]. The limitations of searching with shorter sequences (~ 100 amino acids residues) can be mitigated, in part, by increasing the E-value threshold for alignments from 0.1–1 (typical default settings) to 100 or 1000. However, this modification will lead to an increase in false or questionable hits. Consequently, rigorous follow-up analysis is then needed to confirm hits with additional query sequences derived from different species [6]. Here, I will discuss the use of BLASTP for detecting distant homologies among neuropeptides and illustrate its utility with the identification of homologous candidates for kisspeptin (KISS) and islet amyloid polypeptide (IAPP)/ amylin in *C. elegans.*

NPY is one of the best characterized neuropeptides in various model systems [9–12]. Searches based on amino acid sequence homology fail to establish a bona fide ortholog of NPY in *C. elegans*, despite the fact that homologous nematode receptors can be activated by human NPY [12]. Furthermore, human NPY receptors restore functional activity in mutants with loss-of-function mutations in the endogenous NPY-like receptor (NPR-1) of *C. elegans* [12]. NPY belongs to a subset of neuropeptides referred to as RFamides based on the sequence RFGKR that is cleaved and posttranslationally modified to the characteristic arginine-phenylalanine amide chemical structure, which constitutes the active form of the neuropeptide [2]. NPY actually includes the conservative substitution of tyrosine for phenylalanine and is thus an RYamide.

Several neuropeptides from *C. elegans* serve as biological counterparts of NPY in functional assays. They belong to the FMRFamide-like peptide or *flp* gene family [11, 12]. Whereas mature NPY has the C-terminal sequence RQRY, the functional counterparts in *C. elegans* variably display RLRY (FLP-34.1), VLRF (FLP-18), RPLRF (FLP-21), and PLRF (FLP-15.2) [11, 12]. It is unlikely that any of these neuropeptides represent true orthologs of NPY in the strictest sense [12]. This example illustrates the challenges of identifying distantly related neuropeptides solely on the basis of their amino acid sequences. Moreover, it demonstrates that more than a single peptide can act at a given neuropeptide receptor [2, 12, 13], which suggests deeper evolutionary connections among neuropeptides. Sequence homology analysis can detect vestiges of ancient common precursors of contemporary neuropeptides. Toward this end, I employed BLASTP to identify *C. elegans* counterparts of KISS and IAPP. The *C. elegans* genome contains two G protein-coupled receptor (GPCR) genes that are strong homologs of the mammalian kisspeptin and calcitonin/amylin receptors [14, D.D., unpublished observation]. Therefore, this species should also express neuropeptide ligands to act via these receptors, even if they are not true orthologs, similar to the situation with NPY counterparts. Distant homology BLASTP searches provided attractive candidates for these neuropeptides in *C. elegans* along with hybrid proteins that revealed evolutionary connections between the RFamide family and peptide hormones such as IAPP and calcitonin.

2 Methods

2.1 Distant Homology BLASTP Search of WormBase

1. Retrieve query protein sequences of interest, such as KISS and IAPP, from the European Molecular Biology Laboratory and European Bioinformatics Institute (EMBL-EBI) database, Ensembl (www.ensembl.org) [15]. For this work, the human gene was entered first and < Orthologues> was opened to obtain sequences from other species with established counterparts. Amino acid sequences can be obtained through links in the transcript table. When multiple transcripts were available, the longest one was selected for BLASTP analysis.
2. Copy amino acid sequence (single-letter code) for inputting into the WormBase search.
3. Access WormBase (www.wormbase.org) [16] and find BLAST/BLAST-like alignment tool (BLAT) under the Tools pull-down menu.
4. Paste the sequence of interest into the Query Sequence box and make sure that BLASTP is selected.
5. Set the E-value threshold in the pull-down menu. The default setting is 1E+0; however, a threshold of 1E+4 is recommended for distant homology searches (*see* **Note 1**).
6. The default database (e.g., WS288), which is the most recent version, is the top choice for use.
7. Submit the search and inspect the results (*see* **Note 2**). Figures 1 and 2 show the results of BLASTP searches (at the top of each figure) using kisspeptin neuropeptides from zebrafish and humans.

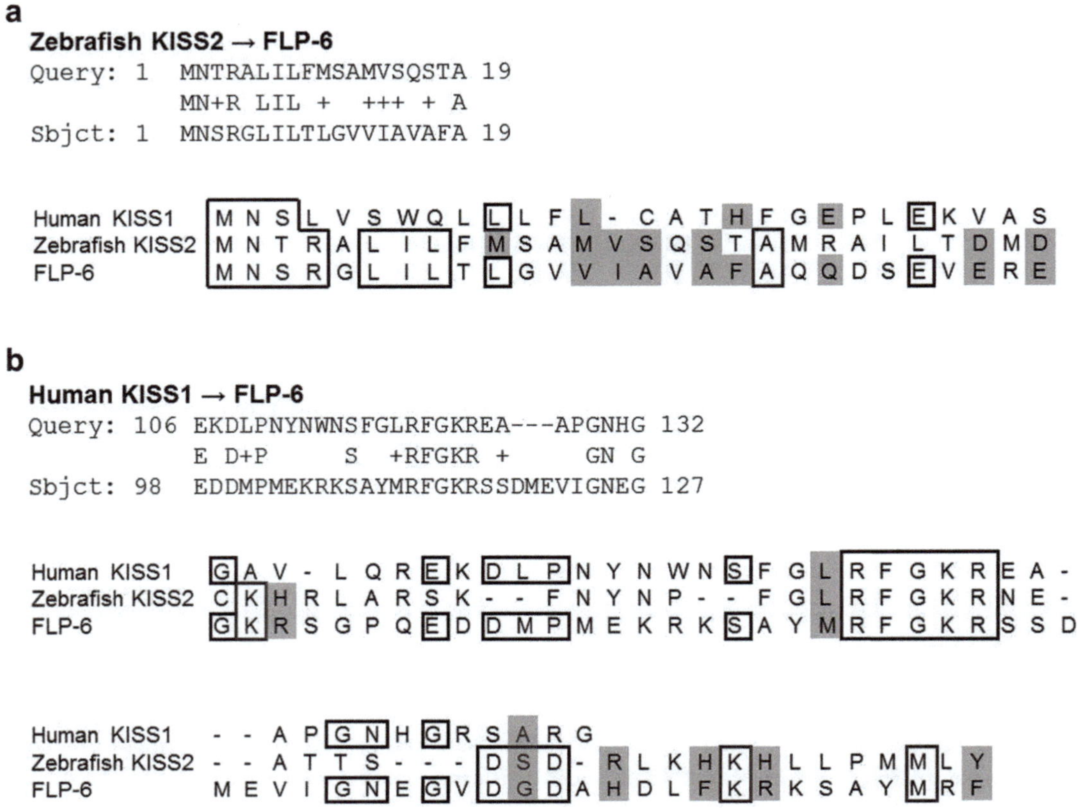

Fig. 1 Results of BLASTP search of WormBase. (**a**) Example search results show that zebrafish kisspeptin-2 detects a region of homology with FLP-6 at the N-terminus of both peptides. In the bottom panel, additional homologous segments were discovered through visual inspection of aligned sequences. Conservative changes are indicated with + signs. (**b**) Search with human KISS1 reveals a homologous segment toward the C-terminus of the peptides that includes the highly conserved RFamide motif. Manual extension of the sequences in both directions yields additional matches. Identical amino acids are outlined or boxed, whereas conserved substitutions are indicated by gray shading

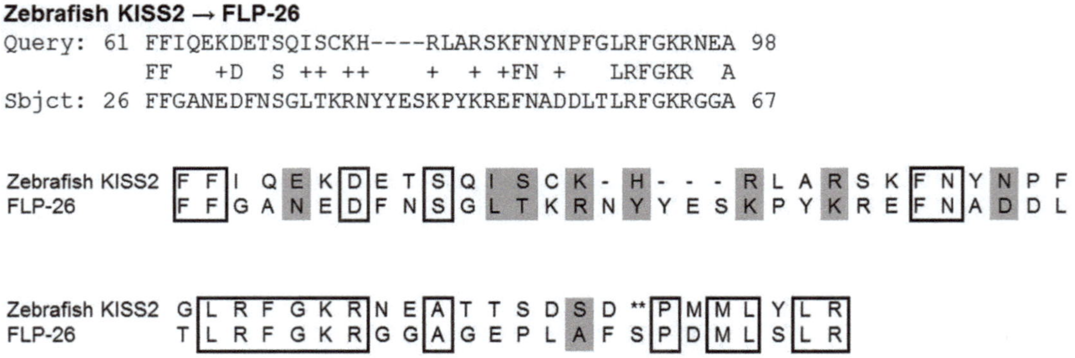

Fig. 2 BLASTP search with zebrafish kisspeptin-2 identifies an extended region of sequence similarity with *C. elegans* FLP-26. Additional significant matches are found beyond the limit of the search results. (** indicates a gap of eight amino acids including a two-residue repeat—KHKH)

Hedgehog IAPP → FLP-12

```
Query: 35 CNTATCATQRL 45
          C  ATCATQ++
Sbjct: 12 CLIATCATQKV 22
```

Hedgehog IAPP → FLP-3

```
Query: 63  TDVGSNTYGKRSAA-----EIVDKEPQNYLPL 89
           T +G+  +GKRSA          + P+N  P
Sbjct: 75  TPLGTMRFGKRSAEPFGTMRFGKRNPENDTPF 106
```

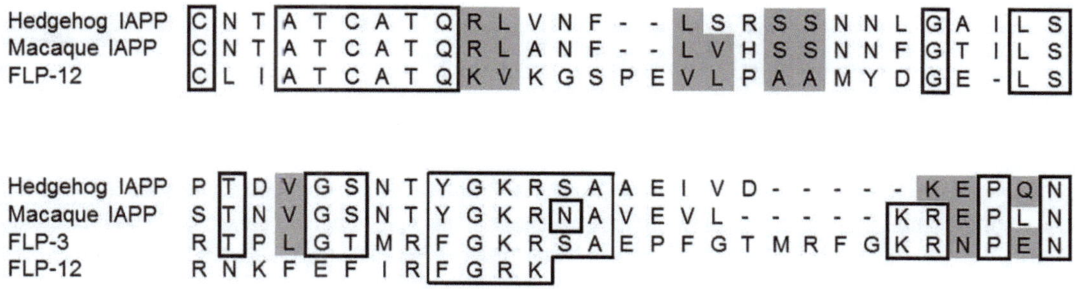

Fig. 3 BLASTP search with the hedgehog IAPP sequence identifies homologous regions in different parts of two separate *C. elegans* FLPs. The top panel shows matches at the N-terminus of FLP-12 and in the C-terminal region of FLP-3. The bottom panel shows multiple sequence alignments of contiguous segments from hedgehog and macaque IAPP with FLP-12 and FLP-3. Extension beyond the original BLASTP search results identifies additional matches. Identical amino acids and conserved substitutions are indicated as before

8. Look for additional matches beyond the initial search results (*see* **Note 3**). Depending on gap penalties and limitations of the search programs to extend alignments, there are often additional sequence matches in the vicinity of the BLASTP results that were missed during seeding. This is illustrated in the bottom part of Figs. 1 and 2.
9. This process can be repeated with additional query sequences of interest. For this case study, hedgehog and macaque IAPP sequences were obtained from Ensembl as described above and used to search for counterparts in WormBase. The results of this search are depicted in Fig. 3. Both FLP-12 and FLP-3 showed clear matches to the query sequences, but involved different regions of the neuropeptides. This type of information can provide novel insights into evolutionary relationships among distantly related peptides and proteins as depicted in Fig. 4 (*see* **Note 4**).
10. Because BLASTP with low-threshold E-values returns more matches—mostly non-orthologous proteins—it is necessary to confirm that an attractive hit merits consideration as a functional counterpart. One strategy relies on obtaining evidence that BLASTP searches with sequences from multiple

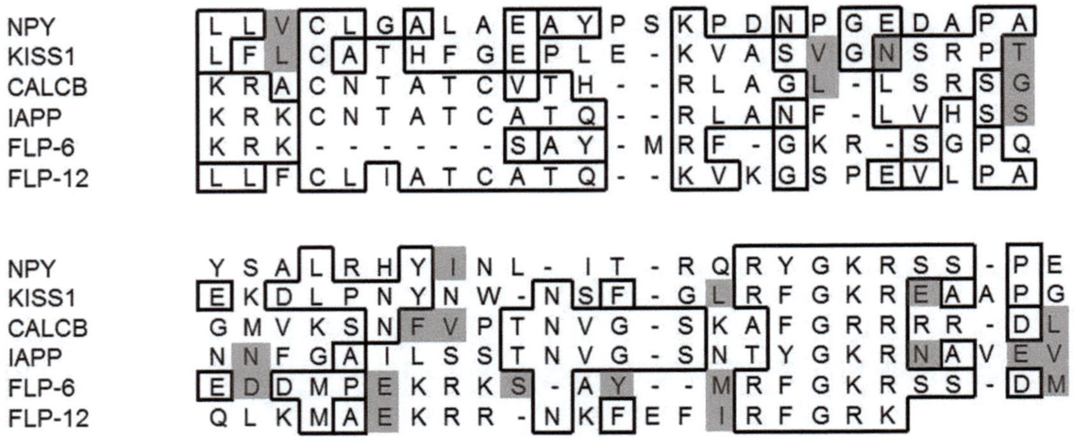

Fig. 4 Shared sequence features of RFamides (NPY and KISS1) and peptide hormones (calcitonin-related polypeptide beta [CALCB] and IAPP) with *C. elegans* FLPs suggest evolutionary connections between the peptide families. The amino acid sequences of CALCB, IAPP, and FLP-6 are contiguous stretches from the top to the bottom panel. There are gaps between the sequences in the top and bottom panels for NPY (6 residues), KISS1 (71 residues), and FLP-12 (34 residues). Cysteines in the N-terminal region are a characteristic feature of peptide hormones (CALCB, IAPP, and others), possibly arising from a common ancestor that resembled FLP-12. The RFamide motif is ancient, and while it remained intact in NPY (except for the F to Y substitution) and KISS1, it may have degenerated in the common precursor of the peptide hormones during evolution, expansion, and specialization of this family

species converge on the same set of candidate hits as outlined in step 9 (*see* **Note 5**). To achieve this goal, steps 1–8 are repeated with an array of different query sequences. Significant insights have been gained with follow-up searches using sequences from zebrafish, spotted gar, golden hamster, hedgehog, zebra finch, Leishan spiny toad, and green anole lizard. The lists of hits from multiple searches are then scanned for recurring members that meet criteria such as similar protein length, similar positioning of matched residues in the overall sequences, and biochemical features consistent with the query protein, such as proteolytic sites in the case of neuropeptides.

2.2 Distant Homology BLASTP Search of Ensembl with *C. elegans* FLPs

1. Confidence in the results of low threshold homology searches can be enhanced by using the sequences of hits from one run to detect the query sequences when searching in the opposite direction.
2. In this case study, the entire set of FLP sequences from *C. elegans* was used to query the Ensembl database for counterparts in representative species (*see* **Note 6**).
3. At WormBase, the <Search directory—for a protein> function is used to obtain the amino acid sequences of individual FLPs for the BLASTP searches. The *C. elegans* genome contains *flp-1-flp-28* and *flp-32-flp-34*. Copy the sequence for use in the Ensembl search.

4. Access the Ensembl website, choose the BLAST/BLAT tool from the menu, and enter/paste the sequence of interest (FLP-1 in this case) in the Sequence data box. Specify "Protein" below the sequence box.
5. The next step is to select the species databases that will be searched for the presence of homologous counterparts. Homo sapiens is the default setting, and other species can be added using the <Change species> function. For this study, a representative sample of species was searched including human, zebra finch, golden hamster, hedgehog, spotted gar, and zebrafish. Once the species menu has been completed, select "Protein database" for the search tool; BLASTP will populate automatically.
6. <Search sensitivity> has three options: near match, normal, and distant homologies, which determine the scoring matrix that will be employed. "Distant homologies" should be selected for expected low homology searches (*see* **Note 7**).
7. General options can be specified next. The default setting for "Maximum E-values for reported alignments" is 1E-1 (0.1). With neuropeptides and shorter sequences, pilot studies found that an E-value of 1000 provided a useful trade-off between speed of searches and the number and fidelity of hits. This setting also necessitates raising the "Maximum number of hits to report," and 1000 is a convenient value here.

   "Word size for seeding alignments" can also be varied, and three is the default setting. This study used two for the word size setting (*see* **Note 8**).
8. Scoring options are adjustable to accommodate distant homology searches of databases. The BLOSUM62 scoring matrix was selected because it shows excellent all-around performance (*see* **Note 9**). Pilot studies indicated that BLOSUM62 worked optimally with the neuropeptide searches. The default "Gap penalties" setting of "Opening" (11) and "Extension" (1) were accepted.
9. Searches were then performed, and the findings were analyzed. The search results are obtained from the BLAST/BLAT dashboard by selecting <View results>. On the webpage navigation menu, the <Find> command was used to search the file for "Gene hits" that included the neuropeptides shown in Table 1: NPY, pancreatic polypeptide (PPY), neuropeptide VF precursor (NPVF), KISS, IAPP, calcitonin-related polypeptide (CALC), adrenomedullin (ADM), and glucagon (GCG). Each species was analyzed in this way, and if the neuropeptide search term was found in the Results file, then that species was added to Table 1 (*see* **Note 10**).

Table 1
Distant homology BLASTP identification of FLP counterparts in vertebrates[a]

Protein	NPY	PPY	NPVF	KISS	IAPP	CALC	ADM	GCG
FLP-1					SG			
FLP-2			F		SG			
FLP-3	SG		HH	Z	SG			
FLP-4	GH, HH							
FLP-5								
FLP-6	GH, HH, SG			Z				
FLP-7								
FLP-8								
FLP-9	F, GH, H, HH							
FLP-10								
FLP-11							HH	
FLP-12					GH, H, HH, SG			
FLP-13	HH		Z					Z
FLP-14								
FLP-15			F, HH					
FLP-16						Z		
FLP-17	F, GH, H, HH, SG		F					
FLP-18	F, GH, HH, SG					Z		
FLP-19								
FLP-20					GH			
FLP-21					SG			
FLP-22	HH				HH			
FLP-23								
FLP-24			GH					
FLP-25	SG							
FLP-26				Z				
FLP-27					SG			
FLP-28	SG					Z		
FLP-32								
FLP-33								SG
FLP-34	F, GH, H, HH, SG, Z	HH						

[a]BLASTP search of Ensembl was conducted with amino acid sequences of FLPs from *C. elegans* as described in the text. When the search yielded a hit from a particular species, it was added to the list. The abbreviations for the different species are as follows: *F* zebra finch, *GH* golden hamster, *H* human, *HH* hedgehog, *SG* spotted gar, and *Z* zebrafish. The peptide abbreviations are as follows: *NPY* neuropeptide Y, *PPY* pancreatic polypeptide, *NPVF* neuropeptide VF precursor (also RFRP), *KISS* kisspeptin, *IAPP* islet amyloid polypeptide, *CALC* calcitonin-related polypeptide (alpha or beta), *ADM* adrenomedullin, and *GCG* glucagon

10. "Big picture" evaluation of the overall data revealed that the NPY and IAPP peptides matched with the largest number of FLPs and across the greatest number of different species (Table 1). This information may reflect the strength of evolutionary relationships among functional counterparts across species and suggest common ancestral origins for the peptides identified (*see* Fig. 4 and **Note 11**).

3 Notes

1. The E-value is a statistics-based measure that determines how many sequence alignments are considered to meet criteria for orthologous matches. A smaller E-value (higher bit score) restricts the number of alignments that qualify and reduces the number of hits obtained in a BLASTP search. Default E-values at different database websites are generally in the range of 1E-1 to 1E+0 (0.1–1). At these default settings, many distantly related counterparts of shorter query proteins are missed. The E-value is a trade-off between search speed and accurate identification of distant relatives. A less stringent E-value, such as 1E+2 (100), will generate a larger list of hits; however, most of these will be false positives. Therefore, additional post-search analysis is required to distinguish high-fidelity ortholog candidates from fortuitously matched, non-orthologous proteins. In the case of neuropeptides (and shorter distantly related proteins), hits that prove to be meaningful often fall into the range of 100–1000 for their E-value. Therefore, relaxed thresholds for BLASTP searches are required to identify functional counterparts of peptides when the evolutionary distance is greater.

   Pilot studies can be useful at the start of a project to evaluate different search settings, including E-values, which allow the identification of sufficient hits so as to include actual distant relatives of the query protein. The only downside of setting the E-value too high is the time required to sift through the more extensive list of hits. If the application of additional exclusion methods to the hit list involves minimal time, then this limitation is not an impediment. Varying E-values in the context of other parameters is a simple process and may yield novel insights during the pilot study.

2. Running BLASTP with relaxed E-values and less stringent conditions will yield more hits as discussed. Follow-up analysis will be needed to identify the most promising functional counterparts of the query protein. It is helpful to establish additional exclusion criteria a priori to facilitate culling the search results. Additional exclusion criteria may include protein size, similar

locations of the matched regions in the protein sequences, conservation of key motifs such as an RFamide site and cysteine residues, similar domains, and absence of an already-established identity for the hit.

3. Despite built-in alignment extension processes, BLASTP tends to ignore limited nearby matches if gaps must be added and scoring parameters are too restrictive. An important adjunct to the automated search is visual inspection of nearby sequences, especially when multiple sequences can be aligned to reveal additional areas of homology. As shown in Figs. 1–3, the number and character of additional matches found manually can be substantial. Multiple sequence alignment tools can be used for this purpose. However, a trained eye can be even more powerful in detecting subtle changes such as short duplications that shift corresponding motifs or aligned sequence-repeat pairs that may indicate similar evolutionary processes.
4. Post-search inspection of sequence matches is important because it can reveal how evolution may have generated homologous proteins. To illustrate this point, two regions of homology between hedgehog IAPP and two different FLPs from *C. elegans* are shown in Fig. 3. The C-terminal region of IAPP matches the C-terminal segment of one of the FLP-3 RFamide peptides. Furthermore, hedgehog IAPP is also homologous to an N-terminal region of FLP-12 that is distinguished by the presence and spacing of conserved cysteine residues. This motif is common to a number of non-RFamide peptide hormones in vertebrates including calcitonin (*see* Fig. 4). Because FLP-12 contains an RFamide segment at its C-terminus, it can be considered a hybrid between RFamides—an ancient neuropeptide system—and more recent peptide hormones such as amylin and calcitonin. The IAPP C-terminal sequence may have diverged from the RFamide motif to further specialize its function. By contrast, NPY, which clearly has functional counterparts in *C. elegans*, and kisspeptin retained RY/Famide activity during evolution. A common ancestor to NPY appears to be the oldest member of this family as suggested by the data in Table 1 (also *see* **Note 10**). Therefore, careful analysis of distant sequence homologies can be used to trace evolutionary relationships of signaling peptides.
5. Distant homology BLASTP searching yields more false positives than actual orthologs. Consequently, search with one sequence from a single species is unlikely to provide adequate confidence in the validity of low homology hits. To overcome this limitation, BLASTP searches should be conducted with sequences from multiple representative species. I recommend the following, but other options may be preferred depending

on the goal of the project: zebrafish, spotted gar, golden hamster, hedgehog, zebra finch, Leishan spiny toad, and green anole lizard. The source of sequences from other species can be weighted toward vertebrates, mammals, or rodents if the evolutionary focus is narrower. The general strategy is to set up queries based on different species and to then analyze the search results for common hits. In our experience, BLASTP with sequences from 5–10 different species will not converge on exactly the same hits, but if the selection of species is broad enough, several strong candidates will typically be shared. Finally, the data with NPY demonstrate that there may be more than a single neuropeptide in *C. elegans* that serves as its functional counterpart.

6. FLP family members are one of the oldest families of neuropeptides that have been identified. Consequently, this family was chosen as the starting point for BLASTP searches aimed at finding whether each one has identifiable counterparts in more recently evolved species. The species used in this study included human, zebra finch, golden hamster, hedgehog, spotted gar, and zebrafish. Findings with this analysis can also help to confirm hits found previously by searching WormBase with sequences from other species. Thus, searching with FLP-6 detected zebrafish kisspeptin-2 as a hit, which confirmed an earlier search of WormBase with the zebrafish sequence as shown in Fig. 1.
7. Neuropeptides have displayed diversification and divergence from ancient common ancestors. Therefore, the expectation is for functional counterparts to show low levels of identical sequences across species. Functional neuropeptides are cleaved from larger precursor proteins with the active peptide often spanning only 20–30 amino acid residues. From a precursor of 100–200 residues, the segment under the most selective pressure is only 10–30% of the total protein. Surrounding sequences of nonessential residues may have substantially diverged during evolution. Consequently, it is necessary to use less stringent conditions for BLASTP searches and "Distant homologies" suits this purpose.
8. Another useful way to relax the search parameters is to decrease the word size for matches. Many search algorithms use a default setting of three for word size. This can be reduced to two if so desired. As a result, more spurious hits will be generated, but with post-search exclusion criteria applied, promising hits can be identified.
9. At the Ensembl website, the BLASTP tool can use several different scoring matrices. When "Near match" is selected from <Search Sensitivity>, the default scoring matrix is

BLOSUM90, whereas for "Distant homologies," it is BLOSUM45. BLOSUM62 is an excellent all-around performer [8] and is the default scoring matrix for WormBase BLASTP searches. Therefore, BLOSUM62 was selected for use in this study. Again, it may be worthwhile to try different settings because needs will vary across projects.

10. The original BLASTP searches of WormBase yielded different peptide hits in *C. elegans* depending on the species that provided the query sequence. Therefore, it made sense to search a representative set of species for the reverse operation, that is, using *C. elegans* FLP sequences as the query source. The species that were searched are listed in Table 1. Moreover, a select group of neuropeptides was chosen for this analysis because they represented a spectrum of two different classes of peptides—the RFamides and non-RFamide peptide hormones such as calcitonin, amylin, and adrenomedullin. Of course, a different set of species and a different group of neuropeptides would provide a different outcome; nevertheless, the trends from this study are informative and illustrate the value of this approach (*see* **Note 11**).

11. Distant homology BLASTP searches revealed several important findings (Table 1). First, certain FLPs matched with a number of different neuropeptides. For example, FLP-3 showed homology to NPY, NPVF, KISS, and IAPP. Second, numerous FLPs were homologous to NPY (11) and IAPP (8) from different species. Finally, other FLPs detected only a single match across all the species examined, for example, FLP-11, or no matches (e.g., FLP-7, FLP-8, and FLP-19). The latter observation provides evidence for the selectivity of the BLASTP searches. Taken together, the data suggest that the neuropeptides NPY and IAPP are closer relatives of ancient common ancestors between deuterostomes and protostomes. A close relationship to common ancestors would be expected to manifest as greater homology with the largest number of present-day family members reflecting their shared evolutionary origin. Other members of the family with fewer hits, for example, PPY or CALC, may have arisen later through gene duplication and/or diverged from the common ancestor more during evolution owing to greater specialization.

    From this type of analysis, it is possible to nominate FLP-3, FLP-6, and FLP-26 as potential kisspeptin counterparts in *C. elegans*. FLP-12 appears to be a solid candidate for a nematode functional equivalent of amylin (IAPP). Based on their more frequent appearance in hit results, FLP-6 and FLP-12 may be closest to the common ancestors of kisspeptin and amylin, respectively (*see* Fig. 4). As an alternative possibility, the lack of clear orthologs of vertebrate neuropeptides and peptide hormones in *C. elegans* may reflect convergent evolution of an adaptable neuropeptide module toward a promiscuous receptor-binding motif. However, divergent evolution can also explain the failure to find unequivocal orthologs in *C. elegans* by simply pushing back the origin of the diverging sequences to the ancient transposable exons or trexons, which gave rise to complex peptides and their receptors [17]. This is consistent with the idea of Jékely who noted that distant neuropeptide families represent deep paralogs [1].

The hybrid structure of FLP-12—part peptide hormone and part RFamide—suggests it may be related to a common ancestral peptide that was duplicated in the genome. One of the duplicates continued the RFamide family with members such as NPY, KISS, and NPVF, whereas the other gave rise to a new family of peptide hormones (IAPP, CALC, etc.). These interesting relationships would have been missed if not for the less stringent threshold for search parameters and careful independent inspection of nearby sequences across multiple species. In conclusion, low homology BLASTP searching is a valuable tool for establishing structural and evolutionary connections among distantly related proteins.

References

1. Jékely G (2013) Global view of the evolution and diversity of metazoan neuropeptide signaling. Proc Natl Acad Sci USA 110:8702–8707
2. Li C, Kim K (2008) Neuropeptides WormBook, pp 1–36
3. Savage-Dunn C, Padgett RW (2017) The TGF-β family in *Caenorhabditis elegans*. Cold Spring Harb Perspect Biol 9:a022178
4. Elphick MR, Mirabeau O, Larhammar D (2018) Evolution of neuropeptide signalling systems. J Exp Biol 221:jeb151092
5. Altschul SF, Madden TL, Schäffer AA, Zhang J, Zhang Z, Miller W et al (1997) Gapped BLAST and PSI-BLAST: a new generation of protein database search programs. Nucl Acids Res 25:3389–3402
6. Franklin C, Dwyer DS (2021) Candidate risk genes for bipolar disorder are highly conserved during evolution and highly interconnected. Bipolar Disord 23:400–408
7. Dwyer DS (2021) Protein receptors evolved from homologous cohesion modules that self-associated and are encoded by interactive networked genes. Life 11:1335
8. Altschul SF, Wootton JC, Gertz M, Agarwala R, Morgulis A, Schäffer A et al (2005) Protein database searches using compositionally adjusted substitution matrices. FEBS J 272:5101–5109
9. Tensen CP, Cox KJA, Burke JF, Leurs R, van der Schors RC, Geraerts WPM et al (1998) Molecular cloning and characterization of an invertebrate homologue of a neuropeptide Y receptor. Eur J Neurosci 10:3409–3416
10. de Bono M, Bargmann CI (1998) Natural variation in a neuropeptide Y receptor homolog modifies social behavior and food response in *C. elegans*. Cell 94:679–689
11. Rogers C, Reale V, Kim K, Chatwin H, Li C, Evans P et al (2003) Inhibition of *Caenorhabditis elegans* social feeding by FMRFamide-related peptide activation of NPR-1. Nat Neurosci 6:1178–1185
12. Gershkovich MM, Gross VE, Kaiser A, Prömel S (2019) Pharmacological and functional similarities of the human neuropeptide Y system in *C. elegans* challenges phylogenetic views on the FLP/NPR system. Cell Commun Signal 17:123
13. Ubuka T, Tsutsui K (2018) Comparative and evolutionary aspects of gonadotropin-inhibitory hormone and FMRFamide-like peptide systems. Front Neurosci 12:747
14. Frooninckx L, Van Rompay L, Temmerman L, Van Sinay E, Beets I, Janssen T et al (2012)

Neuropeptide GPCRs in *C. elegans*. Front Endocrinol 3:167

15. Cunningham F, Allen JE, Allen J, Alvarez-Jarreta J, Amode MR, Armean IM et al (2022) Ensembl 2022. Nucl Acids Res 50: D988–D995
16. Harris TW, Antoshechkin I, Bieri T, Blasiar D, Chan J, Chen WJ et al (2010) WormBase: a comprehensive resource for nematode research. Nucl Acids Res 38:D463–D467
17. Dwyer DS (1998) Assembly of exons from unitary transposable genetic elements: implications for the evolution of protein-protein interactions. J Theor Biol 194:11–27

Chapter 2

Predicting Protein Folding Dynamics Using Sequence Information

Ezequiel A. Galpern, Federico Caamaño, and Diego U. Ferreiro

Abstract

Natural protein sequences somehow encode the structural forms that these molecules adopt. Recent developments in structure prediction are agnostic to the mechanisms by which proteins fold and represent them as static objects. However, the amino acid sequences also encode information about how the folding process can happen and how variations in the sequences impact on the populations of the distinct structural forms that proteins acquire. Here, we present a method to infer protein folding dynamics based only on sequence information. For this, we will rely first on the obtention of a precise "evolutionary field" from the observed variations in the sequences of homologous proteins. We then show how to map the energetics to a coarse-grained folding model where the protein is treated as a string of foldons that interact. We then describe how, for any given protein sequence of a family, the equilibrium folding curve can be computed and how the emergence of protein-folding subdomains can be identified. We finally present protocols to analyze how mutations perturb both the folding stability and the cooperativity, which represent predictions for a deep mutational scan of a protein of interest.

A Google Colab implementation of the method is available at https://colab.research.google.com/github/eagalpern/folding-ising-globular/blob/master/notebooks/custom_potts_simulation_colab.ipynb.

Key words Folding pathway, Folding mechanism, Ising, Evolution, DCA, Potts

1 Introduction

Natural protein molecules are truly amazing objects. Over the last century, it has become clear that these molecules perform their various biological tasks by folding the polypeptide chains into dynamic ensembles of three-dimensional structures. Today, protein folding is understood in the conceptual framework of the energy landscape theory, which posits that evolved proteins are minimally frustrated heteropolymers which navigate physical landscapes that have the overall form of a rough funnel, for which there is a fundamental correlation between structure and energy [1]. Precise, fast, and robust folding is not a general property of most amino acid

Shahid M. Khan and Florencio Pazos (eds.), *Protein Evolution: Methods and Protocols*, Methods in Molecular Biology, vol. 2979,
https://doi.org/10.1007/978-1-0716-4828-5_2,

chains, but results from the selection of specific sequences that, when folded, minimize the energetic conflicts between its parts [2]. The amazing variety of structures that natural protein molecules display is somehow coded in the linear strings of amino acids [3, 4]. Statistical analysis of natural sequence information must then allow us to reverse-engineer the basic aspects of the folding codes [5]. Recently, massive machine learning schemes have been proven successful to predict a three-dimensional structure for a given sequence input [6, 7], apparently solving the structure prediction problem. However, these black boxes tell us nothing about the mechanisms by which the actual polypeptides acquire the structural forms, how broad the structural ensembles are, or how sequence variations impact the folding dynamics.

From a biological standpoint, protein folding appears to be a major driving force in protein sequence evolution [8–10]. As such, the patterns observed when aligning multiple sequences of evolutionary related proteins must somehow account for the energetics of amino acid interactions. Mutations can be interpreted as de facto perturbations of the folding landscape that have passed the sieve of evolution. This accounts for the theoretical underpinning of the now widely used amino acid substitution matrices, domain recognition programs, and protein phylogeny reconstructions [11–13]. Moreover, recent developments of statistical models to account for natural sequence variations have been proved useful for analyzing the energetic effect of single-point mutants and pair-mutations couplings [14–16].

It was discovered when studying the folding of single-domain proteins that, provided there are strong enough native interactions within them to compensate for their entropy loss, distinct protein sections may fold at different periods almost independently. Panchenko et al. have dubbed these units "foldons" as they can fold in a single coordinated step [17]. The exon–foldon correlation has been supported by recent evidence that conserved exons exhibit a strong independent foldability [18]. Thus, it may be possible to use this primary structure partitioning to identify common foldons for each family and examine their folding dynamics, drawing on the annotation of intron–exon borders in several genomes.

We present here a simple model to predict folding dynamics by leveraging the sequence information learned by inverse statistical evolutionary models and a coarse-grained description of protein structure. We summarize how by assuming that protein stability has been the most relevant evolutionary constraint (*see* **Note 1**), it is possible to infer folding curves, free-energy profiles, and subdomain emergence for any given protein sequence.

2 Materials

The essential components for an Ising coarse-grained folding simulation include:

1. A target protein sequence.
2. A multiple sequence alignment of proteins homologous to the target sequence.
3. A partition of the sequence into contiguous folding elements, often referred to as *foldons*.

We propose using a sequence-based energy function that evaluates the probability of observing a protein sequence of a given length within an equilibrated ensemble. These "evolutionary" energy parameters can be computationally derived from a multiple sequence alignment (MSA) using methods such as direct coupling analysis (DCA) [19, 20].

1. Target Protein Sequence

   A target amino acid sequence is required to initiate the folding simulation. This sequence can represent the full length of a protein or a fragment (e.g., a domain) that is assumed to fold independently of the rest of the chain. Variations in folding dynamics are expected across protein families, particularly for protein topologies rich in alpha-helices (*see* **Note 2**). Notably, the impact of all possible single-site mutations can be estimated directly from the wild-type simulation without requiring additional simulations [21].

2. Multiple Sequence Alignment

   To derive a sequence-based evolutionary energy, a set of homologous and aligned sequences is necessary. A multiple sequence alignment (MSA) can be obtained from the Pfam database [22] (now hosted by InterPro [23]) by searching for the protein family containing the target sequence. Alternatively, an MSA can be constructed using the UniParc database [24], where homologous sequences are identified and aligned to the target sequence using tools like *jackhmmer* [25]. Both approaches rely on scoring single-site similarities and penalizing gaps, parameters that one must be careful to tune for obtaining a meaningful MSA. For Pfam alignments, to ensure the MSA aligns with the target sequence, only positions present in the target sequence should be retained. The quality of the evolutionary field to be obtained relies on the depth, coverage, and representativeness of the MSA used for the learning. In other words, a large enough, diverse, and clean MSA is required. We describe some relevant considerations to fulfill these requirements. Protein sequences present in databases represent actual proteins (or predicted sequences from genomic analysis) that the community has put together in databases for diverse reasons, and as such, they are biased by the organisms studied and the phylogenetic relationships among them. To minimize phylogenetic bias, MSA sequences can be globally analyzed in order to detect overall similarity and, assuming that they come from closely related organisms, downgrade the relative contribution. This can be done clustering the complete, unaligned sequences with CD-hit [26] at a 90% cutoff. This cutoff is not stringent but has been proven sufficient for large datasets. A faster alternative is to cluster the aligned sequences using as similarity the normalized Hamming distance between them and applying the same cutoff. A weight defined as *1/n* is assigned to each sequence i, where n_i is the number of sequences in the *i*th cluster, so the sum of clusters is the effective number of sequences. All statistical analyses should be performed using these sequence weights. The likelihood of accurately learning the Potts model parameters increases with the alignment depth, the number of effective sequences of the MSA or simply the sum of sequence weights. For reliable inference of Potts model parameters, it is generally recommended to have an effective number of sequences that is at least 5–10 times the sequence length L [27]. For instance, a protein of length $L = 100$ would require at least 500–1000 effective sequences, while $L = 200$ would require 1000–2000. We highlight that many sequences in raw alignments often do not cover the N-terminus or C-terminus regions (and sometimes in between regions), harming the accuracy of prediction that a model can reach. Certain positions present a very high fraction of gaps, increasing the noise-to-signal ratio during the Potts model learning, being an usual practice to remove from the MSA sequences that have a frequency of gaps higher than a given threshold, that is, 25%.

3. Folding Elements (Foldons)

   The protein sequence must be divided into distinct, non-overlapping folding elements or foldons, which are groups of amino acids that fold and unfold cooperatively. While a common foldon assignment for each protein family (or MSA) is recommended, the method can accommodate sequence-specific partitions. We describe here some heuristics to assign foldons, according to the protein architecture and the hypothesis to test.

   (a) Repeat proteins:

      For periodic or repeat proteins, foldons can be defined as individual repeats or segments of repeats. The periodic nature of these proteins simplifies foldon assignment, and the folding dynamics can be studied by treating each repeat as a single foldon. Also, a periodic partition of the protein can be obtained by separating the repeats into multiple folding units [28, 29] or grouping them into larger foldons.

(b) Secondary structure elements:
For nonsymmetric globular proteins, secondary structure elements (e.g., alpha-helices and beta-strands) can serve as foldons. This approach leverages the natural structural organization of the protein to define folding units.

(c) Exon-based partitioning:
Highly conserved exon boundaries within a protein family can be used to define foldons, as there is often a correspondence between exons and folding units [18].

(d) Arbitrary segments:
Foldons can also be defined based on known structural segments or arbitrary partitions for testing specific hypotheses. This flexibility allows researchers to explore the impact of different folding unit definitions on the simulation results.

(e) Neutral model:
Alternatively, foldons can be assigned using a neutral model, where segment sizes are sampled from a geometric distribution. Multiple neutral partitions can be generated, and results can be averaged across them to account for variability in foldon definitions. This approach is useful for exploring the robustness of the results to different partitioning schemes.

3 Methods

The general procedure for predicting protein folding dynamics using sequence information consists of six steps (Box 1).

1. Learn the Potts model parameters from a MSA.
2. Estimate the selection temperature.
3. Assign the coarse-grained folding Hamiltonian.
4. Set up the Monte Carlo parameters and run simulations.
5. Visualize and analyze results for a single protein.
6. Extra: Exploit family results for protein design.

Model definition and methods pipeline are summarized in Fig. 1.

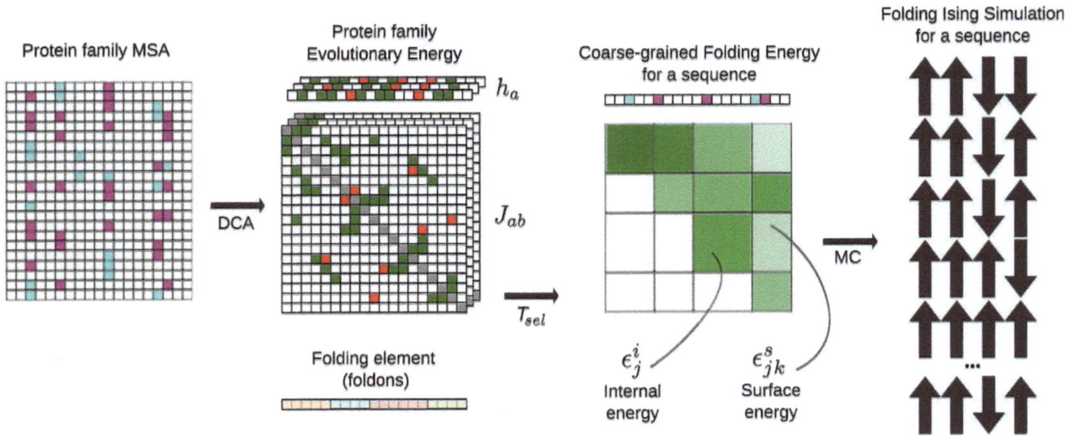

Fig. 1 Model definition. We learn the evolutionary energy field parameters from a multiple sequence alignment (MSA) using a direct coupling analysis (DCA). Given a set of folding units or foldons and the selection temperature of the family, we extract the coarse-grained folding energy for each sequence to input a finite-chain Ising model. The folding dynamics of the sequence is computed using a Monte Carlo simulation

1. Learn the Potts Model Parameters from an MSA

The evolutionary energy parameters can be inferred from a MSA through a direct couple analysis (DCA). A Potts model is defined, containing local fields and pairwise couplings, to account for the occurrence of a given amino acid in a given position and for the co-occurrence of amino acids in pairs of positions. The energy function of the model for a given sequence σ is.

$$E(\sigma) = -\sum_{a=1}^{L} h_a(\sigma_a) - \sum_{a<b}^{L} J_{a,b}(\sigma_a, \sigma_b),$$

where the parameters to learn are the local fields h_a and the pairwise couplings J_{ab}. The exact inference of the model parameters requires the calculation of the partition function, which is computationally intractable. Several approximation strategies have been developed, giving rise to many DCA implementations, including Boltzmann machine learning (bmDCA) [30, 31], pseudolikelihood maximization (plmDCA) [32], and mean field approximation (mfDCA) [19]. A comprehensive review can be consulted in [33]. For first-time DCA users, we recommend starting with plmDCA, implemented in the *pydca* library available for Python. However, we highlight that the folding model may benefit from using more accurate learning methods as bmDCA (*see* **Note 3** for other alternatives). The *pydca* library requires a MSA in FASTA format, and it computes sequence weights via a similarity clustering as explained in Materials, using a cutoff defined by the user. Also, the learning uses two regularization parameters for avoiding overfitting. One of these parameters is the local fields' regularization term, and it was observed that the outputs were robust to it, so a default value for this parameter of 1 can be used [32]. The other parameter, which is the pairwise coupling regularization term, seems to be more critical, therefore requiring a scan of values to determine a precise choice of this parameter. Based on previous observations, a default value of 20 was set for this parameter and used throughout experiments. The accuracy of the learned Potts Model can be estimated by looking at how many contacts of a native reference structure can be predicted as strong couplings in the Potts model. In addition, a simple validation step is recommended to compute the total energy of the target sequence and ensure it is distinguishable from scrambled sequences. A more stringent test is to use the energetic field to obtain artificial sequences. This can be done with Monte Carlo simulations of sequences given the energetic field. One can then compare the single-site and pair-site amino acid frequencies of the simulated sequences versus a test set of natural sequences and their measured energies. The artificial sequences must be statistically indistinguishable from the natural ensemble of the MSA if the Potts field is perfectly inferred.

Also, the mean energy of random sequences must be set to zero. This condition is automatically satisfied when the zero-sum gauge [33] is imposed, as most DCA algorithms do by default. Alternatively, a pretrained Potts model can be used if it matches the target sequence's family. Minor alignment or size discrepancies can be corrected, but the sequence must belong to the family used to train the model. A validation using the described tests is also recommended for pretrained models.

2. Estimate the Selection Temperature

The selection temperature T_{sel} is the apparent temperature at which sequences of a particular family were selected by nature and quantifies how strong the folding constraints have been during evolution [34]. If experimental folding free-energy changes upon mutations $\Delta\Delta G$ are available for at least a protein of the studied family, $k_B T_{sel}$ can be explicitly calculated as the proportionality constant between $\Delta\Delta G$ and the corresponding changes scores as the energy difference of the wild-type and mutant according to Potts field. Alternatively, assuming that the standard deviation of $\Delta\Delta G$ is nearly constant irrespectively of protein families, T_{sel} can be estimated in a relative scale to a known reference, using the ratio of the standard deviation of evolutionary energy changes by single mutations [35]. The reference should be a T_{sel} obtained from $\Delta\Delta G$ experimental data, for instance for PDZ family, T_{sel} will be estimated as.

$$T_{sel}\,\sigma(\Delta E) = T_{sel}^{PDZ}\,\sigma\left(\Delta E^{PDZ}\right),$$

where $\sigma\,(\Delta E)$ denotes the standard deviation of evolutionary energy differences upon point mutations averaged over homologous sequences.

3. Assign the Coarse-Grained Folding Hamiltonian

The protein is modeled as an array of interacting folding elements (foldons) that can be either folded (F) or unfolded (U), as two-state spin variables. The system is represented as a finite-size Ising chain of N elements, where the energy of a coarse-grained configuration, the Hamiltonian, is given by the free energy of the corresponding ensemble of microstates:

$$H = -\sum_{j=1}^{N}\left[T\,s_j(1-\delta_{j,F}) + \epsilon_j^i\,\delta_{j,F}\right] - \sum_{j=1}^{N-1}\sum_{k>j}\epsilon_{jk}^s\,\delta_{j,F}\,\delta_{k,F},$$

where T is the temperature and $\delta_{j,\,F}$ is the Kroeneker symbol taking value one if element j is folded (F) and zero otherwise. If the element j is folded, it has a specific internal folding free energy (averaged over the solvent) ϵ_j^i. If two elements j and k are both folded, we consider also a surface energy ϵ_{jk}^s, describing a specific interaction between the two foldons. If the element j is unfolded, we set the energetic contributions to zero, but there is an explicit entropic contribution given by the entropy s_j of the available spatial configurations of the foldon. Hence, within this model, a protein can unfold as a result of an increase in temperature T.

The energetic $\left(\epsilon_j^i, \epsilon_{jk}^s\right)$ parameters are functions of the amino acid sequence σ. We then map the sequence-based evolutionary energy residue–residue couplings $J_{ab}(\sigma_a, \sigma_b)$ and local fields $h_a(\sigma_a)$ to the specific coarse-grained folding free energy terms for each sequence,

$$\epsilon_j^i = \epsilon^i\left(\sigma_j\right) = k_B\,T_{sel}\left[\sum_{a\in j} h_a(\sigma_a) + \sum_{a,\,b\in j} J_{ab}(\sigma_a, \sigma_b)\right],$$

$$\epsilon_{kj}^s = \epsilon^s\left(\sigma_j, \sigma_k\right) = k_B\,T_{sel}\left[\sum_{a\in j,\,b\in k} J_{ab}(\sigma_a, \sigma_b)\right],$$

where k_B is the Boltzmann constant and T_{sel} is the selection temperature.

To compute the entropic terms (s_j), we recommend to approximate s_j to be independent of amino acid identity and strictly additive and to use an average entropy per residue of $s = 5$ cal mol$^{-1}K^{-1}$res^{-1} that was empirically fitted [29]. Therefore, $s_j = L_j\,s$, being L_j the sequence length of the foldon. Alternatively, if experimental temperature denaturation scans of the protein of interest (or a protein of the family) and mutants are known, the entropic term can be fitted by minimizing the difference between the simulated and the experimental data [29].

4. Set Up the Monte Carlo Parameters and Run Simulations

Once the Hamiltonian has been set for a sequence, the metropolis Monte Carlo simulation of the Ising model must be set up. To describe the folding dynamics, the simulation must be repeated for several temperatures *T*, such as for the lower *T*, the protein is completely folded and, for the higher one, is completely unfolded. The temperature range can be chosen manually or using the automatic range set that ensures the completely folding and unfolding condition. Simulation total time, transient time, and equilibration time parameters may be sensitive to the Ising chain size (number of foldons per protein) which we assume to be constant for each family. An autocorrelation analysis is recommended to choose the optimal parameters for each *T*. As the system presents a critical temperature (at the folding temperature if it has two states) at which the Monte Carlo parameters could diverge, the simulation algorithm is organized in a two-round scheme. During the first round, the simulation is done at a list of given temperatures (typically distributed in a given range) and the critical temperatures are detected according to fluctuations. The simulation is repeated in a second round only at critical temperatures but using an equilibration and transient time multiplied by an extra parameter (critical points factor) that can be chosen according to the autocorrelation analysis.

5. Visualize and Analyze Results for a Single Protein

As the temperature increases, the folding elements are expected to transition from the folded to the unfolded state. For each element, the algorithm outputs the simulated thermal unfolding curves and single-element folding temperatures T_f^j (*see* Fig. 2a). For visualization, the protein tertiary structure is colored according to element T_f^j (*see* Fig. 2b). Such structure and its alignment with the MSA used should be provided. Our implementation simply requires the starting position of the sequence in the Protein Data Bank (PDB) file.

According to the energetics derived from the particular sequence, elements can unfold together at close temperatures. We define the emergence of an apparent subdomain between elements *j*, *k* if $| T_f^j - T_f^k |< 5$. Overlapping domains are separated into the minimum number of nonoverlapping ones. If more than one separation is possible, temperature differences between domains are maximized. The subdomains are shown in a matrix plot (*see* Fig. 2d).

For the complete protein, the fraction of elements folded at each temperature $m(T)$ is calculated taking all the foldons together. This curve can be contrasted to experimental thermal unfolding curves [29]. The protein folding temperature T_f is estimated using a sigmoidalc fit, approximating.

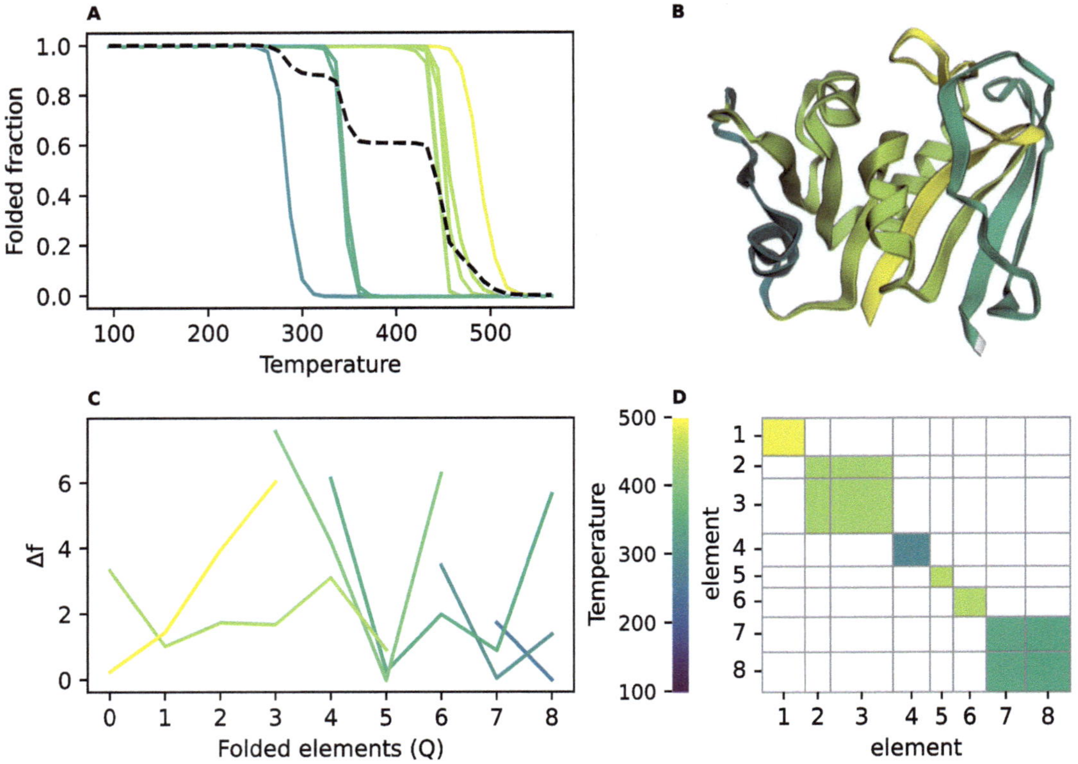

Fig. 2 Simulation results for a dihydrofolate reductase protein sequence (UniProt ID A0A2D5XDZ3). (**a**) Simulated thermal unfolding curves, for the complete protein (black dashed line) and for each element with solid lines, colors identify the folding temperature of each one (the same as B, C, and D). Yellow elements are the most stable ones. Protein folding temperature is $T_f = 411$ *K*. (**b**) The structure (PDB ID 7DFR) is colored according to the folding temperature of each element. (**c**) Free-energy profiles, colored by temperature, with the number of folded elements Q as reaction coordinate. Cooperativity score is $\rho = 5/7$. (**d**) Apparent domain matrix and secondary structure colored by element folding temperatures

$$m(T) = \frac{m_{\max}}{1 + e^{a\left(T - T_f\right)}},$$

where $m_{\max} \epsilon\ [0, 1]$, taking out of the fitting isolated, extremely unstable elements.

A free-energy profile is estimated from the probability of states s with Q folded elements with the metropolis Monte Carlo sampling (*see* Fig. 2c). We considered together sampled states for simulations performed in a window of the 10 closest temperatures. The free-energy profiles are calculated as.

$$\Delta f(Q) = -k_B T \log \left(\frac{\sum_{s|Q} N(s)}{\sum_{s} N(s)} \right),$$

where T is the average temperature, $N(s)$ are the counts of state s, and $s \mid Q$ are the states with Q folded elements.

As a measure of the folding cooperativity, we defined a cooperativity score $\rho = Q_{\text{barrier}}/(N-1)$, the fraction of intermediary Q that were not a minimum of $\Delta F(Q)$ for any T in a protein with N foldons. Together, the protein folding temperature, the cooperativity score, and the domain emergence summarize the simulated folding dynamics.

6. Extra: Exploit Family Results for Protein Design

Once the Ising model has been set up for computing the folding dynamic of a given sequence, it is straightforward to also apply it for many other sequences of the same protein family. The evolutionary Potts model, the foldon assignment, and Monte Carlo parameters are usually shared for the same family. Protein folding temperature and cooperativity are expected to vary according to the evolutionary energy assignment of the sequences [21]. However, depending on the tertiary structure topology, some protein families can present much conserved mechanisms. By construction, sequences with a more favorable total energy will be more stable, so just computing and ranking the overall energy of sequence is enough to pick natural sequences that are expected to be more thermally stable, with potential biotechnological applications.

The cooperativity score is expected to decrease with foldon energetic heterogeneity and increase with the average interaction strength (*see* Fig. 3). We have seen that using a linear fit, useful predictions can be made to estimate folding temperature and cooperativity changes upon single-site mutations (*see* Fig. 4). The folding element assignment modulates the effect of local stability changes. For instance, a destabilizing mutation in a highly stable element reduces the energetic heterogeneity, producing a more cooperative folding (red vertical stripes in Fig. 4a). On top of that, the mutation impact on inter-element interactions also affects predictions. We have seen that the cooperativity variance of natural sequences within a protein family presents a positive correlation with the predicted cooperativity variance for point mutations [21]. Therefore, it is expected that α-protein families are the most sensitive to engineer single variants with different folding cooperativity.

Moreover, the evolutionary Potts model that is being used to input the folding model can also be leveraged to generate an arbitrary amount of new protein sequences for the family using a Monte Carlo simulation [33]. All these generated synthetic sequences can be located in the heterogeneity-interaction phase space (*see* Fig. 3) to predict their folding cooperativity even before computing a folding simulation, without any additional computational cost.

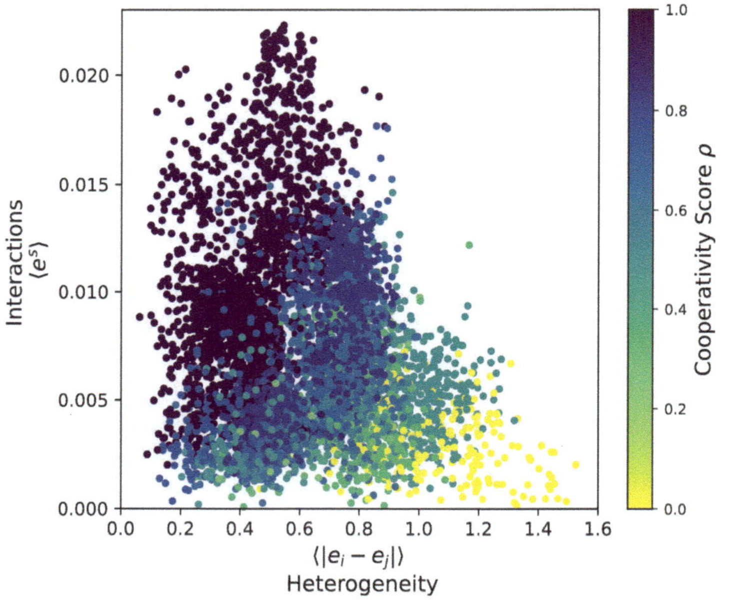

Fig. 3 Cooperativity score for 7490 sequences of 15 different protein families, in a color scale on the space defined by the energetic heterogeneity of foldons and their average interaction strength

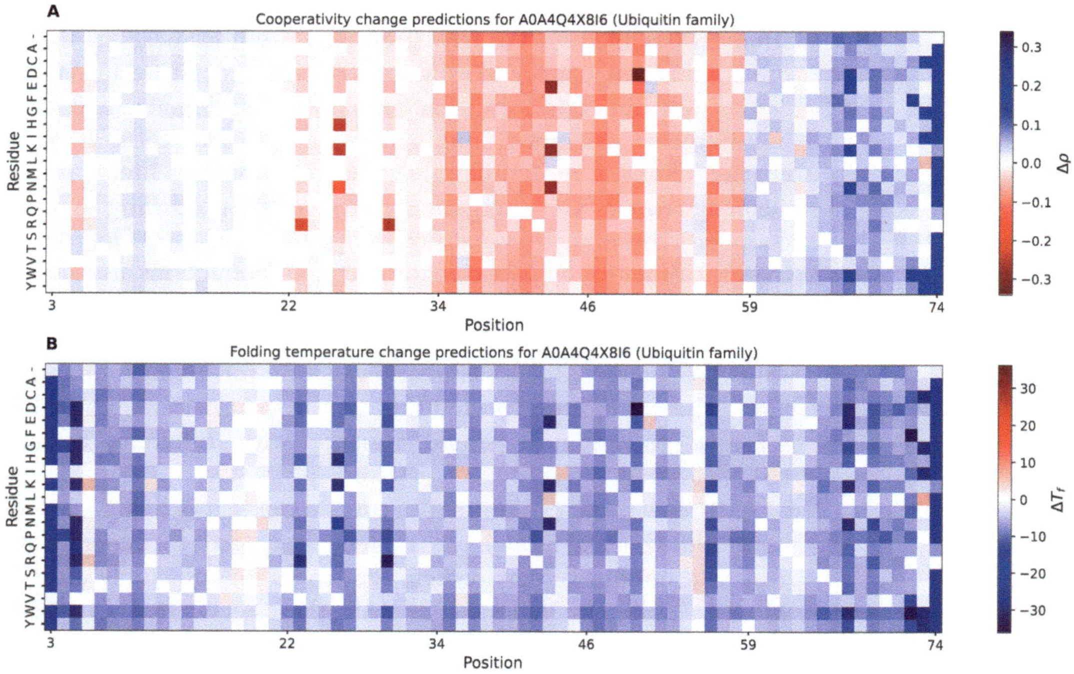

Fig. 4 Single-site mutant predictions. (**a**) Changes in the cooperativity score upon single-point mutations, for a natural sequence of the ubiquitin family (UniProt ID A0A4Q4X8I6). Foldon boundaries are indicated following reference PDB numbering (1ubq). (**b**) For the same sequence, the changes in the folding temperature upon all single-point mutations

4 Notes

1. We highlight that the proposed framework assumes that folding stability is locally the main evolutionary constraint, an approximation in line with the minimal frustration principle. Therefore, sequence positions strongly conserved and conditioned by other selection forces besides folding may affect local stability and some cooperativity predictions, locally frustrating the folding landscape [36].
2. Folding dynamics variability can emerge within protein family members or not according to the protein topology. As a rule of thumb, elongated alpha proteins allow different folding mechanisms depending on the sequence, while compact beta containing proteins do not. For the latter, a vanilla model that uses only contact positions and a binarized secondary structure (alpha/beta or coil/loop) may be enough to get the folding cooperativity, without requiring a sequence-sensitive Potts model [21].
3. Alternatively, Potts model parameters (local fields and couplings) can be also obtained from other evolutionary sequence-based models, often faster to train or already trained and available. For instance, the weights of a Gaussian restricted Boltzmann machine (RBM) model can be exactly mapped to the Potts model parameters [37]. Furthermore, it has been tested that protein large language models as Earth system model 2 (ESM2) [38] "logits" can be leveraged to calculate a categorical Jacobian, a useful approximation for Potts couplings [39].

Acknowledgments

This work was supported by the Consejo de Investigaciones Científicas y Técnicas (CONICET) (DUF is a CONICET researcher and EAG is a postdoctoral fellow), CONICET Grant PIP2022-2024—11220210100704CO, and Universidad de Buenos Aires Grant UBACyT 20020220200106BA. We call the attention of the international scientific community about the catastrophic erosion of Argentina's strong scientific tradition due to current funding constraints and the sudden termination of long-term policies.

References

1. Onuchic JN, Luthey-Schulten Z, Wolynes PG (1997) Theory of protein folding: the energy landscape perspective. Annu Rev Phys Chem 48(1):545–600. https://doi.org/10.1146/annurev.physchem.48.1.545
2. Bryngelson JD, Wolynes PG (1987) Spin glasses and the statistical mechanics of protein

folding. Proc Natl Acad Sci USA 84(21): 7524–7528. https://doi.org/10.1073/pnas.84.21.7524

3. Ferreiro DU, Komives EA, Wolynes PG (2014) Frustration in biomolecules. Q Rev Biophys 47(4):285–363. https://doi.org/10.1017/S0033583514000092
4. Sánchez IE, Galpern EA, Garibaldi MM, Ferreiro DU (2022) Molecular information theory meets protein folding. J Phys Chem B 126(43):8655–8668. https://doi.org/10.1021/acs.jpcb.2c04532
5. Wolynes PG (2015) Evolution, energy landscapes and the paradoxes of protein folding. Biochimie 119:218–230. https://doi.org/10.1016/j.biochi.2014.12.007
6. Akdel M et al (2022) A structural biology community assessment of AlphaFold2 applications. Nat Struct Mol Biol 29(11):1056–1067. https://doi.org/10.1038/s41594-022-00849-w
7. Baek M et al (2021) Accurate prediction of protein structures and interactions using a three-track neural network. Science 373(6557):871–876. https://doi.org/10.1126/science.abj8754
8. Lindquist S (2009) Protein folding sculpting evolutionary change. Cold Spring Harb Symp Quant Biol 74:103–108. https://doi.org/10.1101/sqb.2009.74.043
9. Shakhnovich E (2006) Protein folding thermodynamics and dynamics: where physics, chemistry, and biology meet. Chem Rev 106(5): 1559–1588. https://doi.org/10.1021/cr040425u
10. Ferreiro DU, Komives EA, Wolynes PG (2018) Frustration, function and folding. Curr Opin Struct Biol 48:68–73
11. Dayhoff MO (1976) The origin and evolution of protein superfamilies. Fed Proc 35(10): 2132–2138
12. Henzy JE (2018) Margaret Dayhoff: catalyst of a quiet revolution. In: Whitaker J, Barton M (eds) Women in microbiology. American Society of Microbiology, pp 29–36. https://doi.org/10.1128/9781555819545.ch4
13. Henikoff S, Henikoff JG (1992) Amino acid substitution matrices from protein blocks. Proc Natl Acad Sci 89(22):10915–10919. https://doi.org/10.1073/pnas.89.22.10915
14. Figliuzzi M, Jacquier H, Schug A, Tenaillon O, Weigt M (2016) Coevolutionary landscape inference and the context-dependence of mutations in beta-lactamase TEM-1. Mol Biol Evol 33(1):268–280
15. Espada R, Parra RG, Mora T, Walczak AM, Ferreiro DU (2017) Inferring repeat-protein energetics from evolutionary information. PLoS Comput Biol 13(6):1–16. https://doi.org/10.1371/journal.pcbi.1005584
16. Notin P et al (2023) ProteinGym: large-scale benchmarks for protein design and fitness prediction. *bioRxiv*, no. NeurIPS, pp 1–48
17. Panchenko AR, Luthey-Schulten Z, Wolynes PG (1996) Foldons, protein structural modules, and exons. Proc Natl Acad Sci 93(5): 2008–2013. https://doi.org/10.1073/pnas.93.5.2008
18. Galpern EA, Jaafari H, Bueno C, Wolynes PG, Ferreiro DU (2024) Reassessing the exon–foldon correspondence using frustration analysis. Proc Natl Acad Sci 121(28):e2400151121. https://doi.org/10.1073/pnas.2400151121
19. Morcos F et al (2011) Direct-coupling analysis of residue coevolution captures native contacts across many protein families. Proc Natl Acad Sci USA 108(49). https://doi.org/10.1073/pnas.1111471108
20. Weigt M, White RA, Szurmant H, Hoch JA, Hwa T (2009) Identification of direct residue contacts in protein-protein interaction by message passing. Proc Natl Acad Sci USA 106(1): 67–72. https://doi.org/10.1073/pnas.0805923106
21. Galpern EA, Roman EA, Ferreiro DU (2025) Inferring protein folding mechanisms from natural sequence diversity. Biophys J 124 (15):2531–2541. https://doi.org/10.1016/j.bpj.2025.06.034
22. Finn RD et al (2016) The Pfam protein families database: towards a more sustainable future. Nucleic Acids Res 44(D1):D279–D285. https://doi.org/10.1093/nar/gkv1344
23. Paysan-Lafosse T et al (2023) InterPro in 2022. Nucleic Acids Res 51(D1):D418–D427. https://doi.org/10.1093/nar/gkac993
24. Leinonen R, Diez FG, Binns D, Fleischmann W, Lopez R, Apweiler R (2004) UniProt archive. Bioinformatics 20(17): 3236–3237. https://doi.org/10.1093/bioinformatics/bth191
25. Prakash A, Jeffryes M, Bateman A, Finn RD (2017) The HMMER web server for protein sequence similarity search. Curr Protoc Bioinforma 60(1). https://doi.org/10.1002/cpbi.40
26. Li W, Godzik A (2006) Cd-hit: a fast program for clustering and comparing large sets of protein or nucleotide sequences. Bioinformatics 22(13):1658–1659. https://doi.org/10.1093/bioinformatics/btl158
27. Haldane A, Levy RM (2019) Influence of multiple-sequence-alignment depth on Potts

statistical models of protein covariation. Phys Rev E 99(3):032405. https://doi.org/10.1103/PhysRevE.99.032405

28. Ferreiro DU, Walczak AM, Komives EA, Wolynes PG (2008) The energy landscapes of repeat-containing proteins: topology, cooperativity, and the folding funnels of one-dimensional architectures. PLoS Comput Biol 4(5). https://doi.org/10.1371/journal.pcbi.1000070
29. Galpern EA, Marchi J, Mora T, Walczak AM, Ferreiro DU (2022) Evolution and folding of repeat proteins. Proc Natl Acad Sci 119(31): e2204131119. https://doi.org/10.1073/pnas.2204131119
30. Figliuzzi M, Barrat-Charlaix P, Weigt M (2018) How pairwise coevolutionary models capture the collective residue variability in proteins? Mol Biol Evol 35(4):1018–1027. https://doi.org/10.1093/molbev/msy007
31. Marchi J, Galpern EA, Espada R, Ferreiro DU, Walczak AM, Mora T (2019) Size and structure of the sequence space of repeat proteins. PLoS Comput Biol 15(8):1–23. https://doi.org/10.1371/journal.pcbi.1007282
32. Ekeberg M, Lövkvist C, Lan Y, Weigt M, Aurell E (2013) Improved contact prediction in proteins: using pseudolikelihoods to infer Potts models. Phys Rev E Stat Nonlinear Soft Matter Phys 87(1):1–16. https://doi.org/10.1103/PhysRevE.87.012707
33. Cocco S, Feinauer C, Figliuzzi M, Monasson R, Weigt M (2018) Inverse statistical physics of protein sequences: a key issues review. Rep Prog Phys 81(3). https://doi.org/10.1088/1361-6633/aa9965
34. Morcos F, Schafer NP, Cheng RR, Onuchic JN, Wolynes PG (2014) Coevolutionary information, protein folding landscapes, and the thermodynamics of natural selection. Proc Natl Acad Sci 111(34):12408–12413. https://doi.org/10.1073/pnas.1413575111
35. Miyazawa S (2017) Selection originating from protein stability/foldability: relationships between protein folding free energy, sequence ensemble, and fitness. J Theor Biol 433:21–38. https://doi.org/10.1016/j.jtbi.2017.08.018
36. Freiberger MI et al (2023) Local energetic frustration conservation in protein families and superfamilies. Nat Commun 14(1):8379. https://doi.org/10.1038/s41467-023-43801-2
37. Tubiana J, Cocco S, Monasson R (2019) Learning protein constitutive motifs from sequence data. elife 8. https://doi.org/10.7554/eLife.39397
38. Lin Z et al (2023) Evolutionary-scale prediction of atomic-level protein structure with a language model. Science 379(6637): 1123–1130. https://doi.org/10.1126/science.ade2574
39. Zhang Z, Wayment-Steele HK, Brixi G, Wang H, Kern D, Ovchinnikov S (2024) Protein language models learn evolutionary statistics of interacting sequence motifs. Proc Natl Acad Sci 121(45):e2406285121. https://doi.org/10.1073/pnas.2406285121

Chapter 3

Predictive Model for Single-Site Mutations That Change Intrinsically Disordered Protein Binding Energetics

Soumyanetra Chandra and Raghavan Varadarajan

Abstract

Intrinsically disordered proteins (IDPs) are characterized by structural mobility and lack of stable tertiary structure. These are important for IDP functions and interactions with other protein partners. IDP binding modes are distinctly divergent from globular protein–protein interactions. Comprehensive studies on effects of mutations in this structurally distinct protein class of IDPs are limited, impeding the development of accurate prediction methods for mutational effects in IDPs. Understanding how mutations affect IDP binding affinity is also crucial to peptide drug design, therapeutics engineering, and research on IDP diseases. Previous deep mutational scanning study of intrinsically disordered domain of control of cell death protein A (CcdA) shed light on residue-specific contribution to cognate partner binding energetics and outlined methodology for building IDP—compatible prediction models of binding affinity. The current chapter summarizes our previously described methodology [1] (*see* **Notes 1, 4–7**) for development of a linear regression-based predictive model, *I*ntrinsically disordered *P*rotein *A*ffinity *pred*iction (IPApred). Our method incorporates structural information at the residue level as well as experimentally derived mutational penalties while leveraging the distinct physicochemical properties of different amino acids to predict how single-site substitutions affect partner binding in IDPs. We describe methods for data generation and collection, and the steps of selecting features, training a simple mathematical model, outlining a protocol for using IPApred, and finally discussing validation and evaluation of the model. The chapter also highlights current limitations of the model and challenges faced in modeling IDP energetics and discusses possible strategies of addressing them. Despite drawbacks, IPApred provides a useful framework for predicting mutational effects in disordered proteins and for improved understanding of related biological processes and disease mechanisms.

Key words Affinity prediction, Protein–protein interaction, Computational model, Penalty, Residue depth, Deep mutational scanning, Next-generation sequencing data

1 Introduction

Intrinsically disordered proteins (IDPs) lack stable canonical three-dimensional structures and remain natively unfolded despite having crucial roles in cellular processes. The intrinsic disorder in such proteins is known to help maintain a flexible interface that can readily bind more than one partner with high specificity [2, 3] as

Shahid M. Khan and Florencio Pazos (eds.), *Protein Evolution: Methods and Protocols*, Methods in Molecular Biology, vol. 2979,
https://doi.org/10.1007/978-1-0716-4828-5_3,

well as facilitate efficient regulation of protein levels through targeted degradation. IDPs are believed to interact with partners in a manner coupled to folding, through an induced-fit mechanism [4, 5]. Owing to their unfolded nature, IDPs can facilitate protein–protein interactions (PPIs) efficiently through a much smaller number of residues and interactions that are predominantly hydrophobic in nature, covering a larger interface area in comparison to the well-folded globular proteins [6]. PPIs play a key role in all biological functions including metabolic pathways, immune responses, signal transduction, etc. Mutations in proteins involved in such pathways can alter partner interactions leading to drastic downstream consequences. It is therefore essential to comprehend and predict mutational effects on such protein–protein interactions. However, experimental determination of binding affinity of a large number of protein mutants to their partners is laborious and cumbersome. This necessitates the development of rapid and reliable high-throughput computational tools to predict affinity changes upon mutations.

Several computational predictors of binding affinity changes use regression models trained on structure and sequence features of protein complexes, while others use machine learning approaches trained on experimental data. FoldX [7] and a Rosetta-based approach [8] are examples of the early binding affinity change predictors that are based primarily on physical bond energies and solvation contributions. BeAtMuSiC [9], on the other hand, is a coarse-grained predictor of mutational effects on binding affinity, based on structure-derived statistical energy. The machine learning tool BindProfX [10] estimates affinity changes using interface profile scores derived from multiple structural alignments combined with FoldX physical potentials. The mutation Cutoff Scanning Matrix for Protein–Protein Interaction (mCSM-PPI) [11] uses machine learning to train on graph-based signatures describing interatomic distances and the atomic environment of residues, to predict effects of missense mutations. MutaBind [12] is another machine learning-based predictor of affinity perturbations that relies on various mechanical force fields, energy potentials, and side-chain optimization algorithms derived from structural and evolutionary conservation features. When predictions were tested for mutants of the intrinsically disordered segment of the bacterial antitoxin control of cell death protein A (CcdA), many of these methods fell short in accurately predicting mutational effects on binding affinity toward its partner protein, control of cell death protein B (CcdB) [1]. We suspect that this inaccuracy might be a result of bias in training datasets consisting primarily of experimental data on well-structured globular proteins. Such shortcomings are underemphasized in the literature. This is surprising, given the significant proportion of disordered proteins constituting the proteome and interactome. It has long been recognized that at least 10–20% of all eukaryotic proteins are either disordered or have disordered regions [13]. A prediction study showed that 79% of cancer-associated proteins contain predicted disordered regions [14]. A broad range of human diseases including cancer, diabetes, Alzheimer's disease, and several neurodegenerative and cardiovascular diseases have been associated with IDPs.

We have previously carried out a quantitative investigation of the mutational effects of all possible single-site residue substitutions on partner binding in intrinsically disordered CcdA protein using a facile, high-throughput approach [1]. The CcdA deep mutational scanning (DMS) data show a clear distinction between the IDP mutational landscape and that of globular proteins. Single-point mutations were observed to have significantly milder energetic effects on partner binding in the IDP CcdA, relative to those usually observed in globular protein mutants [1, 15]. The well-studied tolerance patterns in globular protein interfaces are different from those in the IDP CcdA emphasizing the need to develop an IDP-focused predictor of mutational effects on binding energetics. Accurate prediction of mutational effects on partner binding and function of intrinsically disordered proteins or regions is essential for elucidating IDP binding modes, uncovering IDP-related disease mechanisms, discovering drugs for IDP mutant diseases, and developing peptide drugs as therapeutics. Based on exhaustive deep mutational scanning data of binding affinity changes in the intrinsically disordered protein CcdA, we have developed a simplistic prediction approach for this, *I*ntrinsically disordered *P*rotein *A*ffinity *pred*iction (IPApred).

Previous work on globular proteins in our group showed a simple additive model incorporating category and residue-specific penalties to perform well in predicting mutational effects [15]. Category penalty describes the average mutational effect for substitution of a particular class of residue to another, while residue penalty is the average mutational effect of each substituted residue. However, this approach performed poorly and hugely overestimated mutational effects on the binding affinity changes in the CcdA variants. Upon testing different versions of our model, we observed that removal of residue-specific penalties produces better correlations between the predicted results and experimental data in the CcdA mutational dataset. Analysis of the CcdA mutational landscape and studying mutational effects on CcdB binding energetics pointed to the need to distinguish between interacting and noninteracting residues in the IDP for building a binding affinity predictor. The average mutational tolerance of residue positions in CcdA correlated well with the residue depth (RD) parameter calculated in CcdA–CcdB complex. This can be calculated from available crystal structures or reliable structural models of the query protein bound to its protein partner [16]. Residue depth is the depth of a residue averaged over all atoms in the residue, from the nearest surface water molecule [17, 18] (*see* **Note 2**). The strong correlation between residue depth and positional mutational tolerance enabled us to create an internal standard curve which allows estimation of overall sensitivity to mutation (positional predicted mutational tolerance score [MTS_{pred}]) for a residue position using the residue depth parameter [1]. To further improve our predictions and to introduce mutation-specific features into the predictor, we used weighted mutational category penalties in our model (*see* Fig. 1). Unsurprisingly, the mutational effects on the partner binding energetics depend on the extent of contacts the residue makes with the partner protein. Therefore, use of an identical, averaged measure of penalty to predict consequences of substitutions at different positions yields erroneous results. For this purpose, mutational penalties were weighted using a parameter related to the residue depth which is a reliable descriptor of the structural contacts each residue forms with its protein partner at the binding interface. The residue depth was normalized to obtain a suitable scaling factor that can be utilized to specify higher penalties at positions with high-residue burial upon binding, and lower penalties at positions displaying lower-residue burial in the partner-bound complex structure. The CcdA (C-terminal CcdB binding domain) DMS data were split into training and test sets for building the prediction model and validating it, respectively (*see* Fig. 1). After testing several normalizations of the residue depth, the best performing mathematical formula was taken forward and further refined using the training set to build the IPApred model [1].

IPApred provides predicted mutational tolerance score (MTS_{pred}) for each mutant of an IDP which serves as a proxy for its binding affinity to the partner protein. This method provides not only arbitrary scores that describe the binding affinity changes upon mutations in IDPs but also allows estimation of change in Gibbs free energy of binding ($\Delta\Delta G^0{}_{bind}$) upon mutations in protein-interacting disordered regions. We have previously reported the validation of IPApred on CcdA mutants and in other IDPs [1]. Briefly, MTS_{pred} scores calculated using IPApred for all CcdA mutants present in the test dataset correlated well ($r = 0.65$) with the experimental mutational tolerance Score from sequencing (MTS_{seq}) values [1]. The ΔMTS_{pred} (= $MTS_{pred}{}^{mutant}$ − $MTS_{pred}{}^{WT}$) values for CcdA mutants also correlate well ($r = -0.65$) with the experimental $\Delta\Delta G°_{bind}$ values for 14 single-site CcdA mutants, further validating IPApred model [1]. IPApred performed better when compared to other algorithms such as BeAtMuSiC and mCSM-PPI [1, 9, 27].

IPApred methodology highlights how deep mutational scanning data can be utilized to build a simple, IDP-focused prediction algorithm of binding energetics that outperforms well-tested, available methods of prediction. Methodology development in the field of IDP function prediction is indispensable to accurately predict

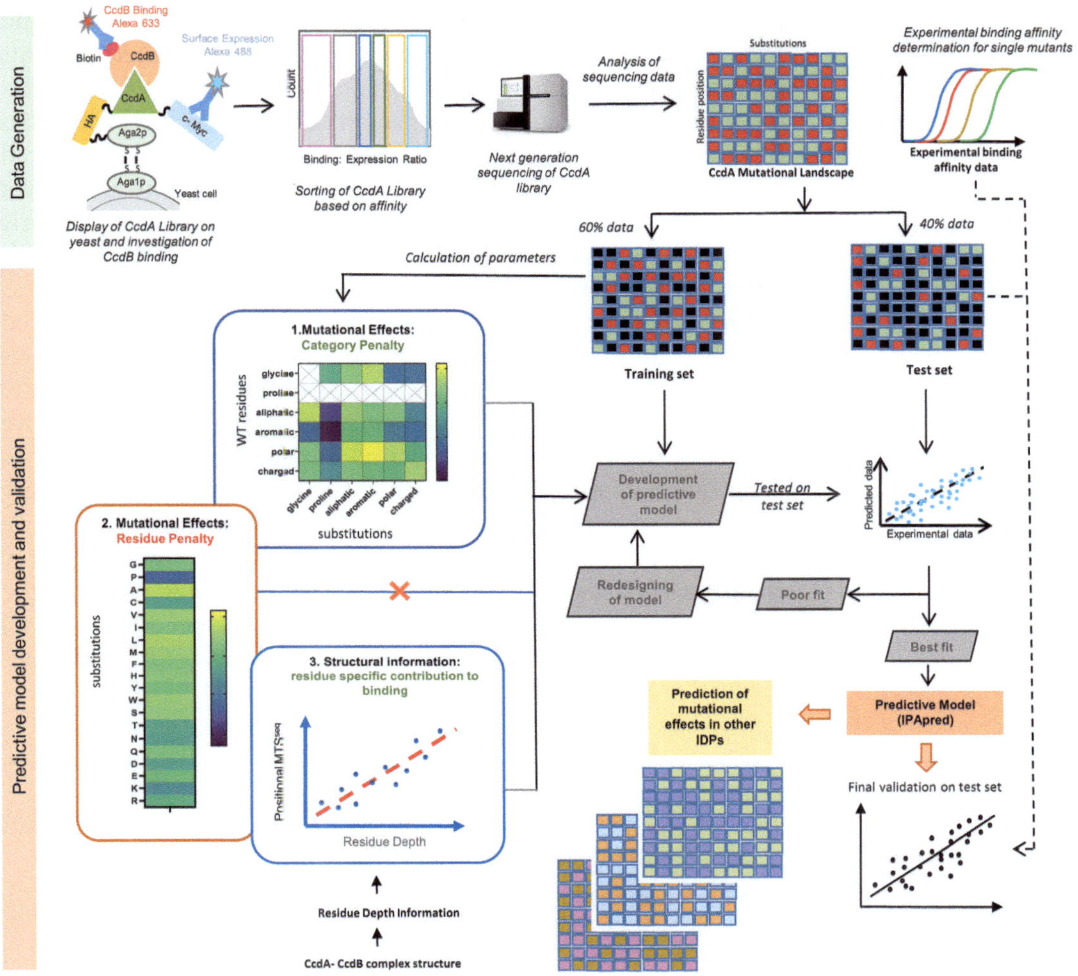

Fig. 1 Schematic describing data generation, intrinsically disordered protein affinity prediction (IPApred) model development, and refinement. Control of cell death protein A (CcdA) single-site mutant library was displayed on the yeast surface and binding to cognate partner, control of cell death protein B (CcdB) was investigated. The ratio of CcdB binding:CcdA expression (using C-terminal c-Myc tag) was used for fluorescence-activated cell sorting (FACS) of the library into different populations with varying binding affinity to CcdB. DNA from these populations was subjected to next-generation sequencing, and the data were analyzed to obtain the complete mutational landscape of CcdA. A small set of individual single mutants of CcdA was subjected to yeast surface display and CcdB titrations to quantify binding affinity parameters K_d and $\Delta G°_{bind}$. For developing IPApred, CcdA (C-terminal domain) mutational data were split into training (60%) and test (40%) sets. The training set was used to derive the category penalties and residue-specific penalties. The residue-specific penalties, however, were found to be redundant for prediction of mutational effects on binding, upon testing on the test set. The observed good agreement of residue depth parameter with mutational effects in the CcdA deep mutational scanning (DMS) data led to incorporation of the complex structure-derived residue depth parameter into the model. Upon several rounds of testing and continuous refinements to improve the predictions (as detailed in Methods), we obtained the best performing model for validation in CcdA test set and subsequently used it to predict mutational effects on partner binding affinity in other intrinsically disordered proteins (IDPs)

mutational effects in disease-related IDPs and better understand how the altered binding energetics of disease-associated mutants give rise to disease phenotypes. A representative example is predicting mutational effects on tumor suppressor protein p53, which has functionally important C- and N-terminal intrinsically disordered regions and where mutations in these disordered regions are known to disrupt interactions ultimately contributing to cancer. Correctly predicting mutational effects on partner binding in IDPs also facilitates peptide drug development. Many peptides developed as drugs are structureless in free form and may form structured stretches upon partner binding. Screening techniques such as phage display, yeast surface display, ribosome display, or affinity selection mass spectrometry are currently employed for extensive screening of better binding mutants of peptides in the process of drug optimization. Establishment of an accurate binding affinity predictor will streamline the screening process, thereby reducing the lengthy drug development timeline, which is one of the key challenges in drug discovery pipelines. In this chapter, we detail the methodology for developing and validating the predictive model and also discuss the model's performance as well as future avenues for further improvement of the predictor.

2 Materials

2.1 Yeast Surface Display

- *Saccharomyces cerevisiae* EBY100.
- pETcon plasmid (Addgene 41,522).
- NdeI (NEB) and XhoI (NEB) enzymes.
- Synthetic Dextrose CasAmino Acid media (SDCAA) (glucose 20 g/L, yeast nitrogen base 6.7 g/L, casamino acid 5 g/L, citrate 4.3 g/L, and sodium citrate dihydrate 14.3 g/L).
- Synthetic Galactose CasAmino Acid media (SGCAA) (galactose 20 g/L, yeast nitrogen base 6.7 g/L, casamino acid 5 g/L, citrate 4.3 g/L, and sodium citrate dihydrate 14.3 g/L).
- Streptavidin-conjugated Alexa Fluor 633 (Invitrogen).
- Chicken anti-cellular myelocytomatosis oncogene (anti-c-Myc) antibody (Invitrogen).
- Anti-chicken antibody conjugated to Alexa Fluor 488 (Invitrogen).
- BD FACSAria III instrument (BD Biosciences).
- EZ-Link™ Sulfo-NHS-SS-Biotin (Thermo Scientific).

2.2 IPApred Model Building and Validation

2.2.1 Protein Data Bank Identifier (PDB ID) of Used Available Structures

The available structures of CcdA–CcdB complex (PDB ID: 3G7Z) [20], cellular Myeloblastosis proto-oncogene protein – CREB-binding protein (c-Myb–CBP) complex (PDB ID: 1SB0) [21], Activator for Thyroid hormone and Retinoid receptors–Nuclear Coactivator Binding Domain (ACTR–NCBD) complex (PDB ID: 1KBH) [22], and Hypoxia-Inducible Factor 1-alpha–CREB-binding protein (HIF-1α–CBP) complex (PDB ID: 1L8C) [23] were analyzed for building and validating the IPApred model..

2.2.2 Calculation of Residue Depth

Residue depth of all positions in the proteins was calculated using the listed complex PDB IDs as inputs from the available DEPTH web server [16–18]. The residue depth (Å), averaged over all atoms for each residue position in the complex structure, was used as input for the predictive model, IPApred.

2.2.3 Mutational Data

Single-mutant experimental binding affinity data were collected for mutants in the disordered c-Myb binding to Kinase-Inducible domain (KIX) of CBP [24], mutants in both partners, disordered activation domain of ACTR and disordered NCBD domain of CBP, binding to each other [25], and mutants in the disordered C-terminal activation domain of HIF-1α involved in binding to transcriptional adaptor zinc-binding 1 (TAZ1) domain of CBP [26].

2.2.4 Available Prediction Methods Used for Comparison

The available mCSM-PPI [27], BeAtMuSiC [9], and Mutabind2 [12] algorithms were used for predicting CcdA mutational effects to test these methods and compare them to IPApred performance.

3 Methods

3.1 Data Generation

3.1.1 Deep Mutational Scanning Data

The mutational tolerance data on the effect of all single-site mutations of CcdA on its binding to CcdB were generated in our yeast surface display-based deep mutational scanning study [1]. Briefly, a single-site saturation mutation library of CcdA was constructed by substituting each residue position with NNK codons (N = any bases and K = G, T bases). The library was then cloned in the yeast surface display plasmid pETcon between Nde1 and Xho1 sites and transformed using a high efficiency lithium acetate/single-stranded deoxyribonucleic acid/polyethylene glycol (LiAc/ssDNA/PEG) chemical transformation method [19] in *S. cerevisiae* EBY100 strain. The CcdA proteins had genetically fused N-terminal HA (hemagglutinin) tag and a C-terminal c-Myc (cellular myelocytomatosis proto-oncogene epitope) tag. CcdA proteins were expressed on the yeast cell surface as a C-terminal fusion to Aga-2p (agglutinin subunit 2 protein) by growing at 30 °C, 250 rpm in SDCAA media till optical density (OD) 2 is reached and then inducing in SGCAA media containing

2% galactose at 30 °C, 250 rpm for 16 h. Yeast cells displaying the CcdA library were subjected to binding to 2 nM-biotinylated CcdB and then labeled with streptavidin-conjugated Alexa Fluor 633 (1: 2000) to probe binding. In parallel, the yeast surface-displayed CcdA library was also labeled with chicken anti-c-Myc antibody (1:400) and anti-chicken antibody conjugated to Alexa Fluor 488 (1:300) to measure the level of CcdA expression on yeast surface. CcdB protein was purified [15] and biotinylated using EZ-Link™ Sulfo-NHS-SS-Biotin following manufacturer's protocol. The binding-to-expression ratio was analyzed using flow cytometry, and the CcdA library was sorted into seven gates with varying binding: expression ratio on a BD FACSAria III instrument [1]. The *ccdA* gene, PCR (polymerase chain reaction) amplified from each gate individually with gate-specific barcoded primers, was subjected to next-generation (amplicon) sequencing using the Illumina NovaSeq 6000 150 PE Platform (Macrogen, South Korea). The sequencing reads were analyzed using a custom-built pipeline developed in-house and were used to estimate the binding/expression mean fluorescence intensity ratio ($\mathrm{MFI}^{\mathrm{ratio}}$):

$$\int \mathrm{MFI}^{\mathrm{ratio}}_{\;x} = \Sigma_i[(D^x{}_i) \times F_i] \tag{1}$$

where x and i represent mutant and gate identity, respectively; F_i is the mean fluorescence for gate i, and $D^x{}_i$ is the normalized fraction of a mutant x in gate i, thus accurately representing the frequency distribution of mutants across gates [1].

A mutational tolerance score derived from deep sequencing, $\mathrm{MTS}_{\mathrm{seq}}$, was assigned to each mutant using the formula below:

$$\mathrm{MTS}_{\mathrm{seq}} = \mathrm{MFI}^{\mathrm{ratio}}\ \text{of mutant}/\mathrm{MFI}^{\mathrm{ratio}}\ \text{of WT} \tag{2}$$

3.1.2 Experimental Determination of Binding Affinity of CcdA Single Mutants

WT CcdA and 20 single-site mutants of CcdA were cloned in a pETcon vector in EBY100. The WT and mutant CcdA proteins were individually expressed on the yeast surface and bound to varying concentrations (0.01 pM to 200 nM) of biotinylated CcdB and subsequently to streptavidin-conjugated Alexa Fluor 633 (1:2000) before flow cytometry analysis. The mean binding fluorescence intensity data were plotted as a function of CcdB concentration, and the data were fit to a single-site ligand binding equation:

$$y = x\ B_{\mathrm{max}}/(x + K_{\mathrm{d}}) \tag{3}$$

where x is the concentration of CcdB, y is the fluorescence intensity, B_{max} is the saturated fluorescence intensity, and K_d is the dissociation constant of binding.

The Gibbs free energy of binding ($\Delta G^{\circ}_{\mathrm{bind}}$) was calculated using the relationship $\Delta G^{\circ}_{\mathrm{bind}} = \mathrm{RT}\ \ln K_{\mathrm{d}}$, where R (ideal gas constant) = 1.9872 * 10^{-3} kcal K^{-1} mol^{-1} and T temperature) = 298.15 K. $\Delta\Delta G^{\circ}_{bind}$ (= $\Delta G^{\circ}_{bind}{}^{mutant}$ - $\Delta G^{\circ}_{bind}{}^{WT}$) values were also calculated using single mutant and WT CcdA values.

3.2 Development of a Predictive Model IPApred

3.2.1 Model Selection

Through systematic analysis of the CcdA mutation dataset, a number of factors including chemistry of wild-type and mutated residue side chains, position of residues in the protein sequence, and extent of interaction of a particular CcdA residue with its cognate partner CcdB in the known CcdA–CcdB complex structure were identified to play an important role in determining the magnitude of mutational effect on CcdB binding. Linear regression models were used to determine the exact relationship among parameters like residue depth, mutational effects, and category penalties. Based on such relationships, a simple predictive empirical model [1] was built to estimate binding affinity changes upon mutation (*see* Fig. 1).

3.2.2 Building a Training Set

The CcdA mutational screening dataset of residues 40–72 comprising the CcdB binding C-terminal domain was split into training (60%) and validation (40%) sets by random sampling method.

3.2.3 Predicting Overall Mutational Effects for Each Residue Position

Analysis of the CcdA dataset clearly showed dependence of overall mutational sensitivity at each residue position on the extent of interaction with partner protein CcdB. After investigation of the relationship of positional mutational tolerance with structure-related parameters like accessible surface area, residue depth, and number of hydrogen (H) bonds and nonbonded interactions, it was found that positional mutational tolerance in the CcdA training set correlates best with structure-derived residue depth. This was used to establish a standard curve for predicting overall mutational effects for each residue position.

$$\text{Positional MTS}_{pred} = 1.315 - 0.0934 \text{ residue depth} \quad (4)$$

3.2.4 Calculation of Category Penalty

To account for mutational effects due to various physicochemical properties of amino acid substitutions, amino acids were classified into six classes, namely charged (D, E, K, and R), polar (S, T, N, and Q), aromatic (F, H, Y, and W), aliphatic (A, C, V, I, L, and M), glycine (G), and proline (P) [15]. Mutations, therefore, can be classified into 36 categories ($6 \times 6 = 36$), of which glycine→glycine and proline→proline substitutions result in wild-type sequence making a total of 34 (36–2 = 34) possible categories of substitutions. CcdA lacks proline in its wild-type sequence, and therefore the CcdA dataset lacks all five categories of substitutions from proline to other classes, making 29 (34–5 = 29) possible categories of CcdA mutations. Each mutation in the training set was assigned to one of the 29 categories, and a mutational penalty was determined for each category of mutation from the CcdA training dataset.

Table 1
Category penalties for different types of substitutions, calculated from the CcdA training set

	Substitutions					
WT residues	Glycine	Proline	Aliphatic	Aromatic	Polar	Charged
Glycine		−0.19357	−0.1016	−0.04086	−0.3409	−0.337
Aliphatic	−0.03267	−0.44349	−0.07483	−0.12905	−0.11374	−0.28175
Aromatic	−0.36094	−0.52549	−0.15981	−0.1578	−0.30887	−0.33961
Polar	−0.12158	−0.36342	−0.01657	0.017241	−0.04185	−0.18169
Charged	−0.1361	−0.27106	−0.12307	−0.15567	−0.16816	−0.05902

These data were generated in previous work on CcdA mutational data and IPApred model building [1]

A category-averaged mutational score (category-averaged MTS_{seq}) was calculated for each of the 29 categories of mutations by averaging the mutational score across all mutations assigned to a particular category.

$$\text{Category averaged } MTS_{seq}(a \rightarrow b) = (1/n)\sum\nolimits_n MTS_{seq}(a \rightarrow b) \tag{5}$$

Category penalties (*see* Table 1) for each of the 29 categories in CcdA were calculated by subtracting the WT MTS_{seq} from the category-averaged MTS_{seq}:

$$\text{Category penalty}(a \rightarrow b) = \text{category averaged } MTS_{seq}(a \rightarrow b)\text{-WT } MTS_{seq} \tag{6}$$

where the value of wild-type (WT) MTS_{seq} is 1.

3.2.5 Weighing of Category Penalty Using Residue Depth

To accurately predict mutational effects of interacting and noninteracting residues, a scaling factor related to the residue depth parameter was used to weigh the calculated penalties. Various scaling factors were tested, and the best results were obtained with a normalized residue depth (nRD):

$$nRD^j = RD^j / RD^{max} \tag{7}$$

where *j* is the residue position, RD^j is the residue depth of the wild-type residue at position *j*, and RD^{max} is the maximum residue depth observed in the protein segment of interest.

Through rigorous testing of the CcdA training dataset and validating on the test set, the best performing mathematical formula was found to be as follows:

$$\text{Predicted score}^j(a \rightarrow b) = \text{positional } MTS_{pred}{}^j + [2 \times nRDj \times \text{Category Penalty}(a \rightarrow b)] \tag{8}$$

The normalization factor of 2 in Eq. 8 was determined based on the best performance on the training set after testing several variations of the formula incorporating the three parameters: positional MTS_{pred}, residue depth (RD), and category penalties.

3.2.6 Correction of Predicted Score to Improve the Performance

The predicted scores were calculated for all CcdA mutants present in the training set using Eq. 8. These predicted scores correlated well with experimental values in the training set. While the well-correlated linear relationship between the predicted and experimental scores validated the method, it also allowed us to improve our prediction by deriving predicted mutational scores using the predicted score through this standard curve.

$$MTS_{pred}{}^{j}(a \rightarrow b) = 0.45 + \left[\text{Predicted score}^{j}(a \rightarrow b)\right] \times 0.62\right] \quad (9)$$

MTS_{pred} thus calculated provides a predicted mutational sensitivity score for any mutation at any residue position of CcdA or other IDPs.

3.2.7 Prediction of Quantitative Change in Binding Affinity

MTS_{pred} was calculated for CcdA mutants for which experimental binding affinity data had already been determined. Good agreement was observed between ΔMTS_{pred} (= $MTS_{pred}{}^{mutant}$ − $MTS_{pred}{}^{WT}$) and the experimental $\Delta\Delta G°_{bind}$ (= $\Delta G°_{bind}{}^{mutant}$ − $\Delta G°_{bind}{}^{WT}$) values for single-site CcdA mutants. This relationship was used as a standard curve, to derive $\Delta\Delta G°_{pred}$ values from the predicted MTS_{pred} values as follows:

$$\Delta\Delta G°{}_{pred}{}^{j}(a \rightarrow b) = -1.31 \times \Delta MTS_{pred}{}^{j}(a \rightarrow b) - 0.01 \quad (10)$$

where $\Delta\Delta G°_{pred}$ (= $\Delta G°_{pred}{}^{mutant}$—$\Delta G°_{pred}{}^{WT}$) is a predicted score obtained from Eq. 10, though the quantitative values of the individual parameters $\Delta G°_{pred}{}^{mutant}$ and $\Delta G°_{pred}{}^{WT}$ are not obtained from IPApred.

3.3 Validation of IPApred Model

3.3.1 Calculation of MTS_{pred} and $\Delta\Delta G_{pred}$

The residue depth was determined as described above for a residue position to be used in prediction. For each substitution, the category of substitution was identified, and respective penalty values previously determined were input into Eq. 9 alongside calculated residue depths to obtain the MTS_{pred}. $\Delta\Delta G°_{pred}$ was then determined using Eq. 10.

3.3.2 Test on CcdA Test Dataset

MTS_{pred} was calculated for all single mutants in the test set (40% of CcdA mutants), and these were correlated with the available MTS_{seq} deep mutational scanning experimental data. $\Delta\Delta G°_{pred}$ was also calculated for all these single mutants in the CcdA test set and then correlated with available MTS_{seq} deep mutational scanning experimental data. Correlation coefficient (r) was calculated for these correlations for investigation of the performance of IPApred.

ΔMTS_{pred} (= $MTS_{pred}{}^{mutant}$ – $MTS_{pred}{}^{WT}$) values were calculated using the IPApred model (*see* Eq. 9) for a subset of single-site mutants of CcdA for which K_d values and the $\Delta\Delta G°_{bind}$ were experimentally determined. Predicted ΔMTS_{pred} values were correlated with the experimental $\Delta\Delta G°_{bind}$ values for this small number of individual CcdA mutants to investigate the performance of IPApred and also to derive the standard curve for $\Delta\Delta G°_{pred}$ calculations (*see* Eq. 10).

3.3.3 Test on Other IDP Mutational Datasets

MTS_{pred} and $\Delta\Delta G^0{}_{pred}$ values were calculated using IPApred for single-site mutants of c-Myb, ACTR, disordered NCBD domain, and HIF-1α proteins for which experimental energetics data were available and previously published. These predicted ΔMTS_{pred} and $\Delta\Delta G^0{}_{pred}$ values were correlated with the experimental $\Delta\Delta G°_{bind}$ values to test the performance of IPApred.

3.4 Protocol for Using IPApred for Single Mutants of IDP

1. Locate the PDB ID for the structure of the IDP–partner complex.
2. Go to DEPTH server: https://cospi.iiserpune.ac.in/depth. Input the structure ID to generate residue depth information of each IDP residue (RD^j for residue j). Also find the residue in the IDP with the maximum value of residue depth in the complex, and assign this value to RD^{max}. Input RD^j and RD^{max} values into Eq. 11 in **step 4**.
3. Identify mutation category (a → b) of IDP mutation for which a prediction is to be made and input the specific category penalty (a → b) from Table 1 into Eq. 11 as described in **step 4**.
4. Calculate $MTS_{pred}{}^{j}$ (a → b) for each IDP mutant for residue position j and mutation category (a → b) following Eq. 11 (derived by combining Eqs. 4, 7, 8, and 9) as follows:

$$MTS_{pred}{}^{j}(a \rightarrow b) = 1.2653 + \{1.24 \times \text{Category Penalty}(a \rightarrow b) \times (RD^{j}/RD^{max})\} - \{0.0579 \times RD^{j}\} \quad (11)$$

5. Calculate $\Delta\Delta G°_{pred}{}^{j}$ (a → b) for each IDP mutant for residue position *j* and mutation category (a → b) following Eq. 10 and MTS_{pred} value obtained in **step 4**.

$$\Delta\Delta G°_{pred}{}^{j}(a \rightarrow b) = -1.31 \times \Delta MTS_{pred}{}^{j}(a \rightarrow b) - 0.01 \quad (12)$$

4 Notes

1. We observed that among various parameters analyzed, residue depth showed the highest correlation with average mutational effects in CcdB binding C-terminal region of CcdA. Other

structural parameters, like accessible surface area change upon binding and number of interactions with CcdB, also showed modest correlations to observed mutational effects, while poor correlation was observed for evolutionary conservation, helix propensity, and intrinsic disorder propensity.

2. While IPApred uses the DEPTH server, alternative methods for calculating residue depth can also be used where suitable [16–18].
3. The current method uses available experimental IDP–protein complex structures. In the absence of an available complex structure, a high-confidence model of the complex structure predicted by state-of-the-art deep learning methods like Alpha-Fold or RoseTTAFold may be used for residue depth calculation, though this has not been tested in the study and the prediction accuracy would be highly dependent on the accuracy of the predicted models [28, 29] (*see* **Note 3**).
4. The $\Delta\Delta G°_{bind}$ mutational effects predicted by available algorithms (namely mCSM-PPI, BeAtMuSiC, and Mutabind2) are significantly larger than those observed experimentally for CcdA. This highlights the drawbacks of using available methods for predicting affinity changes caused by mutations in IDPs like CcdA, where it is evident that individual mutational effects are less drastic than that commonly seen for globular protein mutants, likely because of the extended nature of the protein–protein interface.
5. The extent of mutational effects on interactions ($\Delta\Delta G°_{bind}$ values) appear to depend on the length of the IDP domain as well as the interface area of the PPI. While predicted MTSpred provides a reliable projection of binding affinity relative to WT, the $\Delta\Delta G°_{pred}$ values predicted by IPApred may in certain cases underestimate or overestimate binding energy values in IDPs with smaller or larger interfaces relative to CcdA–CcdB complex.
6. In case of availability of experimental $\Delta G°_{bind}$ values for a small set of mutants of IDP of interest, first the MTS_{pred} values can be predicted using Eq. 11 of IPApred, and then an independent standard curve may be derived from the relationship between the $\Delta\Delta G°_{bind}$ (mutant–WT) values and ΔMTS_{pred} (mutant–WT) values for the small set of mutants of IDP, providing a new equation $y = mx + c$, where y is $\Delta\Delta G°_{pred}$ and x is ΔMTS_{pred}. For mutants of interest for which no experimental $\Delta G°_{bind}$ values are available, MTS_{pred} values can be predicted using Eq. 11 of IPApred, and then, using the new equation, $\Delta\Delta G°_{pred}$ may be predicted instead of using Eq. 10 of IPApred.
7. The CcdA mutational dataset lacks mutants that significantly improve partner binding relative to wild-type sequence.

Therefore, IPApred currently trained solely on CcdA training set is unable to predict stabilizing mutational effects on binding energetics. The CcdA WT sequence lacks proline, and therefore the mutational dataset lacks information on proline-to-other-residue substitutions, limiting IPApred usage in case of mutations of wild-type proline in IDPs.

8. We believe significant improvement to IPApred can be made by incorporating a machine learning model trained on diverse IDP binding energetics landscapes.

Acknowledgments

We would like to acknowledge the funders who supported the original work (published in Chandra et al 2023, Protein Science) on which the current protocol paper is based on. The work was funded in part by a grant to RV from the Department of Biotechnology, grant number-BT/COE/34/SP15219/2015, DT.20/11/2015), Government of India, Science and Engineering Research Board (SERB), Government of India (grant number EMR/2017/0040S4 and SR/S2/JCB-10/2007 (JC Bose Fellowship))and grants from DST and MHRD. We also acknowledge funding for infrastructural support from the following programs of the Government of India: DST FIST, UGC Centre for Advanced study, Ministry of Human Resource Development (MHRD), and the DBT IISc Partnership Program. The funders had no role in study design, data collection and interpretation, or the decision to submit the work for publication.

References

1. Chandra S, Manjunath K, Asok A et al (2023) Mutational scan inferred binding energetics and structure in intrinsically disordered protein CcdA. Protein Sci 32:e4580
2. Dyson HJ, Wright PE (2005) Intrinsically unstructured proteins and their functions. Nat Rev Mol Cell Biol 6:197–208
3. Uversky VN (2002) Natively unfolded proteins: a point where biology waits for physics. Protein Sci 11:739–756
4. Sugase K, Dyson HJ, Wright PE (2007) Mechanism of coupled folding and binding of an intrinsically disordered protein. Nature 447: 1021–1025
5. Arai M (2018) Unified understanding of folding and binding mechanisms of globular and intrinsically disordered proteins. Biophys Rev 10:163–181
6. Dyson HJ (2016) Making sense of intrinsically disordered proteins. Biophys J 110:1013–1016
7. Guerois R, Nielsen JE, Serrano L (2002) Predicting changes in the stability of proteins and protein complexes: a study of more than 1000 mutations. J Mol Biol 320:369–387
8. Kortemme T, Baker D (2002) A simple physical model for binding energy hot spots in protein–protein complexes. Proc Natl Acad Sci 99:14116–14121
9. Dehouck Y, Kwasigroch JM, Rooman M et al (2013) BeAtMuSiC: prediction of changes in protein-protein binding affinity on mutations. Nucleic Acids Res 41:W333–W339
10. Xiong P, Zhang C, Zheng W et al (2017) BindProfX: assessing mutation-induced binding affinity change by protein Interface profiles with pseudo-counts. J Mol Biol 429:426–434

11. Pires DE V, Ascher DB, Blundell TL (2014) mCSM: predicting the effects of mutations in proteins using graph-based signatures. Bioinformatics 30:335–342
12. Li M, Simonetti FL, Goncearenco A et al (2016) MutaBind estimates and interprets the effects of sequence variants on protein-protein interactions. Nucleic Acids Res 44:W494–W501
13. Xue B, Dunker AK, Uversky VN (2012) Orderly order in protein intrinsic disorder distribution: disorder in 3500 proteomes from viruses and the three domains of life. J Biomol Struct Dyn 30:137–149
14. Iakoucheva LM, Brown CJ, Lawson JD et al (2002) Intrinsic disorder in cell-signaling and cancer-associated proteins. J Mol Biol 323: 573–584
15. Tripathi A, Gupta K, Khare S et al (2016) Molecular determinants of mutant phenotypes, inferred from saturation mutagenesis data. Mol Biol Evol 33:2960–2975
16. Tan KP, Nguyen TB, Patel S et al (2013) Depth: a web server to compute depth, cavity sizes, detect potential small-molecule ligand-binding cavities and predict the pKa of ionizable residues in proteins. Nucleic Acids Res 41: W314–W321
17. Chakravarty S, Varadarajan R (1999) Residue depth: a novel parameter for the analysis of protein structure and stability. Structure 7: 723–732
18. Tan KP, Varadarajan R, Madhusudhan MS (2011) DEPTH: a web server to compute depth and predict small-molecule binding cavities in proteins. Nucleic Acids Res 39:W242–W248
19. Gietz RD, Schiestl RH (2008) High-efficiency yeast transformation using the LiAc/SS carrier DNA/PEG method. Nat Protoc 2:31–35
20. De Jonge N, Garcia-Pino A, Buts L et al (2009) Rejuvenation of CcdB-poisoned gyrase by an intrinsically disordered protein domain. Mol Cell 35:154–163
21. Zor T, De Guzman RN, Dyson HJ et al (2004) Solution structure of the KIX domain of CBP bound to the transactivation domain of c-Myb. J Mol Biol 337:521–534
22. Demarest SJ, Martinez-Yamout M, Chung J et al (2002) Mutual synergistic folding in recruitment of CBP/p300 by p160 nuclear receptor coactivators. Nature 415:549–553
23. Dames SA, Martinez-Yamout M, De Guzman RN et al (2002) Structural basis for Hif-1 alpha /CBP recognition in the cellular hypoxic response. Proc Natl Acad Sci USA 99:5271–5276
24. Giri R, Morrone A, Toto A et al (2013) Structure of the transition state for the binding of c-Myb and KIX highlights an unexpected order for a disordered system. Proc Natl Acad Sci 110:14942–14947
25. Dogan J, Mu X, Engström Å et al (2013) The transition state structure for coupled binding and folding of disordered protein domains. Sci Rep 3:2076
26. Lindström I, Andersson E, Dogan J (2018) The transition state structure for binding between TAZ1 of CBP and the disordered Hif-1α CAD. Sci Rep 8:7872
27. Rodrigues CHM, Myung Y, Pires DE V et al (2019) mCSM-PPI2: predicting the effects of mutations on protein–protein interactions. Nucleic Acids Res 47:W338–W344
28. Abramson J, Adler J et al (2024) Accurate structure prediction of biomolecular interactions with AlphaFold 3. Nature 630(8016): 493–500
29. Baek M, DiMaio F et al (2021) Accurate prediction of protein structures and interactions using a three-track neural network. Science (New York, N.Y.) 373(6557):871–876. https://doi.org/10.1126/science.abj8754

Chapter 4

Detection of Specificity-Determining Positions in Protein Families

Borja Pitarch, David Juan, and Florencio Pazos

Abstract

The determination of a protein's functional residues is key for understanding its function at the molecular level and devising ways of modifying it to our benefit. A plethora of computational methods exist for predicting these sites, and most of them make use of the information contained in a multiple sequence alignment of homologous proteins. Nowadays, these approaches are mature enough, both in terms of reliability and ease of use, to become part of the toolbox of any biologist. This chapter describes a simple protocol for generating a multiple sequence alignment for a protein of interest and detecting different types of functional positions on it, all using freely available web and desktop tools with graphical user interfaces.

Key words Multiple sequence alignment, Functional residue, Specificity-determining position, Protein family, Protein function

1 Introduction

Nowadays, it is relatively cheap and fast to obtain the amino acid sequences of large collections of proteins and entire proteomes by sequencing and translating the corresponding genomic sequences [1]. Metagenomic projects (the organism-independent sequencing of deoxyribonucleic acid [DNA] in environmental samples) are also increasing the number of sequenced proteins and protein fragments [2]. A different issue is the experimental characterization of these proteins, either at the whole-protein level (i.e., to determine protein molecular function and biological role) or at the residue level (functional residues). Experimental methods for inferring functional features cannot cope with the deluge of sequenced proteins, as they are expensive and time-consuming. As a result, most proteins characterized at the sequence level lack associated experimental information [3].

Knowing which residues of a protein are functional is important not only for understanding the protein's function at the

Shahid M. Khan and Florencio Pazos (eds.), *Protein Evolution: Methods and Protocols*, Methods in Molecular Biology, vol. 2979,
https://doi.org/10.1007/978-1-0716-4828-5_4,

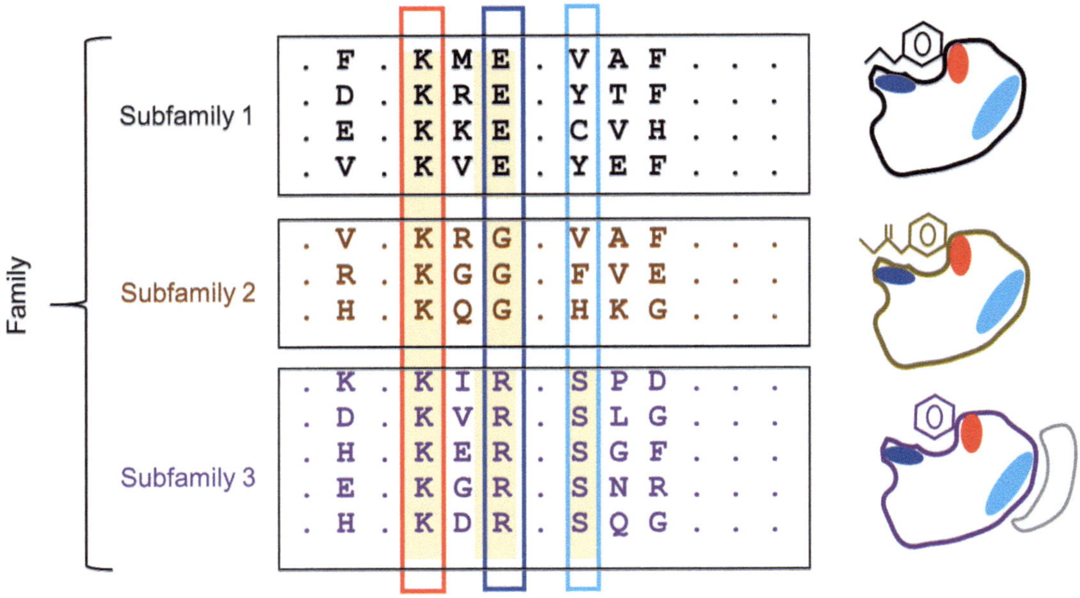

Fig. 1 Some functional features extracted from multiple sequence alignments (MSA). A schematic MSA with three subfamilies is depicted. Conservation (highlighted in yellow) is apparent in different forms: full conservation (red) and family-dependent conservation (dark and light blue). The type of protein regions that map these types of conservation patterns are also illustrated on the left

molecular level, but also to conceive ways for modifying that function to our benefit (i.e., for biomedical or biotechnological purposes). This importance, together with the difficulty in determining them experimentally, prompted the development of computational approaches for the in silico prediction of protein residues and regions involved in function. These computational methods are usually combined with experimental approaches, generally restricting the number of candidates to test (mutate). There are many illustrative examples of this interplay between computational and experimental methods, where in silico approaches generate a set of candidates that are subsequently mutated to check their involvement in function or, more interestingly, to fine-tune protein functions or design new ones (*see* examples in [4]).

Although the landscape of approaches for predicting protein functional sites is large and very diverse, most methods take advantage of the sequence deluge commented earlier to compile evolutionary information for the target protein by retrieving homologous sequences and aligning them in a multiple sequence alignment (MSA) (*see* **Note 1**). A position (pos.) of an MSA (column in its standard layout, *see* Fig. 1) represents equivalent residues in the set of homologous proteins from a structural, evolutionary, and functional point of view [5]. Hence, an MSA position can be regarded as a representation of the amino acids allowed by the evolution at that equivalent residue of the homologous proteins (*see* **Note 2**). For this reason, MSAs are rich sources of structural and functional information at the residue level [6].

While the most obvious function-related feature in an MSA is full conservation, there are other patterns of sequence variation that point to particular functional aspects (*see* Fig. 1). When the MSA of homologous proteins can be subdivided into subfamilies with subtle variations of the common function of the family (e.g., different substrate specificities), a position variation pattern emerges: Some positions are differentially conserved in these subfamilies (either being conserved only in certain subfamilies or the conserved amino acid being different) (*see* Fig. 1). These positions are called "specificity-determining positions" (SDPs) as they are generally involved in that functional aspect [6]. Detecting these positions is important because they complement fully conserved positions as predictors of functionality. But, more importantly, these positions are key for understanding functional specificity, and by playing with them, it is possible to change that specificity and fine-tune protein functional aspects [4]. There are many different methodologies for detecting SDPs in MSAs (for recent reviews, *see* [3, 7]).

In this chapter, we describe a step-by-step protocol for detecting SDPs in a family of interest, starting from a single representative sequence. The protocol involves web-accessible tools and desktop software with user graphical interfaces freely available for all operating systems.

2 Methods

The protocol described here uses web-accessible tools and desktop software with user graphical interfaces freely available for all operating systems. The protocol is instantiated here for the human Harvey rat sarcoma viral oncogene homolog (HRas) protein (UniProt accession P01112) as an example, whose sequence is available at https://rest.uniprot.org/uniprotkb/P01112.fasta.

2.1 Homolog Retrieval

The first step is to look for sequences homologous to that of interest in protein databases. For that, we use the basic local alignment search tool (*BLAST*) [8] which can scan large collections of proteins for sequence matches against one of interest in a very fast way. We use it through a web interface available at the National Center for Biotechnology Information: https://blast.ncbi.nlm.nih.gov/ (*see* **Note 3**). On the main page, we select "Protein BLAST."

In the input form, we paste the sequence of interest in the "Enter query sequence" box and select "UniProt/Swissprot" as "Database." We expand "Algorithm parameters" and select 5000 as "Max target sequences" and 0.001 in "Expected threshold." Finally, press "BLAST" to start the search (*see* Fig. 2 and *see* **Note 4**).

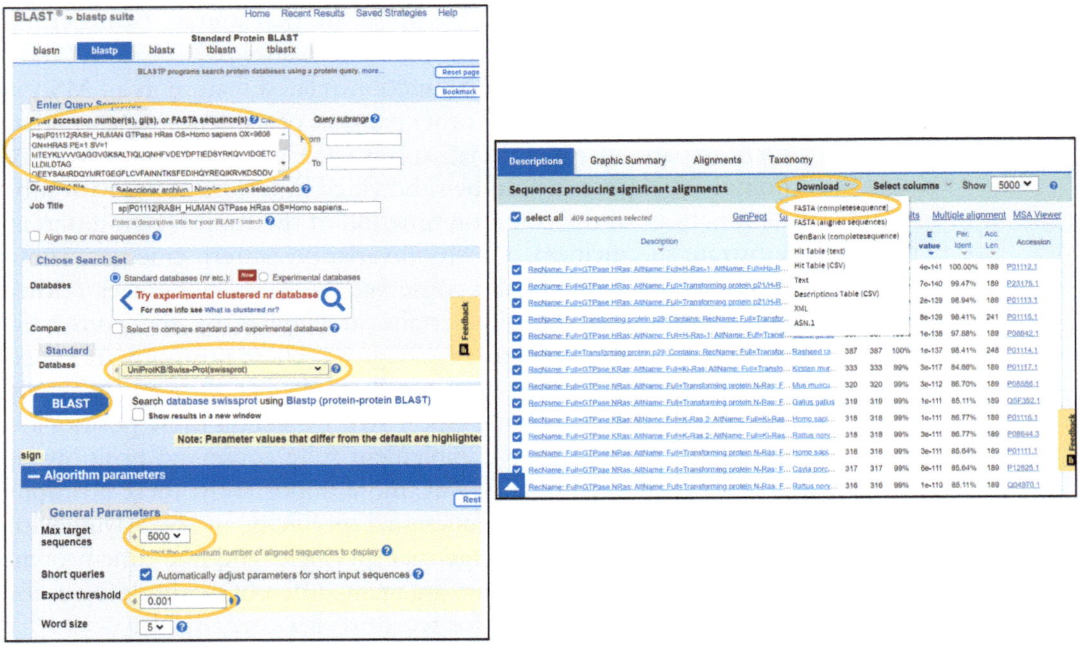

Fig. 2 Left: web input form of the *BLAST* tool, highlighting the fields and parameters touched in this example. Right: the same for the results page

In a few seconds, the list of found homologs is shown, by default sorted by *BLAST*'s main score ("expected value" or E-value). At the top of this list, there is a "Download" menu with an option to retrieve the sequences of the found homologs as a FASTA file to our local disk ("FASTA completesequence") (*see* Fig. 2).

2.2 Generation of a Multiple Sequence Alignment with the Homologs

Although *BLAST* reports pairwise alignments of all hits against the query sequence, these might not be the optimal alignments, as the method is optimized for speed. To generate an MSA with the set of homologs found by *BLAST*, we use an alignment program which takes as input the file with the homologous sequences downloaded in the previous step. We are going to use the *Clustal Omega* program [9] available through a web interface at the European Bioinformatics Institute: https://www.ebi.ac.uk/jdispatcher/msa/clustalo [10] (*see* **Note 5**).

In the input form of this tool, we upload the FASTA file generated in the previous step; select "Pearson/FASTA" in "OUTPUT FORMAT." Open "More options" and select "input" in "order" so that our query sequence (the top hit in the BLAST search) is also the first one (master) in the MSA (*see* **Note 6**). Finally, press "Submit" (*see* Fig. 3).

When the alignment job is finished, follow the "View Results" link and press the "Download" button to obtain another FASTA file with our sequences aligned in an MSA (*see* **Note 7**).

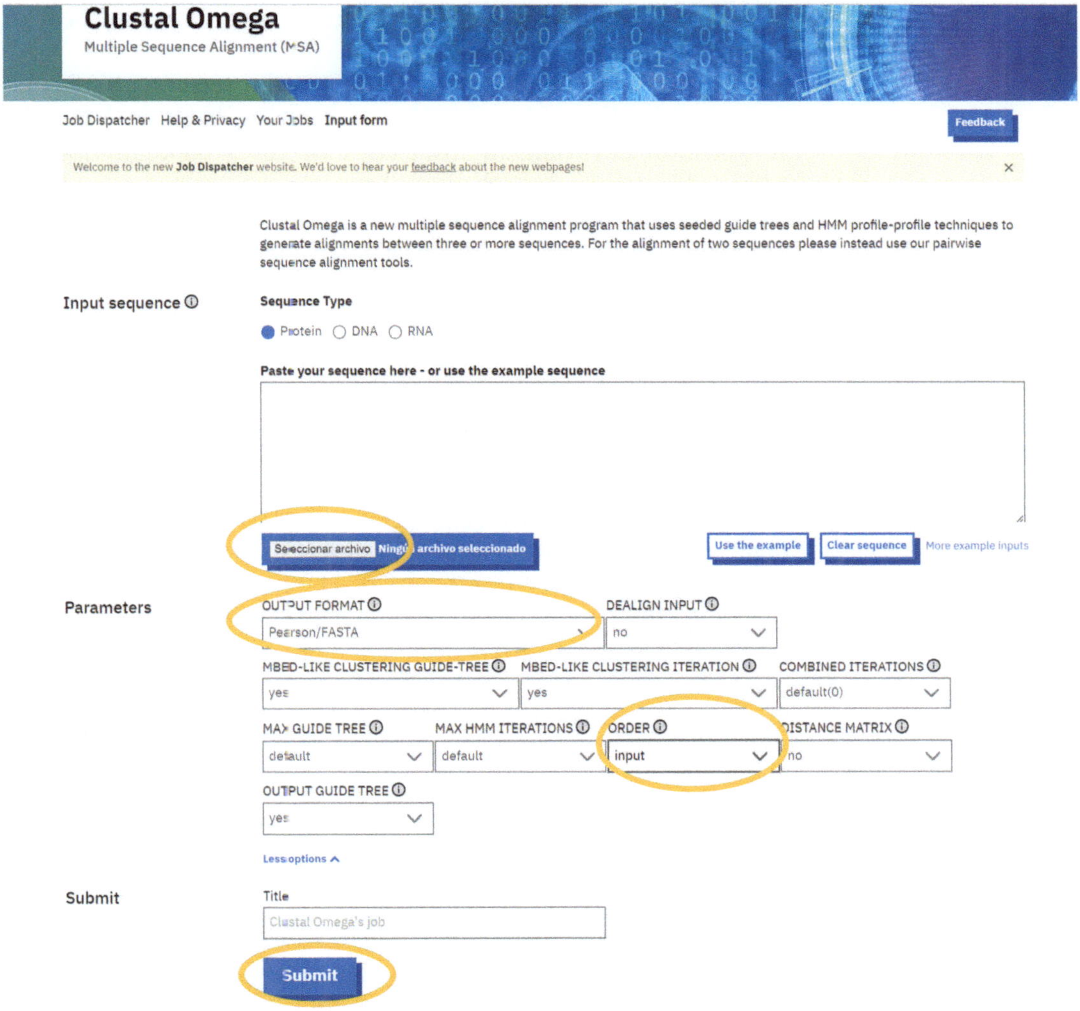

Fig. 3 Input form of the *Clustal Omega* web server, highlighting the fields and parameters modified in this example

2.3 Using JDet for Inspecting the Multiple Sequence Alignment and Detecting SDPs and Conserved Positions

JDet [11] is a free multiplatform desktop software for the interactive calculation and visualization of function-related conservation patterns in multiple sequence alignments and structures (https://csbg.cnb.csic.es/JDet/).

2.4 Installing JDet

Detailed installation instructions can be found at the program's help page (https://csbg.cnb.csic.es/JDet/JDet_help.html#install). In short, you have to download the .zip file with the latest version from the main page above and uncompress it. The only requirement of the program is to have Java (1.5 or higher) installed in your system.

If you plan to use one of the programs for detecting SDPs that comes with the package (*S3Det*), you also have to have the *R* statistical environment installed and edit a configuration file (detailed instructions in the link above).

To run the program, just double-click the "jdet-x.x.jar" file in the main folder of the distribution. Take a look at some operating-system-dependent instructions in the link above (*see* **Note 8**).

2.5 Filtering the Multiple Sequence Alignment

At the program's main interface, use "File > Load alignment" to import the MSA file generated in the previous steps.

Next, we remove redundancy, outliers ("divergent" sequences), and MSA positions without equivalent in the master sequence with the "File > Filter alignment" option. In the filter dialog, we set "Remove sequences with %ID with master lower than" to 25, "Remove redundancy at %ID" to 97, "Fragments, % coverage with master" to 60, and check "remove positions with a gap in the master sequence" (*see* Fig. 4 and *see* **Note 9**).

At this point, we have a filtered alignment with (hopefully) a comprehensive and balanced representation of the sequences of our protein family of interest (*see* **Note 10**).

JDet incorporates two methods for detecting SDPs: *Xdet* [12] and *S3Det* [13]. *Xdet* also calculates sequence conservation, quantified

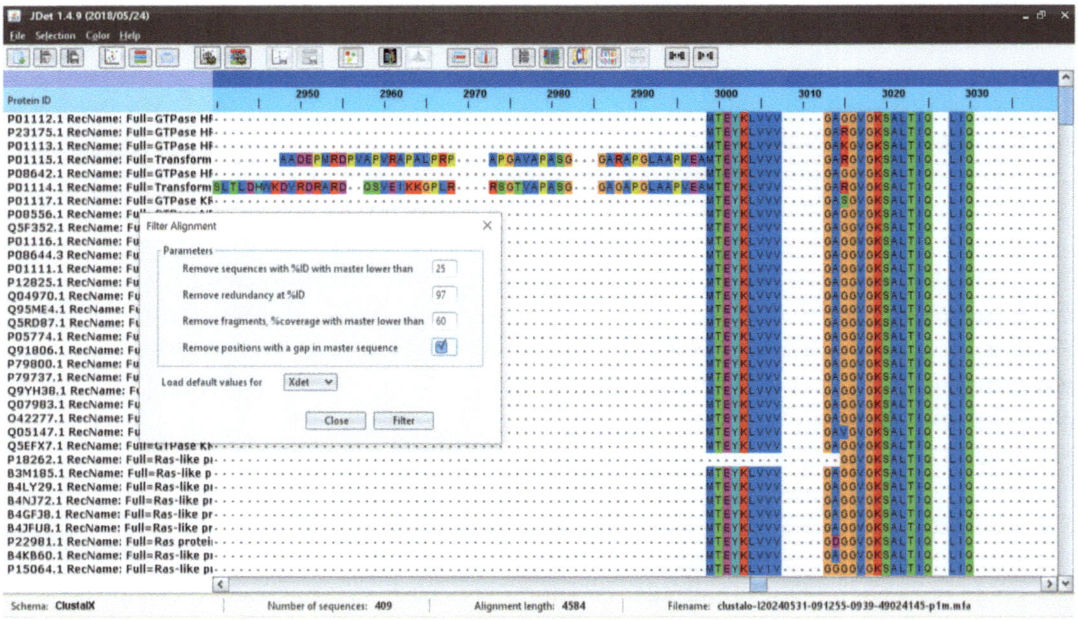

Fig. 4 *JDet* main interface with a multiple sequence alignment just loaded, and the "Filter alignment" dialog box

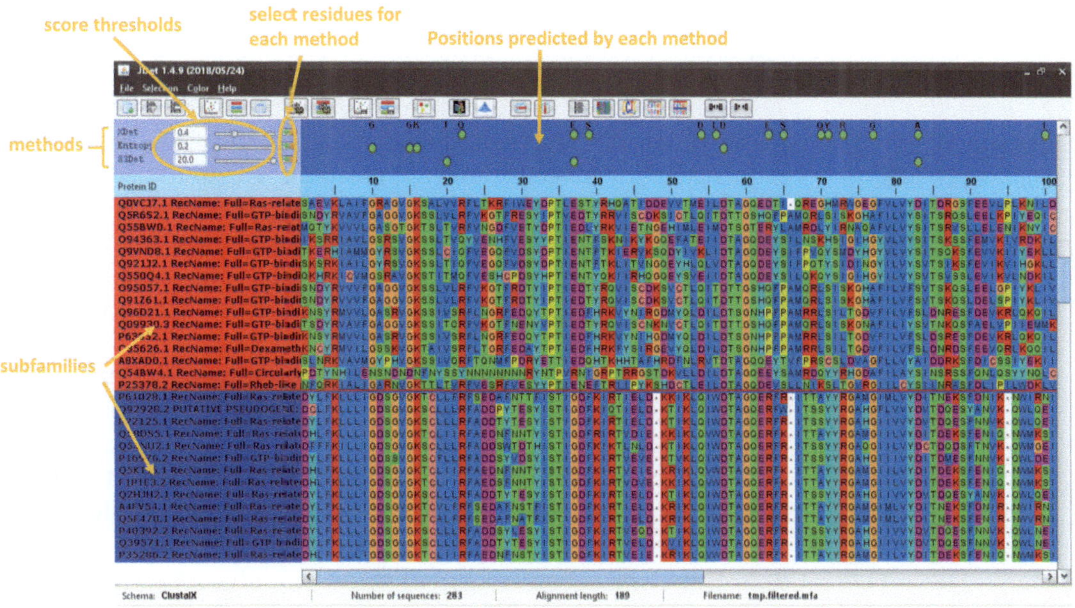

Fig. 5 *JDet* interface after running some methods for detecting functional sites and subfamilies

2.6 Detecting Subfamilies, SDPs, and Conserved Positions

as sequence entropy. To run these methods on our alignment, go to "File > Run method > Xdet" and "File > Run method > S3Det." It can take some time for these methods to run in desktop computers for a large number of sequences such as that in our example (*see* **Note 11**). Once the methods have run, it is recommended to save their results in local files to avoid forthcoming runs for the same alignments. This can be done with "File > Save method results > Xdet" and "File > Save method results > S3Det." *JDet* has some features for importing other methods' results (*see* **Note 12**).

S3Det automatically splits the MSA into potential subfamilies based on its sequence distribution, besides reporting the associated SDPs. These subfamilies are labeled with different colors (*see* Fig. 5). In this example, *S3Det* finds five subfamilies which are in good agreement with what is known about the HRas superfamily (*see* **Note 13**).

The positions reported by the two SDP detection approaches, as well as those conserved ("Entropy"), are marked with green dots in three rows above the alignment (one for each method) (*see* Fig. 5). Each method has a sliding bar to increase or decrease the score thresholds of the positions to show. These thresholds can also be changed with the corresponding text boxes. For *Xdet* and *S3Det*, the higher the score, the higher the probability the position is an SDP, while for Entropy, it is the other way around: A value of 0.0 represents a perfect conservation, and the position becomes more variable as we increase the value. Let us set these thresholds to 0.4 (*Xdet*), 20 (*S3Det*), and 0.2 (*Entropy*). We see a number of conserved positions and SDPs spread through the protein (*see* Fig. 5).

2.7 Inspecting the Sequence Compositions of SDPs and Conserved Positions

JDet can generate "sequence logos" for a set of selected positions to get an overview of their sequence composition. To generate the sequence logos for the conserved positions, we go to "Selection > From method > Entropy" (or press the green "+" button in the row of this method) in order to select the positions currently shown for this method, and then go to "Selection > Sequence logo." In this example, we see the sequence logo for the six more conserved positions (*see* Fig. 6).

In order to do the same for the SDP positions, we first have to unselect the previously selected conserved positions ("Selection > Unselect residues") and then press the green "+" for *Xdet* and *S3Det* in order to generate a cumulative selection with the SDPs of both methods. Then, we go to "Selection > subfamilies logo" (*see* Fig. 6 and *see* **Note 14**). For this example, we can see, in the logos, the typical tendency of SDPs to be differentially conserved within subfamilies (*see* **Note 15**).

2.8 Mapping the Positions in Protein Three-Dimensional Structures

In order to better understand the functional role of the MSA positions, *JDet* allows mapping them in protein three-dimensional (3D) structures if available for the family of interest.

JDet can load a protein structure from a local Protein Data Bank (PDB) file or retrieve any structure from the web, providing its PDB four-character identifier (ID).

To check whether our protein of interest (or a close homolog) has an experimental 3D structure, we repeat the same database

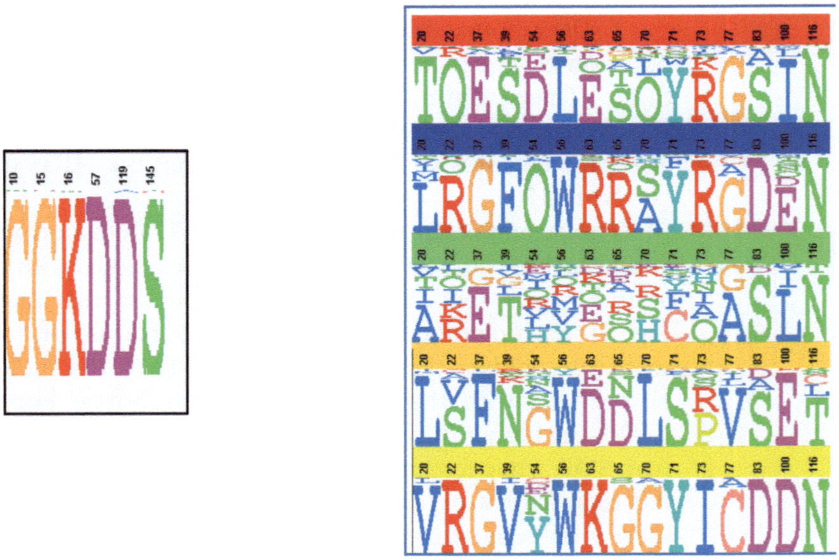

Fig. 6 Sequence logos for conserved positions (left) and SDPs (right)

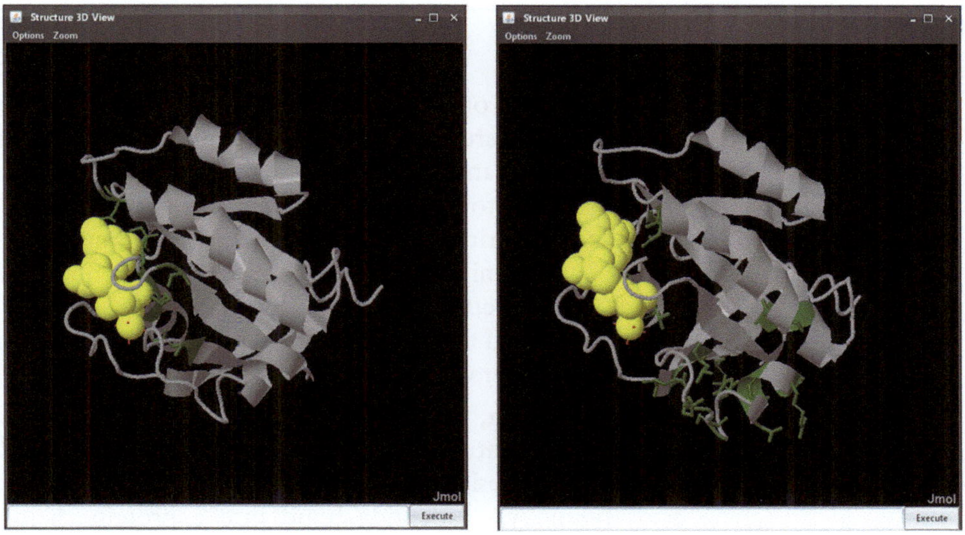

Fig. 7 Conserved positions (left) and SDPs (right) mapped into the three-dimensional structure of a representative protein

sequence search of **step 1** for HRas (UniProt accession P01112) but against the database of experimental structures ("pdb"), instead of UniProt/Swiss-Prot. We can see that the top hit reported by *BLAST*(PDB ID 4Q21, chain A) is exactly our protein (100% sequence ID), and hence we can use that structure (*see* **Note 16**).

To load that structure in *JDet*, we go to "File > Load structure (PDB)." In the dialog, we make sure that in "Alignment sequence," the MSA sequence for which we are going to import a structure is that for which we searched in PDB in the previous step (P01112). We enter the PDB ID (4Q21) in "4-letter PDB-code," "A" in "chain," and press "Load from server." When "status" shows "Download complete," we press "Load." The alignment between our sequence and that in the PDB file is shown to double-check that it is a suitable structure. Pressing "Ok" in this dialog opens the *Jmol* molecular viewer (https://jmol.sourceforge.net/) integrated in *JDet* with the structure (*see* Fig. 7 and *see* **Note 17**).

The residues in the structure corresponding to the positions selected in *JDet* are marked in green. If the structure contains hetero groups, these are shown in yellow. To map in this structure the conserved positions, we select them with the procedure above (making sure to unselect any previously selected positions), and they will show up in the molecular viewer (*see* Fig. 7). We repeat the same procedure for selecting the SDPs so that they become marked in the 3D structure (*see* Fig. 7 and *see* **Note 18**).

3 Notes

1. Homologous proteins are those sharing a common ancestor. Homologs share a similar overall three-dimensional (3D) structure and many functional features, both inherited from that ancestor. In general, homology can be detected by sequence similarity, although some remote homologs might have lost any similarity easily detectable at the sequence level. Profile-based methods, able to capture subtler sequence patterns, are used in these cases, together with structural information or any other clue indicative of common ancestry.
2. In the following, we will use the term "amino acid" for each one of the 20 natural amino acids that are the building blocks for proteins, "residue" for an instance of one of these amino acids bound via a peptide bond into a protein chain, and "position" (of an MSA) for the set of evolutionary equivalent residues in a family of proteins (corresponding to a column in the standard representation of an MSA).
3. There are many web interfaces for the *BLAST* set of tools hosted at different places worldwide. They can be found with a web search. Although they share the same main features, they have different side functionalities that could be worth exploring. For example, some servers include a functionality to automatically generate an MSA with the found hits, a procedure that here is performed in a dedicated server (*see* Subheading 2). Expert users can also install the free *BLAST* package locally for a better control of its functionalities.
4. These parameters are fine for most searches intended to detect SDPs. If we do not obtain enough sequences with them, we can try another database with more sequences or a program able to find more remote homologs:
   - "Database": For this illustrative example, we select "Swissprot" as it contains enough sequences of our protein of interest linked to functional information which might eventually help us in a forthcoming interpretation of the results, but, in other cases, it could be necessary to scan the "whole" set of nonredundant protein sequences ("nr"). Another common scenario is to search against the set of proteins with experimentally determined structures ("pdb").
   - "Algorithm": The original *BLAST* algorithm is optimized for speed, but it might not be sensible enough to detect subtle sequence relationships, such as those linking remote homologs (*see* **Note 1**). While for our particular example *BLAST* is enough, variations of this method exist, such as position-specific iterated *BLAST* (*PSI-BLAST*), a profile-based method able to detect more remote homologs. If

the standard *BLAST* search does not find enough homologs, we can try the same procedure with *PSI-BLAST*.

5. There are many programs for multiple aligning a set of protein sequences. In this particular example, the set of homologs is close (no remote homologs), and almost any alignment algorithm will do a good job. For other cases, including distant homologs, more sophisticated algorithms, such as *MUSCLE* [14] or *T-Coffee* [15], could be required. Similarly, different web interfaces hosting these programs can have different side features besides the main alignment functionality, and hence the user is advised to check some of them. To pipe other alignment methods in the workflow described here, make sure that you choose "aligned FASTA" or "multiple FASTA" as output format.
6. The information contained in the MSA is the same regardless of the order of the sequences. Nevertheless, it is a good practice to have our protein of interest as the first one ("master") for easily locating it in alignment viewers, matching alignment position numbering with residue numbers in this sequence, etc.
7. This results page contains some simple viewers for inspecting the MSA and the resulting phylogenetic tree representing the evolutionary relationships between our sequences. Although we are going to use a desktop MSA viewer in the next step, the features available at this results page can be used for a quick inspection of the alignment.
8. *JDet*'s distribution comes with documentation, examples, and a guided tutorial, all of which are also available online. It is advised to read the documentation and follow the tutorial to get familiar with the package and its complete set of features, not described here.
9. The performance of SDP detection programs predicting functional positions depends on the quality of the input MSA. Although there are no fixed rules for generating a "good" MSA, some minimal tips should be followed. The MSA should contain a balanced representation of the sequence space around our query protein. Consequently, redundancy has to be removed to avoid large groups of almost identical or very similar proteins. On the other hand, very divergent sequences that might fall outside our family of interest have to be removed too. Short fragments, common, for example, if we include metagenomic sequences in our initial sequence search, should also be removed as they introduce artificial gaps and "missing" regions in the alignment.
10. All the previous steps were intended to generate a filtered MSA of homologs for a generic protein. There are resources with precompiled collections of homologs and MSAs generated

with different tools and criteria, and some of them could be useful for detecting SDPs. For example, the *Pfam* resource [16] contains pregenerated alignments for protein domains. Its web interface has an option ("SEQUENCE SEARCH") to look for those matching our sequence of interest. If this sequence contains more than one domain, we will obtain different *Pfam* alignments, and the following steps of the workflow described here have to be carried out for each independently. From *Pfam*, download the"full alignment" in FASTA format, which might be redundant and require filtering. Another resource with pregenerated alignments for protein domains is *PANTHER* [17].

11. If the number of sequences makes it unfeasible to run these methods in desktop computers, one option is to apply a more restrictive redundancy filter to the alignment (in case we still have enough sequences). Another possibility is to run the methods outside *JDet*'s interface, in a high-performance computer, and then import the result files in *JDet* with "File > Load method results." *JDet* includes binaries of the methods for the main operating systems that can be run as standalone programs outside it.

12. *JDet* can import the results of any program that generates predictions at the position level, providing they are converted to a simple format. This can be done with the option "File > Load method results > User defined." The dialog box explains the file format required, which basically contains information, for every position of the MSA, on the position number, the residue in the master sequence, and an associated score, as well as whether the score increases or decreases for the confident predictions.

13. The Ras superfamily has been extensively studied in many publications due to the importance of this protein in key cellular processes and its involvement in cancer. Some of these works deal with its subfamily composition and the different functions/specificities of these subfamilies [18–20]. All members of the superfamily act as molecular switches in a functional cycle driven by adenosine triphosphate (ATP) hydrolysis/adenosine diphosphate (ADP) release (common function for all members). Within this common function, the subfamilies differ in the downstream processes that the switch is coupled to and the different effectors controlling the ATP/ADP cycle, functions that are mediated by the differential interaction with different proteins. The first family found by *S3Det* (red) contains part of the Ras family, formed by proteins Ras, Rhe, Rap, DiRas, Rit, and Ral, involved in "gene expression and regulation of cellproliferation, differentiation, and survival" [20]. The second family (dark blue)

comprises mainly the Rab family, containing proteins Rab, Ypt, and Sec, involved in regulating intracellular vesicular transport. The third family (green) contains other part of the Ras family, formed by proteins RHEBL, RasL, and REM, as well as the Arf family, involved in regulation of vesicular transport. The fourth family (orange) is the Rho family, consisting of the proteins Rho, RAC, and ARAC, involved in "cell polarity, cell movement, cell shape, and cell-cell and cell-matrix interactions, as well as in regulation of endocytosis and exocytosis" [20]. Finally, the last family (yellow) represents the Ran family, formed by proteins Ran and GSP, involved in nucleocytoplasmic transport of both ribonucleic acid (RNA) and proteins.

14. As SDPs are differentially conserved within particular subfamilies, the sequence logos for the whole alignment ("Selection > Sequence logo") are not very informative and do not reflect their subfamily-dependent behavior. The option "Selection > family logo" generates the sequence logos independently for each subfamily to better reflect this subfamily differential conservation.

15. We can see cases where the position is variable except in one or two subfamilies where it is conserved (e.g., pos. 54), and cases where it is conserved but with different amino acids in different families (e.g., pos. 37 and 39). *JDet* has different color schemas for the amino acids, available under the "Color" menu, which may help getting insight into the molecular differences associated to these differential conservations. Due to the importance of the HRas family (*see* **Note 13**), the involvement of many of these positions in controlling functional specificity has been experimentally corroborated [21]. In some cases, it was experimentally shown that by mutating only one or a couple of these positions, it was possible to switch the interaction specificities of the corresponding proteins. Thisis the case of position 37 (Glu in Ras, Rhe, Rap, DiRas, Rit, Ral; Gly in Rab; Gly in RAD and RasL; Glu in RHEBL, REM, and Arf; Phe in Rho; and Gly in Ran) (*see* Fig. 6).

16. It is not necessary to find a perfect hit (100% identity). Any structure within a trustable range of sequence identity for homology modeling (> ~ 30%) can be used to map residues from our protein as it is going to have the same overall 3D structure and their alignment (to map residue equivalences) is going to be correct.

17. If we do not find a suitable experimental structure with this procedure (roughly speaking, %ID >30%), we can generate a 3D model with the recently developed *AlphaFold* system [22], or download it in PDB format from the database of pregenerated models (https://alphafold.ebi.ac.uk/). Once this model is

available as a local PDB file, we load itwith the "Enter PDB file / Browse" option of the dialog.

18. As expected, fully conserved positions map around the bound guanosine triphosphate (GTP) (shown in yellow), as that is the function common to all members of the superfamily. On the contrary, SDPs are concentrated in the interfaces conferring interaction specificity (*see* **Note 15**).

Acknowledgments

This work was partially funded by the Spanish Ministry of Economy and Competitiveness with European Regional Development Fund (grant number PID2022-140047OB-C22 to F.P.).

References

1. van Dijk EL, Auger H, Jaszczyszyn Y, Thermes C (2014) Ten years of next- generation sequencing technology. Trends Genet 30: 418–426
2. Segata N, Waldron L, Ballarini A et al (2012) Metagenomic microbial community profiling using unique clade-specific marker genes. Nat Methods 9:811–814. https://doi.org/10.1038/nmeth.2066
3. Pazos F (2022) Computational prediction of protein functional sites-applications in biotechnology and biomedicine. Adv Protein Chem Struct Biol 130:39–57. https://doi.org/10.1016/bs.apcsb.2021.12.001
4. Chagoyen M, Garcia-Martin JA, Pazos F (2016) Practical analysis of specificity- determining residues in protein families. Brief Bioinform 17:255–261. https://doi.org/10.1093/bib/bbv045
5. Pazos F, Sanchez-Pulido L (2014) Protein superfamilies. In: eLS. Wiley, Chichester. https://doi.org/10.1002/9780470015902.a0025587
6. de Juan D, Pazos F, Valencia A (2013) Emerging methods in protein co-evolution. Nat Rev Genet 14:249–261
7. Chakraborty A, Chakrabarti S (2015) A survey on prediction of specificity- determining sites in proteins. Brief Bioinform 16:71–88
8. Altschul SF, Madden TL, Schaffer AA et al (1997) Gapped BLAST and PSI- BLAST: a new generation of protein database search programs. Nucl Acids Res 25:3389–3402
9. Sievers F, Wilm A, Dineen D et al (2011) Fast, scalable generation of high-quality protein multiple sequence alignments using Clustal omega. Mol Syst Biol 7:539. https://doi.org/10.1038/msb.2011.75
10. Madeira F, Pearce M, Tivey ARN et al (2022) Search and sequence analysis tools services from EMBL-EBI in 2022. Nucleic Acids Res 50:W276–W279. https://doi.org/10.1093/nar/gkac240
11. Muth T, García-Martín JA, Rausell A et al (2012) JDet: interactive calculation and visualization of function-related conservation patterns in multiple sequence alignments and structures. Bioinformatics 28:584–586
12. Pazos F, Rausell A, Valencia A (2006) Phylogeny-independent detection of functional residues. Bioinformatics 22:1440–1448
13. Rausell A, Juan D, Pazos F, Valencia A (2010) Protein interactions and ligand binding: from protein subfamilies to functional specificity. Proc Natl Acad Sci USA 107:1995–2000
14. Edgar RC (2004) MUSCLE: multiple sequence alignment with high accuracy and high throughput. Nucleic Acids Res 32:1792–1797
15. Notredame C, Higgins DG, Heringa J (2000) T-coffee: a novel method for fast and accurate multiple sequence alignment. J Mol Biol 302: 205–217
16. Finn RD, Bateman A, Clements J et al (2014) Pfam: the protein families database. Nucleic Acids Res 42:D222–D230
17. Mi H, Muruganujan A, Thomas PD (2013) PANTHER in 2013: modeling the evolution of gene function, and other gene attributes, in the context of phylogenetic trees. Nucleic Acids Res 41:D377–D386. https://doi.org/10.1093/nar/gks1118

18. Rojas AM, Fuentes G, Rausell A, Valencia A (2012) The Ras protein superfamily: evolutionary tree and role of conserved amino acids. J Cell Biol 196:189–201
19. Takai Y, Sasaki T, Matozaki T (2001) Small GTP-binding proteins. Physiol Rev 81:153–208. https://doi.org/10.1152/physrev.2001.81.1.153
20. Wennerberg K, Rossman KL, Der CJ (2005) The Ras superfamily at a glance. J Cell Sci 118:843–846. https://doi.org/10.1242/jcs.01660
21. Bauer B, Mirey G, Vetter IR et al (1999) Effector recognition by the small GTP- binding proteins Ras and Ral. J Biol Chem 274:17763–17770
22. Jumper J, Evans R, Pritzel A et al (2021) Highly accurate protein structure prediction with AlphaFold. Nature 596:583–589. https://doi.org/10.1038/s41586-021-03819-2

Chapter 5

Co-Var, a Measure to Determine Intra-protein, Inter-protein, Protein–DNA, or Protein–RNA Coevolution

Ishita Mukherjee and Saikat Chakrabarti

Abstract

Interacting biomolecules tend to coevolve, and compensatory or coordinated residue changes may occur to preserve functional interactions. Predicting coevolutionary pairings could be important for determining structurally and functionally important residues in complexes or interfaces. Correlated variation (Co-Var) aims to determine both intramolecular and intermolecular coevolving residues in proteins, protein–protein, protein–DNA, or protein–RNA complexes. It utilizes two statistical parameters namely mutual information and Bhattacharyya coefficient. In this article, we have described how to utilize the Co-Var method to identify coevolutionary parings within and/or across biomolecules.

Key words Mutual information, Intermolecular coevolution, Sequence homology, Coevolving residues, Sequence–structure relationships

1 Introduction

Coevolution detects those position pairs within multiple sequence alignment (MSA) that exhibit interdependent amino acid frequencies or similar amino acid substitution patterns. For this purpose, homologous protein or nucleic acid sequences from both the protein/gene families are utilized [1]. Determining coevolving residues can provide insights into structure, function, or interaction mechanisms in proteins or complexes [2]. This may help in studying different biological processes, pathways, or interactions important for cellular function [3]. Applications of coevolution analysis include predicting protein contact or protein interface, predicting protein–protein interaction, predicting amino acid residue positions crucial for function, and identifying mutations that disrupt protein activity [4, 5].

Intramolecular coevolution studies suggest that coevolving positions may occur in two different categories. The first category comprises positions that coevolve with one or two other positions

Shahid M. Khan and Florencio Pazos (eds.), *Protein Evolution: Methods and Protocols*, Methods in Molecular Biology, vol. 2979, https://doi.org/10.1007/978-1-0716-4828-5_5,

and often share direct amino acid side-chain interactions in proximity. The second category includes positions that coevolve with multiple other positions which are predominantly located in active sites or regions involved in intermolecular interactions and recognition and thus play crucial roles in protein function [6, 7]. Intramolecular coevolution analysis approaches include mutual information of amino acid frequencies, coevolution analysis using protein sequences (CAPS), MSA global statistical models like direct coupling analysis (DCA) or protein sparse inverse covariance (PSICOV), and substitution correlations like McLachlan-based substitution correlation (McBASC) [8–12]. In addition, statistical coupling analysis (SCA) can be used to detect protein folds or domains [2].

Intermolecular coevolution analysis of protein complexes determines coevolving residues, which could be important for functional interactions. Methods such as CAPS, EV complex, and MirrorTree can be used to analyze protein–protein coevolution. MirrorTree quantifies similarities between the MSA-derived trees of orthologous set of sequences of the two protein families by extracting inter-ortholog distance matrices and calculating the linear correlation [13]. In contrast, CAPS identifies amino acid covariation with the help of BLOSUM-corrected amino acid distances and phylogenetic sequence relationships [9]. However, EV complex predicts inter-evolutionary couplings using a global probability model of sequence coevolution utilizing pseudolikelihood maximization (PLM)-based calculation of coupling parameters on the alignment of concatenated sequences [11].

Correlated variation (Co-Var) is a covariation measure based on mutual information and Bhattacharyya coefficient that aims to determine coevolving residues in proteins, protein–protein, and protein–DNA or protein–RNA complexes [1]. Following sections briefly describe the materials and methods to identify coevolving positions within single protein and also between protein–protein, protein–DNA, and protein–RNA complexes using the Co-Var package.

2 Materials

1. *Install software packages* on your Linux or Windows computer. This method can be used locally on a desktop/workstation or online on the web server from a Linux or Microsoft Windows device. The computation time would depend on the number of sequences that one uses in the alignments. On the web server, around 50 sequences would result in a computation time of 2–3 hours. The prerequisites for the local installation include a homology detection tool like basic local alignment search tool (BLAST) [14] and its databases, an alignment tool like multiple alignment using fast Fourier transform (MAFFT) [15], and Co-Var local version. The Co-Var package uses R; R packages ggplot2, plot3D, and forcats; and Python package NetworkX [16]. Users are advised to refer to individual web pages for installation instructions. To calculate coevolutionary pairings with Co-Var, the following software and packages should be installed on the local machine.
   (a) Install R using a package manager. The software and installation instructions can be found at "https://www.r-project.org/" (tested on R 3.6.3). Additional packages like ggplot2, plot3D, and forcats should also be installed (https://bioconductor.org/install/).
   (b) The Python package NetworkX (https://networkx.org/documentation/stable/install.html) is also a prerequisite for Co-Var.
   (c) The alignment tool like MAFFT (https://mafft.cbrc.jp/alignment/software/source.html) should be installed if the user is not submitting an alignment file (optional step).
   (d) After downloading the Co-Var package, place the scripts in a local folder (note: Co-Var uses Perl, Python, and R packages).
2. *Input dataset generation.* Depending on the question, one can curate a list of proteins and complexes for analysis. Previous studies on coevolution provide a good reference for coevolving complexes [1, 17, 18]. FASTA format of sequences of these biomolecules needs to be collected from UniProt or National Center for Biotechnology Information (NCBI) gene databases [19–21] (*see* **Notes 1** and **2**). For protein complexes, Mirror-Tree can be utilized to determine whether a pair of protein families coevolve or not using alignments of homologous sequences of the query proteins. Additionally, the conserved domain database (CDD) contains carefully curated protein sequence alignments that can be used for this analysis as well [22].

3. *Protein Data Bank (PDB) structure for mapping* the coevolving residues and subsequently drawing their functional relevance.
   (a) Construct a query (phrases) to search for protein structures or complexes in the Protein Data Bank (PDB) [23] using the instructions provided at the link https://www.rcsb.org/docs/search-and-browse/basic-search. The UniProt entry for the protein can also provide a link to the associated PDB ID or record. The PDB ID can also be used to search for the relevant complex and download the relevant structure file. Only X-ray crystallography structures are compatible for this analysis.
4. *Collection of homologous sequences.* Close homolog/orthologs of query proteins or biomolecules should be collected with the help of database searches using sequence similarity search programs or algorithms. Sequence similarity search methods include basic local alignment search tool (BLAST) [14] for sequence–sequence comparisons (*see* **Note 3**), hidden Markov model scan (HMMscan) [24] for sequence–profile comparisons, and HMM-HMM-based lightning-fast iterative sequence search (HHblits) [25, 26] for profile–profile comparisons. Examples of sequence databases include nonredundant protein [27] and nucleotide database, protein families database (Pfam), and conserved domain database (CDD) [28]. This step can be performed online on the respective web portals or locally depending on the expertise of the user (*see* **Notes 4–10**).
5. *Alignment generation* (optional). Multiple sequence alignment (MSA) representative of each sequence family can be prepared using "multiple alignment using fast Fourier transform" (MAFFT) [15] or using any other equivalent tools available to the user (*see* **Notes 1, 11,** and **12**). A number of progressive MSA algorithms like CLUSTALW, DIALIGN-TX, MUSCLE, T-Coffee, Probcons, Probalign, or CLUSTAL OMEGA with global or local pairwise alignment options provide differing levels of accuracy or speed [29].
6. *Co-Var online version or local setup.* An online version of the Co-Var methodology can be accessed at http://www.hpppi.iicb.res.in/ishi/covar/index.html. This method can be used to identify coevolutionary pairings in intra-protein, inter-protein, protein–DNA, or protein–RNA complexes. A local version is available for download at Co-Var web server (http://www.hpppi.iicb.res.in/ishi/covar/download/covar-loc.zip) and at the GitHub repository (https://github.com/Ishita2690/Co-Var).

3 Methods

A phenomenon where a change in one locus possibly affects the selection pressure at another locus and results in reciprocal change (s) is referred to as molecular coevolution. Direct evolutionary interaction may occur between sites within a single protein referred to as intra-protein coevolution or between different proteins in which case it is referred to as inter-protein coevolution [30].

3.1 Identify Coevolving Residues Online Using Co-Var Web Server

Co-Var Web server identifies coevolutionary pairings in intra-protein, inter-protein, protein–DNA, or protein–RNA complexes. This Web server is available at http://www.hpppi.iicb.res.in/ishi/covar/index.html.

3.1.1 Intra-protein Coevolutionary Analysis

Coevolving protein fragments can be informative about key mutations with roles in genetic diseases, protein conformational changes, folding intermediates, peptide assembly, subfamily-dependent motifs, and active-site functional residues or interacting residues [31]. In order to study intra-protein coevolution, the submission form available at http://www.hpppi.iicb.res.in/ishi/covar/IntraProt.html can be used for job submission. Example files are provided so that the user can get an idea about the required file format.

1. The input files have to be uploaded and the user can decide whether to start with sequences or an alignment. In case a set of homologous sequences is provided, the default alignment steps and thresholds included in the portal will be utilized to optimize it.
2. The reference structure provided is used for mapping and assigning relevance to the coevolved residues predicted. However, this is optional. To provide a reference structure, one needs to download the corresponding PDB file from the PDB database and upload it here and add the reference chain information. The first sequence in the alignment is considered as the reference, and ideally, the same protein in PDB format should be used.
3. Select a z-score threshold for the analysis. The recommended threshold is −1, and lower values usually provide more stringent results.
4. In order to select interface residues that are coevolving, a distance threshold can be specified.
5. An email address and a job name should be specified before submission. A link to the results will be provided to the user on the given email address after the completion of the job.

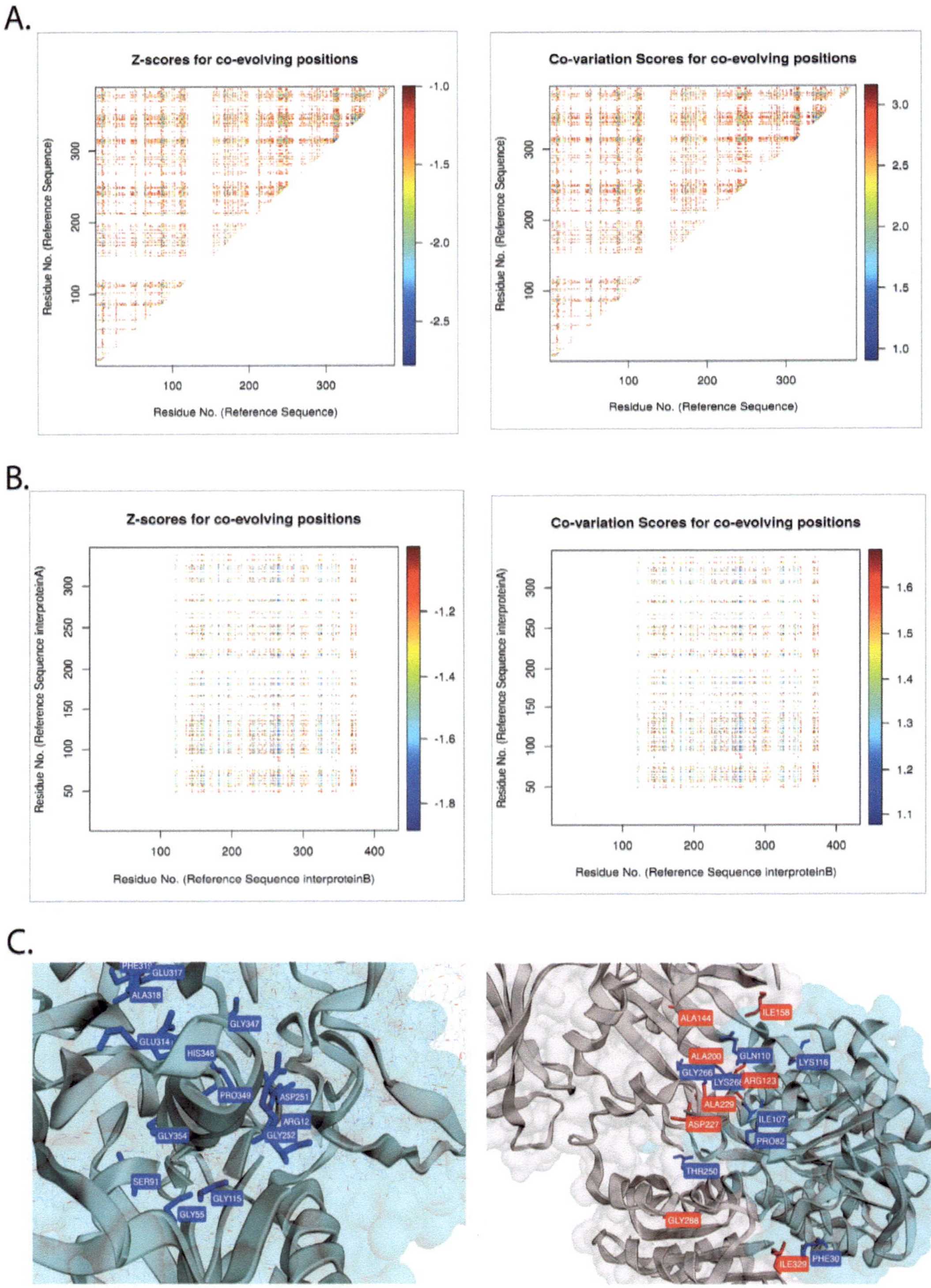

Fig. 1 (**a**) Coevolving residues in intra-protein coevolution with z-scores (left) and Co-Var scores (right). (**b**) Coevolving residues in inter-protein coevolution with z-scores (left) and Co-Var scores (right). (**c**) Structure-mapped coevolved positions in intra-protein coevolution (left) and inter-protein coevolution (right)

Covarying positions selected based on a z-score threshold are reported considering the reference sequence, which is the first sequence in alignment or input sequences and structure mapped positions. A sample result page is available at http://www.hpppi.iicb.res.in/ishi/covar/chjkP2/getIntraRes.html. Reference sequence mapped coevolving positions can be viewed on the plots and patterns within functional domains may appear (*see* Fig. 1a). The result file can be downloaded, and an alternate threshold can be used to select coevolved pairs for subsequent analysis. Depending on the interacting residues' distance threshold, the selected high-scoring coevolved residues can be visualized on the PDB structure along with amino acid, number, and chain (*see* Fig. 1c, left panel). The result file with pairing information considering sequence and structure positions, distance, and Co-Var score along with z-scores can also be downloaded. An overview of distances between coevolving positions is also provided.

3.1.2 Inter-protein Coevolutionary Analysis

Residues involved in coevolutionary pairings may be structurally or functionally important in protein–protein complex interaction and cellular process or signaling. Previous studies suggest that interface residues in obligate complexes coevolve among interacting partners. However, interface residues in transient protein interaction complexes exhibit an increased rate of substitution [18]. In addition, coevolutionary pairings were observed in functional domains and interface residues in intercellular protein complexes involved in metastasis [1]. Thus, Co-Var can be used to find residue position interactions that can be modified to alter or supplement the complex functionality. A protein–protein coevolution analysis job can be submitted using this submission form (http://www.hpppi.iicb.res.in/ishi/covar/protPROT.html). Steps for job submission include:

1. Select option for analysis using sequences or alignment.
2. Prepare sequence and alignment files in the format specified in the example files and upload them.
3. Select and upload a reference structure containing the reference sequences from protein family A and B (optional). Provide the PDB code and respective chain identifiers.
4. Select significance thresholds for coevolving positions and coevolving interface residues.
5. Submit the job with an email address and job name.

Once the analysis is complete, a result file similar to this sample result (http://www.hpppi.iicb.res.in/ishi/covar/kmn/ResInterProt.html) will be available. In addition to the plots and visualizations for sequence mapped and structure mapped positions as mentioned above (*see* Fig. 1b, c, right panels), a list of high-degree coevolutionary pairings is also provided for download. A high-degree coevolving position in one protein is one which has a tendency to coevolve with a large number of residue positions in the coevolving protein partner.

3.1.3 Protein–DNA or Protein–RNA Coevolution Analysis

An example of protein–DNA interactions would be transcription factor–DNA interactions, and Co-Var can be utilized to study coevolution in such complexes. Coevolution analysis of transcription factor binding specificity or binding site sequence organization can provide insights into gene regulation and transcriptional regulatory network of gene expression [32]. RNA binding proteins mediate co-transcriptional and posttranscriptional regulation to mediate gene expression, and Co-Var determines coevolutionary pairings in such protein–RNA complexes for insights into RNA binding protein (RBP) evolutionary signatures or RNA–protein interaction networks mediating cellular function [33]. In order to study coevolution in protein–DNA or protein–RNA complexes, user should utilize the submission form available at http://www.hpppi.iicb.res.in/ishi/covar/protDNA.html or http://www.hpppi.iicb.res.in/ishi/covar/protRNA.html on the Web server, respectively. The steps are similar to the previous cases.

1. Select a complex for study and prepare relevant input files according to the format specified in the example files and the homology detection steps with the sequence or structure databases. One can select the gene sequence region depending on the question at hand and their requirements (*see* **Notes 7–10, 13** and **14**). The first sequence in alignment or input sequences is considered as the reference to report the covarying positions.
2. Select mode for analysis (sequence/alignment).
3. Upload the sequence or alignment files and enter a threshold to select coevolved positions [note: negative z-scores are indicative of positions more likely to coevolve].
4. Provide a job name and email address before submitting the job.
5. Once the job is complete, a link to the results will be emailed to the user and the user can download and analyze the result files.

The results file for protein–DNA and protein–RNA coevolution includes only sequence-mapped coevolved positions and a plot to depict the same as shown here http://www.hpppi.iicb.res.in/ishi/covar/ppDn/result-interDNA.html or http://www.hpppi.iicb.res.in/ishi/covar/ppR2/result-interRNA.html.

3.2 Command Line Analysis of Coevolving Positions with Co-Var

A local version of the Co-Var method to determine coevolutionary pairings within and across biomolecules is available for download from the Co-Var Web server (http://www.hpppi.iicb.res.in/ishi/

covar/download/covar-loc.zip) and the GitHub repository (https://github.com/Ishita2690/Co-Var). Please refer to the GitHub directory and links to packages for installation instructions.

3.2.1 Steps to Analyze Intra-protein Coevolution

In order to understand how sequence variation affects protein evolution and function, study the sequence-mapped and structure-mapped coevolving positions. For the protein of interest, following steps should be performed:

1. Prepare input sequences or alignment file as outlined above.
2. Identify coevolving positions mapped onto the reference sequence with the help of the following command:

```
./wrapper-combi-Intra [sequence/alignment file] [input type]
[threshold] [output dir]
```

Here, [sequence/alignment file] refers to the path to the input file (sequences/alignment) (e.g., "/path to workingDir/filename.fa"). Mention the input type as "sequence" or "msa" for sequence file or alignment file, respectively. Provide a z-score threshold (e.g., −1) and the name of the output directory. These arguments must be provided in order.

3. Determine three-dimensional structure-mapped coevolved positions with the following command:

```
./strucresAnal [PDB file (Reference structure)] [chain ID
(Reference structure)] [sequence/alignment file] [threshold]
[output dir] [PDB ID] [Distance threshold]
```

Here, the following arguments should be provided in given order:

(i) PDB file (e.g.," /path to workingDir/file.pdb").
(ii) Chain ID provided in the reference structure.
(iii) Input file used in the previous step.
(iv) Significance threshold for coevolved positions (an integer, e.g., −1).
(v) Output directory name.
(vi) PDB structure ID.
(vii) Distance threshold for interface residues.

Intra-protein coevolution analysis often identifies functionally important residues as coevolving [7]. Further, less variable positions within a protein family or non-proximal clusters of positions close to active sites and binding regions may also coevolve

[6, 34]. Once the analysis is complete, results are generated in the output directory specified. A summary of results can be viewed in the browser by opening the file '*getIntraRes.html*'. Here, it would be interesting to observe the pattern of coevolved residues and the intrachain residue–residue distances or active-site residues.

3.2.2 Protein–Protein Coevolution with Co-Var

Disease causing missense mutations are often observed among residues involved in enzymatic function, ligand-binding interaction sites, and protein–protein interaction interface core region [35, 36]. Coevolutionary pairings appeared in interface and non-interface regions which contained missense mutations in some ligand–receptor complexes involved in protein–protein signaling and cancer metastasis [1]. Coevolved positions participate in native pathway interactions, and as such, this information is helpful for studying complex functionality and designing strategies for modifying the interactions toward therapeutic options.

1. Prepare input sequences or alignment files as outlined above.
2. Determine reference sequences mapped coevolved positions using the following command:

```
./wrapper-combi-Inter [sequence/alignment file (Family A)]
[sequence/alignment file (Family B)] [input type] [threshold]
[output dir]
```

Here, [sequence/alignment file (Family A)] refers to the path to the input file 1 (sequences/alignment) (e.g., "/path to workingDir/filename.fa") followed by file 2. Next, mention the input type as "sequence" or "msa" for sequence file or alignment file respectively and provide a z-score threshold (e.g., −1) along with the name of the output directory.

3. In order to calculate, reference structure-mapped coevolved positions run the command as below:

```
./strucresAnal-IP [Reference structure PDB file] [chain ID of
reference sequence A] [chain ID of reference sequence B] [input
sequence/alignment file (family A)] [input sequence/alignment
file (family B)] [threshold] [output dir] [PDB id] [distance
threshold]
```

Here, the following arguments should be provided in given order:

(i) PDB file (e.g.," /path to workingDir/file.pdb").
(ii) Chain ID provided in the reference structure for family A followed by B.

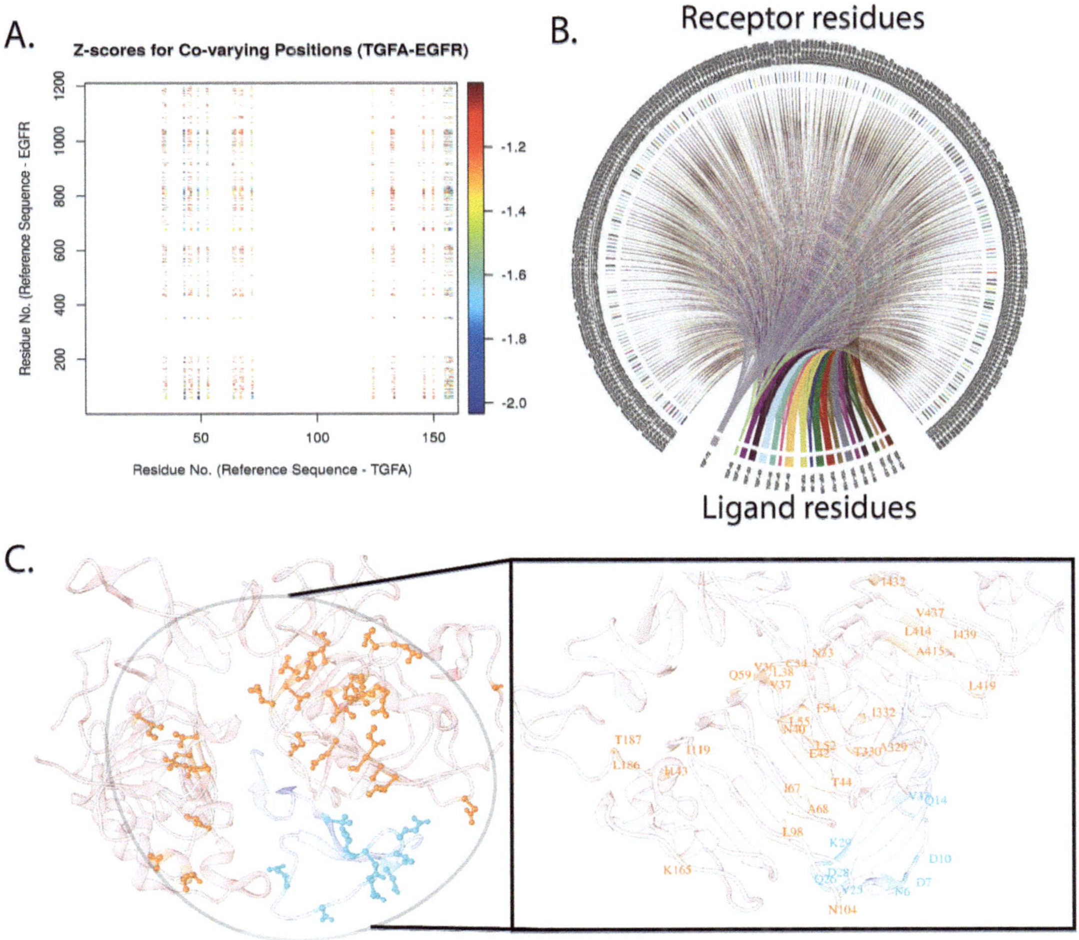

Fig. 2 (**a**) Sequence-mapped coevolved positions and scores in TGFA-EGFR complex obtained from inter-protein coevolution analysis. (**b**) High-degree coevolved positions in a ligand–receptor complex represented as a chord diagram. (**c**) Structure-mapped high-degree coevolved positions in inter-protein coevolution [color code—blue: ligand (TGFA); orange: receptor (EGFR)]

(iii) Input file used in the previous step (family A and B).

(iv) Significance threshold for coevolved positions (an integer, e.g., −1).

(v) Output directory name.

(vi) PDB structure ID.

(vii) Distance threshold for interface residues.

For inter-protein analysis, results are generated in the output directory specified and the results summary *'ResInterProt.html'* can be viewed in the browser. A set of coevolved positions is provided, and sequence-mapped results for transforming growth factor alpha (TGFA)-epidermal growth factor receptor (EGFR) complex are shown as an example in Fig. 2a. A number of high-degree coevolving positions occur in interface and non-interface regions but often in regions critical for protein function [1]. Such pairings can be studied with the help of network diagrams or chord diagram to study residue pairing among ligand and receptor positions (*see* Fig. 2b). Significant coevolved positions when mapped onto the complex structure help study interface properties (*see* Fig. 2c), and inter-residue distances can also help assign relevance (data reproduced from Mukherjee and Chakrabarti, 2021). As such, these residues could be important for structure-based drug design to modify altered protein signaling in diseases such as cancer.

3.2.3 Protein–DNA or Protein–RNA Coevolution Analysis

Transcription factors regulate the expression of multiple genes and have a DNA-binding preference, while RNA-binding proteins also have specific binding domains for RNA to mediate gene regulation. Some examples include tryptophan repressor protein, gal transcriptional activator, hormone nuclear receptor, and embryonic lethal abnormal vision (ELAV)-like protein 1. Analyzing coevolution in these complexes may highlight how the protein or its family mediates regulation of multiple targets and their binding specificity or variations observed across different species.

1. Prepare input sequences or alignment files as outlined above. The neighboring DNA region and (+/−) 150 nucleotides to (+/−) 600 nucleotides can be included with the binding site sequence to improve alignment during the input generation step (for TF-DNA coevolution).
2. Determine reference sequence-mapped coevolved positions (protein–DNA):

   In the directory where the Co-Var scripts have been saved, type the following command:

```
./wrapper-combi-InterDNA [sequence/alignment file (Family A)]
[sequence/alignment file (Family B)] [input type] [threshold]
[output dir]
```

3. Determine reference sequence-mapped coevolved positions (protein–RNA):

```
./wrapper-combi-InterRNA [sequence/alignment file (Family A)]
[sequence/alignment file (Family B)] [input type] [threshold]
[output dir]
```

Here, the following arguments are required: protein sequence or alignment file "/path to workingDir /filename.fa") and RNA or DNA sequence or alignment file. Indicate the input type as "sequence" or "msa" for sequence or alignment, respectively. Provide a z-score significance threshold (an integer) and an output directory.

Coevolution can be studied with the help of prokaryotic or eukaryotic transcription factors such as Lex-A (locus for X-ray sensitivity A), a repressor protein responsible for SOS regulon

(SOS) mechanism or DNA repair. An example alignment for proteins and DNA-binding site used for a sample analysis is shown in Fig. 3a. The sequence-mapped coevolved positions obtained from Co-Var are depicted in Fig. 3b (data reproduced from Mukherjee and Chakrabarti, 2021). These mainly occurred within the binding domain and were visualized on a reference structure using UCSF Chimera (*see* Fig. 3c) [37].

4 Notes

1. It is ideal to use a minimum of 50 homologous sequences for the multiple sequence alignments.
2. A set of protein sequence alignments that can be used to study protein–protein coevolution (positive set) is available from the Dryad data repository [1].
3. In order to determine protein orthologs, domain enhanced lookup time accelerated BLAST (DELTA-BLAST) [38] can be used and significant sequences can be selected with the following parameters: E-value = 70%, and sequence identity > = 45%. Taxonomy filtered non-redundant sequences can then be used for the alignment.
4. Homology searches from protein or nucleotide databases may suggest "unnamed," "partial," "hypothetical," "putative," or "low-quality protein" as homologous. These sequences can be omitted before preparing the alignment, since it helps prune the alignment to be more specific.
5. Tools such as cluster database at high identity with tolerance (CD-HIT) [39] can be utilized to check redundancy among sequences, and only representative sequences should be retained for the alignment generation step.
6. Different algorithms for protein homology detection can be explored. BLAST for protein (BLASTp) compares the query to a protein database, whereas position-specific iterated BLAST (PSI-BLAST) allows iterative comparisons of the query to a database using a user defined position-specific scoring matrix (PSSM). Additionally, pattern-hit initiated BLAST (PHI-BLAST) is guided by pattern matching between the query and alignments and domain enhanced lookup time accelerated BLAST (DELTA-BLAST) searches a sequence database with the help of a profile from CDD search [38, 40, 41].
7. For DNA or RNA sequences, it is better to use sequence bases upstream and downstream from the binding site motif alignment matches obtained from the BLAST search.
8. In order to determine DNA or RNA orthologs, MegaBLASTn (Mega Basic Local Alignment Search Tool Nucleotide) [43]

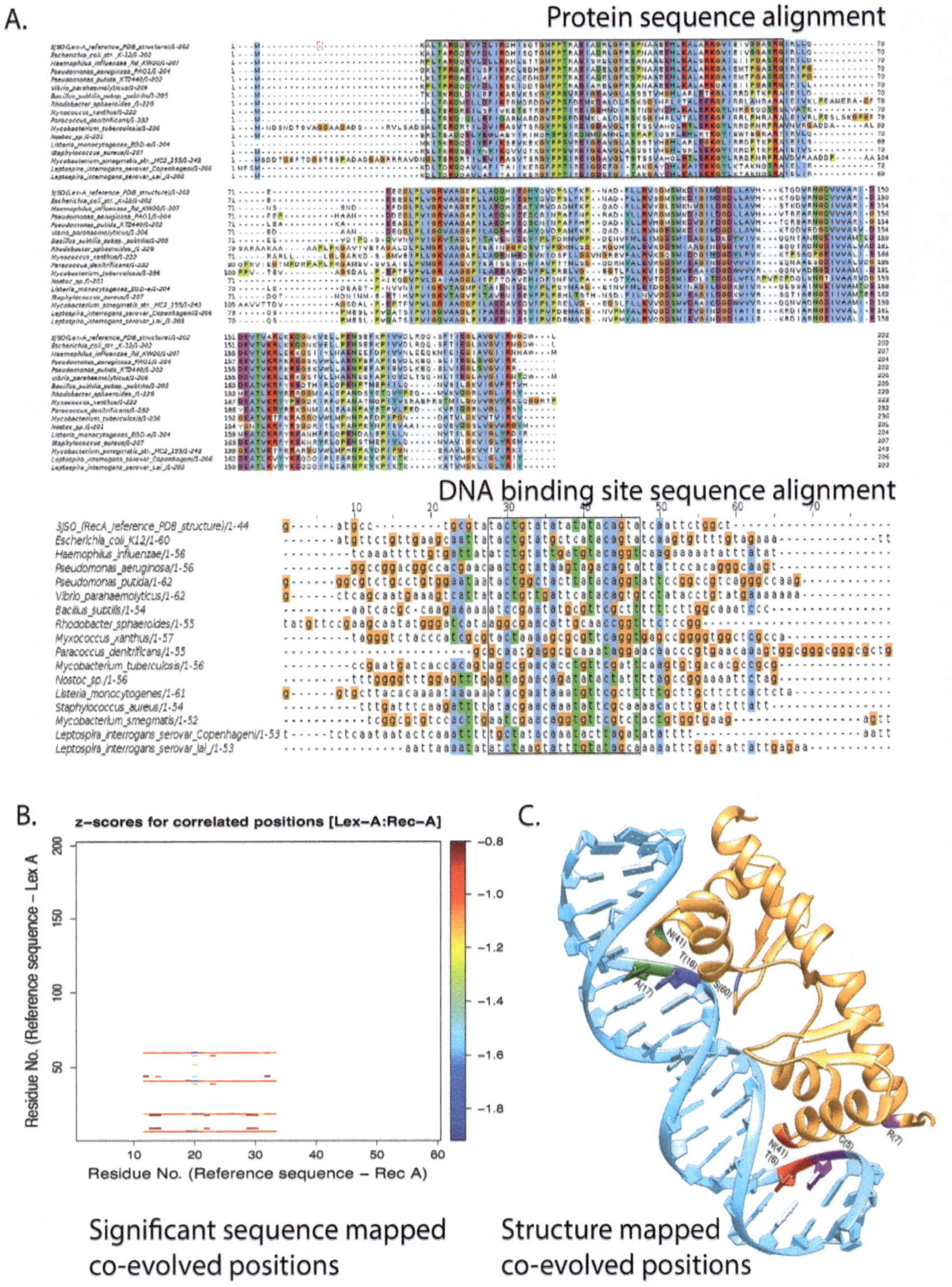

Fig. 3 (**a**) Representative protein sequences and DNA-binding site alignment for Lex-A and Rec-A. (**b**) Sequence-mapped coevolved positions in protein–DNA complex. (**c**) Structure-mapped coevolved positions in Lex-A::Rec-A complex

may be used and significant sequences can be selected with the following parameters: E-value = 60%, and sequence identity > = 60%. Taxonomy-filtered non-redundant sequences can then be used for the alignment.

9. DNA or RNA sequence homologs can be filtered to remove some sequences that are "uncultured," "synthetic," "clone," "scaffold," or "contig" to improve alignment quality.
10. Different algorithms can be used for nucleotide BLAST including BLASTn, Mega Basic Local Alignment Search Tool (MegaBLAST), or discontinuous MegaBLAST depending upon the user requirement and scientific question.
11. In case alignment generation is performed with MAFFT, different algorithms can be explored to obtain the best possible alignment. Different method types include progressive methods (FFT-NS1 and FFT-NS2), iterative refinement methods (FFT-NS-i and NW-NS-i), and iterative refinement methods using weighted sum-of-pairs score and consistency scores (L-INS-i, E-INS-i, and G-INS-i). Different algorithms may work better depending upon the speed or accuracy required; in our experience, L-INS-i, E-INS-i (local alignment), and G-INS-i (global alignment) work better [15, 42].
12. The alignment can be viewed in an interactive viewer like Jalview [45] for studying the representative domain or binding region alignment and alignment in conserved or non-conserved regions.
13. An additional motif scan in regulatory sequence analysis tool (RSAT) [44] or a similar tool can help prune sequences with similar binding or interaction domains.
14. CLIP db (Cross-linking and Immunoprecipitation Database) provides a good resource for users to browse and download binding site information of RNA-binding proteins as obtained by CLIP-seq (Cross-linking and Immunoprecipitation Sequencing). Users can query protein-coding genes or non-coding RNA of interest [33].

References

1. Mukherjee I, Chakrabarti S (2021) Co-evolutionary landscape at the Interface and non-interface regions of protein-protein interaction complexes. Comput Struct Biotechnol J 19:3779–3795. https://doi.org/10.1016/j.csbj.2021.06.039
2. Salinas VH, Ranganathan R (2018) Co-evolution-based inference of amino acid interactions underlying protein function. elife 7: e34300. https://doi.org/10.7554/eLife.34300
3. Tillier ER, Charlebois RL (2009) The human protein co-evolution network. Genome Res 10:1861–1871. https://doi.org/10.1101/gr.092452.109
4. Green AG, Elhabashy H, Brock KP et al (2021) Large-scale discovery of protein interactions at residue resolution using co-evolution calculated from genomic sequences. Nat Commun 12:1396. https://doi.org/10.1038/s41467-021-21636-z

5. Colavin A, Atolia E, Bitbol AF et al (2022) Extracting phylogenetic dimensions of co-evolution reveals hidden functional signals. Sci Rep 12:820. https://doi.org/10.1038/s41598-021-04260-1
6. Gloor GB, Martin LC, Wahl LM, Dunn SD (2005) Mutual information in protein multiple sequence alignments reveals two classes of coevolving positions. Biochemistry 44:7156–7165
7. Chakrabarti S, Panchenko AR (2009) Co-evolution in defining the functional specificity. Proteins 75:231–240. https://doi.org/10.1002/prot.22239
8. Korber BT, Farber RM, Wolpert DH, Lapedes AS (1993) Covariation of mutations in the V3 loop of human immunodeficiency virus type 1 envelope protein: an information theoretic analysis. Proc Natl Acad Sci USA 90:7176–7180
9. Fares MA, McNally D (2006) CAPS: co-evolution analysis using protein sequences. Bioinformatics 22(22):2821–2822
10. Fares MA, Travers SA (2006) A novel method for detecting intramolecular co-evolution: adding a further dimension to selective constraints analyses. Genetics 173(1):9–23. https://doi.org/10.1534/genetics.105.053249
11. Hopf TA, Schärfe CP, Rodrigues JP, Green AG, Kohlbacher O, Sander C, Bonvin AM et al (2014) Sequence co-evolution gives 3D contacts and structures of protein complexes. elife 3:e03430. https://doi.org/10.7554/eLife.03430
12. Jones DT, Buchan DWA, Cozzetto D, Pontil M (2012) PSICOV: precise structural contact prediction using sparse inverse covariance estimation on large multiple sequence alignments. Bioinformatics 28:184–190
13. Ochoa D, Pazos F (2010) Studying the co-evolution of protein families with the Mirrortree web server. Bioinformatics 26(10):1370–1371. https://doi.org/10.1093/bioinformatics/btq137
14. Altschul SF, Gish W, Miller W, Myers EW, Lipman DJ (1990) Basic local alignment search tool. J Mol Biol 215(3):403–410
15. Katoh K, Misawa K, Kuma K, Miyata T (2002) MAFFT: a novel method for rapid multiple sequence alignment based on fast Fourier transform. Nucleic Acids Res 30(14):3059–3066
16. R Core Team (2016) R: a language and environment for statistical computing. https://www.R-project.org
17. Sowmya G, Breen EJ, Ranganathan S (2015) Linking structural features of protein complexes and biological function. Protein Sci 24(9):1486–1494
18. Mintseris J, Weng Z (2005) Structure, function, and evolution of transient and obligate protein-protein interactions. Proc Natl Acad Sci USA 102(31):10930–10935
19. Breuza L, Poux S, Estreicher A, Famiglietti ML, Magrane M, Tognolli M et al (2016) The UniProtKB guide to the human proteome. Database 2016:bav120. https://doi.org/10.1093/database/bav120
20. The UniProt Consortium (2015) UniProt: a hub for protein information. Nucleic Acids Res 43:D204-12
21. The UniProt Consortium (2019) UniProt: a worldwide hub of protein knowledge. Nucleic Acids Res 47(D1):D506–D515. https://doi.org/10.1093/nar/gky1049
22. Marchler-Bauer A, Derbyshire MK, Gonzales NR, Lu S, Chitsaz F, Geer LY, Geer RC, He J, Gwadz M, Hurwitz DI et al (2015) CDD: NCBI'S conserved domain database. Nucleic Acids Res 43(D1):D222–D226. https://doi.org/10.1093/nar/gku1221
23. Rose PW, Prlić A, Bi C, Bluhm WF, Christie CH, Dutta S et al (2015) The RCSB protein data bank: views of structural biology for basic and applied research and education. Nucleic Acids Res 43:D345–D356
24. Eddy SR (1998) Profile hidden Markov models. Bioinformatics 14(9):755–763
25. Söding J (2005) Protein homology detection by HMM-HMM comparison. Bioinformatics 21(7):951–960
26. Remmert M, Biegert A, Hauser A, Söding J (2011) HHblits: lightning-fast iterative protein sequence searching by HMM-HMM alignment. Nat Methods 9(2):173–175
27. Pruitt KD, Tatusova T, Maglott DR (2007) NCBI reference sequences (RefSeq): a curated non-redundant sequence database of genomes, transcripts and proteins. Nucleic Acids Res 35:D61–D65
28. Finn RD, Bateman A, Clements J, Coggill P, Eberhardt RY, Eddy SR et al (2014) The Pfam protein families database. Nucleic Acids Res 42(D1):D222–D230
29. Pais FS, Ruy PC, Oliveira G, Coimbra RS (2014) Assessing the efficiency of multiple sequence alignment programs. Algorithms Mol Biol 9(1):4. https://doi.org/10.1186/1748-7188-9-4
30. Lovell SC, Robertson DL (2010) An integrated view of molecular co-evolution in protein–protein interactions. Mol Biol Evol 27(11):2567–2575. https://doi.org/10.1093/molbev/msq144

31. Oteri F, Sarti E, Nadalin F, Carbone A (2022) iBIS2Analyzer: a web server for a phylogeny-driven co-evolution analysis of protein families. Nucleic Acids Res 50(W1):W412–W419. https://doi.org/10.1093/nar/gkac481
32. Rosanova A, Colliva A, Osella M et al (2017) Modelling the evolution of transcription factor binding preferences in complex eukaryotes. Sci Rep 7:7596. https://doi.org/10.1038/s41598-017-07761-0
33. Yang YCT, Di C, Hu B et al (2015) CLIPdb: a CLIP-seq database for protein-RNA interactions. BMC Genomics 16:51. https://doi.org/10.1186/s12864-015-1273-2
34. Mandloi S, Chakrabarti S (2017) Protein sites with more co-evolutionary connections tend to evolve slower, while more variable protein families acquire higher co-evolutionary connections. F1000Res 6:453. https://doi.org/10.12688/f1000research.11251.2
35. Gao M, Zhou H, Skolnick J (2015) Insights into disease-associated mutations in the human proteome through protein structural analysis. Structure 23(7):1362–1369. https://doi.org/10.1016/j.str.2015.03.028
36. David A, Sternberg MJ (2015) The contribution of missense mutations in core and rim residues of protein-protein interfaces to human disease. J Mol Biol 427(17):2886–2898. https://doi.org/10.1016/j.jmb.2015.07.004
37. Pettersen EF et al (2004) UCSF chimera – a visualization system for exploratory research and analysis. J Comput Chem 13:1605–1612
38. Boratyn GM, Schäffer AA, Agarwala R, Altschul SF, Lipman DJ, Madden TL (2012) Domain enhanced lookup time accelerated BLAST. Biol Direct 7:12. https://doi.org/10.1186/1745-6150-7-12
39. Li W, Godzik A (2006) Cd-hit: a fast program for clustering and comparing large sets of protein or nucleotide sequences. Bioinformatics 22(13):1658–1659. https://doi.org/10.1093/bioinformatics/btl158
40. Altschul SF, Madden TL, Schäffer AA, Zhang J, Zhang Z, Miller W, Lipman DJ (1997) Gapped BLAST and PSI-BLAST: a new generation of protein database search programs. Nucleic Acids Res 25(17):3389–3402
41. Zhang Z, Miller W, Schäffer AA, Madden TL, Lipman DJ, Koonin EV, Altschul SF (1998) Protein sequence similarity searches using patterns as seeds. Nucleic Acids Res 26(17):3986–3990. https://doi.org/10.1093/nar/26.17.3986
42. Katoh K, Kuma K, Toh H, Miyata T (2005) MAFFT version 5: improvement in accuracy of multiple sequence alignment. Nucleic Acids Res 33(2):511–518. https://doi.org/10.1093/nar/gki198
43. Chen Y, Ye W, Zhang Y, Xu Y (2015) High speed BLASTN: an accelerated MegaBLAST search tool. Nucleic Acids Res 43(16):7762–7768. https://doi.org/10.1093/nar/gkv784
44. Santana-Garcia W, Castro-Mondragon JA, Padilla-Gálvez M, Nguyen NTT, Elizondo-Salas A, Ksouri N, Gerbes F, Thieffry D, Vincens P, Contreras-Moreira B, van Helden J, Thomas-Chollier M, Medina-Rivera A (2022) RSAT 2022: regulatory sequence analysis tools. Nucleic Acids Res 50(W1):W670–W676. https://doi.org/10.1093/nar/gkac312
45. Waterhouse AM, Procter JB, Martin DM, Clamp M, Barton GJ (2009) Jalview Version 2 – a multiple sequence alignment editor and analysis workbench. Bioinformatics 25(9):1189–1191. https://doi.org/10.1093/bioinformatics/btp033

Chapter 6

adabmDCA 2.0—A Flexible but Easy-to-Use Package for Direct Coupling Analysis

Lorenzo Rosset, Roberto Netti, Anna Paola Muntoni, Martin Weigt, and Francesco Zamponi

Abstract

In this methods article, we provide a flexible but easy-to-use implementation of direct coupling analysis (DCA) based on Boltzmann machine learning, together with a tutorial on how to use it. The package adabmDCA 2.0 is available in different programming languages (C++, Julia, Python) usable on different architectures (single-core and multicore CPU, GPU) using a common front-end interface. In addition to several learning protocols for dense and sparse generative DCA models, it allows to directly address common downstream tasks like residue-residue contact prediction, mutational-effect prediction, scoring of sequence libraries, and generation of artificial sequences for sequence design. It is readily applicable to protein and RNA sequence data.

Key words Direct coupling analysis, Generative probabilistic models, Protein-sequence modeling

1 Introduction

Over the past decade, the increase in sequence data availability has significantly propelled the application of machine learning in biomolecular studies. Generative models, in particular, have emerged as highly effective tools. Even though the field of bioinformatics is moving toward the adoption of larger and larger deep neural network models, notably with the advent of large language models (LLMs), direct coupling analysis (DCA) methods [1] provide a simple, user-friendly, energy-efficient, and interpretable tool that can be successfully adopted for generating functional protein sequences [2], predicting contacts in the ternary structure [3], predicting mutational effects in proteins [4], and unveiling coevolutionary signals in MSAs of homologous protein or RNA families [5].

This tutorial presents a new version of adabmDCA [6]. The package comes in three different languages: C++ (single-core

Shahid M. Khan and Florencio Pazos (eds.), *Protein Evolution: Methods and Protocols*, Methods in Molecular Biology, vol. 2979, https://doi.org/10.1007/978-1-0716-4828-5_6,

CPU), Julia (multicore CPU), and Python (GPU-oriented). They share the same front-end interface from the terminal allowing the user to install and use one of the three equivalent versions based on hardware or software constraints.

We provide three different training routines:

1. **bmDCA**: Trains a fully connected DCA model [7].
2. **eaDCA**: Trains a DCA model on a sparse coupling network by progressively adding couplings during the training [8].
3. **edDCA**: Starts from a trained bmDCA model and iteratively removes the less informative couplings until the target sparsity is reached [9].

Additionally, we provide several routines for sampling and analyzing the generated sequences once a DCA model is trained, for constructing and evaluating—according to a DCA model—a single-mutant library from a given wild type, and finally, for computing the pairwise contact scores, in terms of average-product-corrected Frobenius norms of the DCA couplings [10].

This tutorial is organized as follows. In the first section, we present the theoretical framework of DCA models that is useful for understanding the parameters of the algorithm and the possible issues that one can encounter during the training. We also describe the typical preprocessing pipeline for preparing the data before using them to train the DCA model. In the second section, we describe the different routines and the most important hyperparameters that need to be set. Finally, in the last section, we summarize the main terminal commands for the different routines, and we summarize all the input parameters. The reader who is already familiar with DCA modeling of biological sequences can skip the first part of the tutorial and go directly to the third section.

2 Boltzmann Learning of Biological Models

2.1 Structure of the Model and Boltzmann Learning

DCA models (Fig. 1) are probabilistic generative models that infer a probability distribution over sequence space. Their objective is to assign high probability values to sequences that are statistically similar to the natural sequences used to train the architecture while assigning low probabilities to those that significantly diverge. The input dataset is a multiple sequence alignment (MSA), where each sequence is represented as a L-dimensional categorical vector $\boldsymbol{a} = (a_1, \ldots, a_L)$ with $a_i \in \{1, \ldots, q\}$, each number representing one of the possible amino acids or nucleotides, or the alignment gap. To simplify the exposition, from here on, we will assume them to be amino acids. The following equation then gives the DCA probability distribution:

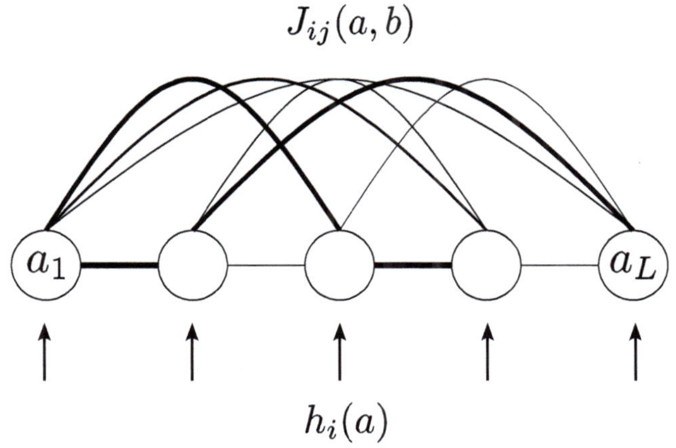

Fig. 1 Schematic representation of a DCA model

$$p(\boldsymbol{a}|\boldsymbol{J},\boldsymbol{h},\mathcal{G}) = \frac{1}{Z(\boldsymbol{J},\boldsymbol{h},\mathcal{G})} e^{-E(a_1,\ldots,a_L)}$$
$$= \frac{1}{Z(\boldsymbol{J},\boldsymbol{h},\mathcal{G})} \exp\left(\sum_{(i,a)\in\mathcal{V}} h_i(a)\delta_{a_i,a} + \sum_{(i,a,j,b)\in} J_{ij}(a,b)\delta_{a_i,a}\delta_{a_j,b} \right). \tag{1}$$

In this expression, Z is the normalization constant and E is the DCA *energy function*. The interaction graph $\mathcal{G}=(\mathcal{V},\mathcal{E})$ is represented in a *one-hot encoding* format: The vertices $\mathcal{V}$ of the graph are the $L \times q$ combinations of all possible symbols on all possible sites, labeled by $(i,a)\in\{1,\cdots,L\}\times\{1,\cdots,q\}$, and the edges $\mathcal{E}$ connect two vertices (i,a) and (j,b). The *bias* (or field) $h_i(a)$ corresponding to the amino acid a on the site i is activated by the Kronecker $\delta_{a_i,a}$ if and only if $a_i = a$, and it encodes the conservation signal of the MSA. The *coupling matrix* $J_{ij}(a,b)$ represents the coevolution (or epistatic) signal between pairs of amino acids at different sites and is activated by the $\delta_{a_i,a}\delta_{a_j,b}$ term if $a_i = a$ and $a_j = b$. Note that $J_{ii}(a,b)=0$ for all a, b to avoid a redundancy with the $h_i(a)$ terms and that $J_{ij}(a,b)=J_{ji}(b,a)$ is imposed by the symmetry structure of the model. The interaction graph $\mathcal{G}$ can be chosen fully connected as in the bmDCA model, or it can be a sparse graph as in eaDCA and edDCA.

Training the Model The training consists of adjusting the biases, the coupling matrix, and the interaction graph to maximize the log likelihood of the model for a given MSA, which can be written as

$$\mathcal{L}(\{\boldsymbol{a}^{(m)}\}|\boldsymbol{J},\boldsymbol{h},\mathcal{G}) = \frac{1}{M_{\text{eff}}}\sum_{m=1}^{M} w^{(m)}\left[\sum_{(i,a)\in\mathcal{V}} h_i(a)\delta_{a_i^{(m)},a} + \sum_{(i,a,j,b)\in\mathcal{E}} J_{ij}(a,b)\delta_{a_i^{(m)},a}\delta_{a_j^{(m)},b}\right] - \log Z(\boldsymbol{J},\boldsymbol{h},\mathcal{G}) = \sum_{(i,a)\in\mathcal{V}} h_i(a)f_i(a) + \sum_{(i,a,j,b)\in\mathcal{E}} J_{ij}(a,b)f_{ij}(a,b) - \log Z(\boldsymbol{J},\boldsymbol{h},\mathcal{G}), \tag{2}$$

where $w^{(m)}$ is the weight of the data sequence m, with $\sum_{m=1}^{M} w^{(m)} = M_{\text{eff}}$, and

$$f_i(a) = \frac{1}{M_{\text{eff}}}\sum_{m=1}^{M} w^{(m)}\delta_{a_i^{(m)},a}, \qquad f_{ij}(a,b) = \frac{1}{M_{\text{eff}}}\sum_{m=1}^{M} w^{(m)}\delta_{a_i^{(m)},a}\delta_{a_j^{(m)},b}, \tag{3}$$

are the empirical single-site and two-site frequencies computed from the data. Roughly speaking, $f_i(a)$ tells us what is the empirical probability of finding the amino acid a in the position i of a training sequence, whereas $f_{ij}(a,b)$ tells us how likely it is to find together the amino acids a and b at positions i and j, respectively, in a training sequence.

For a fixed graph $\mathcal{G}$, we can maximize the log likelihood by iteratively updating the parameters of the model in the direction of the gradient of the log likelihood, meaning

$$h_i(a) \leftarrow h_i(a) + \gamma\frac{\partial\mathcal{L}(\{\boldsymbol{a}^{(m)}\}|\boldsymbol{J},\boldsymbol{h},\mathcal{G})}{\partial h_i(a)}, \qquad J_{ij}(a,b) \leftarrow J_{ij}(a,b) + \gamma\frac{\partial\mathcal{L}(\{\boldsymbol{a}^{(m)}\}|\boldsymbol{J},\boldsymbol{h},)}{\partial J_{ij}(a,b)}, \tag{4}$$

where γ is a small rescaling parameter called *learning rate*. By differentiating the log likelihood (2), we find the update rule for Boltzmann learning:

$$h_i(a) \leftarrow h_i(a) + \gamma(f_i(a) - p_i(a)), \qquad J_{ij}(a,b) \leftarrow J_{ij}(a,b) + \gamma(f_{ij}(a,b) - p_{ij}(a,b)), \tag{5}$$

where $p_i(a) = \langle\delta_{a_i,a}\rangle$ and $p_{ij}(a,b) = \langle\delta_{a_i,a}\delta_{a_j,b}\rangle$ are the one-site and two-site marginals of the model (1). Notice that the convergence of the algorithm is reached when $p_i(a) = f_i(a)$ and $p_{ij}(a,b) = f_{ij}(a,b)$.

Monte Carlo Estimation of the Gradient The difficult part of the algorithm consists of estimating $p_i(a)$ and $p_{ij}(a,b)$, because computing the normalization Z of Eq. 1 is computationally intractable, preventing us from directly computing the probability of any sequence. To tackle this issue, we estimate the first two moments of the distribution through a Monte Carlo simulation. This consists of sampling a certain number of fairly independent sequences from the probability distribution (1) and using them to estimate $p_i(a)$ and $p_{ij}(a,b)$ at each learning epoch. There exist several equivalent strategies to deal with it. Samples from the model (1) can be obtained via Markov Chain Monte Carlo (MCMC) simulations either at equilibrium or out of equilibrium, where we start from N_c configurations (we refer to them as *chains*), chosen uniformly at random, from the data or the last configurations of the previous learning epoch, and update them using Gibbs or Metropolis-Hastings sampling steps up to a certain number of MCMC sweeps. It has been shown in [6] for Boltzmann machines and, in general, for energy-based models [11] that under certain conditions, training the model estimating the marginals from an out-of-equilibrium sampling leads to good enough generative models. Depending on the training set, these models can even be fairly close to what is achieved with perfectly equilibrated training. For this reason, in `adabmDCA 2.0`, we implement a persistent contrastive divergence (PCD) scheme, in which the computation of the gradient is done with chains that may be slightly out of equilibrium. In particular, chains are *persistent*, i.e., they are initialized at each learning epoch using the last sampled configurations of the previous epoch. This is done because sampling from configurations that are already close to the stationary state of the model at the current training epoch is much more convenient. Furthermore, the number of sweeps to be performed should be large enough to ensure that the updated chains are close enough to an equilibrium sample of the probability (1). In practice, this is done by fixing the number of sweeps to a convenient value, k, that trades off between a reasonable training time and a fair mixing of the chains.

Convergence Criterium To decide when to terminate the training, we monitor the two-site connected correlation functions of the data and of the model, which are defined as

$$C_{ij}^{\text{data}}(a,b) = f_{ij}(a,b) - f_i(a)f_j(b) \,, \qquad C_{ij}^{\text{model}}(a,b) = p_{ij}(a,b) - p_i(a)p_j(b) \,. \quad (6)$$

When the Pearson correlation coefficient between the two reaches a target value, set by default at 0.95, the training stops.

Once the model is trained, we can generate new sequences by sampling from the probability distribution (1), infer contacts on the tertiary structure by analyzing the coupling matrix, or propose and assess mutations through the DCA energy function E (*see* Subheading 4). While during the training we keep the number of Monte Carlo sweeps fixed, we underline that whenever we want to sample new sequences from a model (i.e., after convergence of training), we compute the mixing time as explained in Subheading 4.1 and we ensure that sequences are sampled in equilibrium.

2.2 Training Sparse Models

What we have described so far is true for all the DCA models considered in this work, but we have not yet discussed how to adjust the topology of the interaction graph $\mathcal{G}$. In the most basic implementation, bmDCA, the graph is assumed to be fully connected (every amino acid of any residue is connected with all the rest), and the learning will only tweak the strength of the connections. This results in a coupling matrix with $L(L-1)q^2/2$ independent parameters, where L is the aligned sequence length, $q=21$ for amino acid and $q=5$ for nucleotide sequences. However, it is well known from the literature that the interaction network of protein families tends to be relatively *sparse*, suggesting that only a few connections should be necessary for reproducing the statistics of biological sequence data. This observation brings us to devising a training routine that produces a DCA model capable of reproducing the one- and two-body statistics of the data with a minimal amount of couplings.

To achieve this, we implemented two different routines: eaDCA that promotes initially inactive (i.e., zero) coupling parameters to active ones starting from a profile model, and edDCA, which iteratively prunes active but negligible coupling parameters starting from a dense fully connected DCA model.

Element Activation DCA In eaDCA (Fig. 2b), we start from an empty interaction graph $\mathcal{E}=\oslash$, meaning that no connection is present. Each training step is divided into two different moments: First, we update the graph, and then we bring the model to convergence once the graph is fixed (a similar pipeline has been proposed in [8]).

To update the graph, we first estimate $p_{ij}(a,b)$ for the current model. Then, we decide how many couplings we want to target as the fraction `factivate` of the number of currently inactive couplings, and from all the possible quadruplets of indices (i,j,a,b), we select those for which $p_{ij}(a,b)$ is "the most distant" (according to the criterion introduced in [8]) from the target statistics $f_{ij}(a,b)$. We then activate the couplings corresponding to these quadruplets, obtaining a new graph $\mathcal{E}' \supseteq \mathcal{E}$. Notice that some of

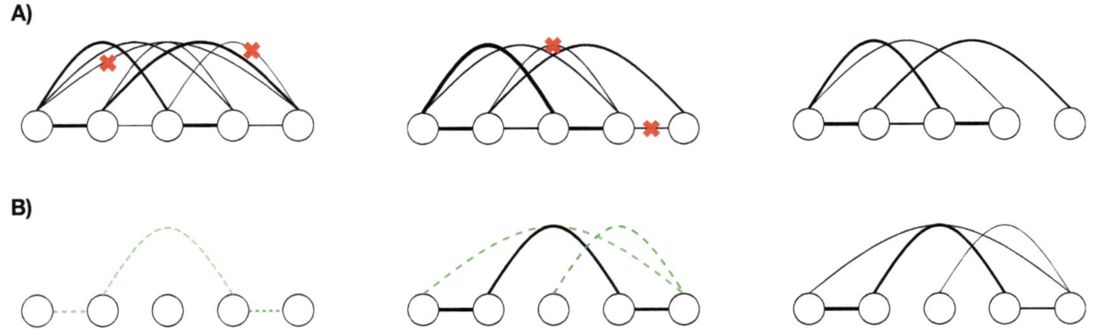

Fig. 2 Schematic representation of the sparse model training. (**a**) edDCA, the sparsification is obtained by progressively pruning contacts from an initial fully connected model. (**b**) eaDCA, the couplings are progressively added during the training

the selected couplings might be already active, so the model partially autoregulates the number of new couplings that have to be activated at each step.

Element Decimation DCA In edDCA (Fig. 2a), we start from a previously trained bmDCA model and its fully connected graph $\mathcal{G}$. We then apply the decimation algorithm, in which we prune connections from the edges $\mathcal{E}$ until a target density of the graph is reached, where the density is defined as the ratio between the number of active couplings and the number of couplings of the fully connected model. Similarly to eaDCA, each iteration consists of two separate moments: graph updating and active parameter updating.

To update the graph, we remove the fraction `drate` of active couplings that, once removed, produce the smallest perturbation on the probability distribution at the current epoch. In particular, for each active coupling, one computes the symmetric Kullback-Leibler distances between the current model and a perturbed one, without that target element. One then removes the `drate` elements which exhibit the smallest distances (*see* [9] for further details).

Parameter Updates in Between Decimations/Activations In both procedures, to bring the model to convergence on the graph, we perform a certain number of parameter updates in between each step of edge activation or decimation, using the update rule (5). Between two subsequent parameter updates, k sweeps are performed to update the Markov chains.

In the case of element activation, we perform a fixed number of parameter updates, specified by the input parameter `gsteps`. Instead, when pruning the graph, we keep updating the parameters with the rule (5) until the Pearson correlation coefficient reaches a target value.

2.3 Input Data and Preprocessing

`adabmDCA 2.0` takes as input a multiple sequence alignment (MSA) of aligned amino acid or nucleotide sequences, usually forming a protein or RNA family. DCA implementations require the data to be saved in FASTA format [12].

`adabmDCA 2.0` implements the three default alphabets shown in Table 1, but the user can specify an ad hoc alphabet as far as it is compatible with the input MSA.

An example of a FASTA file format is shown in Fig. 3. In particular, `adabmDCA 2.0` correctly handles FASTA files in which line breaks within a sequence are present.

Preprocessing Pipeline The `adabmDCA 2.0` code applies the following preprocessing pipeline to the input MSA:

Table 1
Default alphabets implemented in `adabmDCA 2.0`, where "-" indicates the alignment gap

Protein	-, A, C, D, E, F, G, H, I, K, L, M, N, P, Q, R, S, T, V, W, Y
RNA	-, A, C, G, U
DNA	-, A, C, G, T

```
>chain 1
GCAAGAGCCAAGGGGUGUUCCGUUUCGGGGCUGAGAUC----------AAACCCUUAGAACCUGAUCCGGUUAAUACCGGCGGAGGAAUGCGUCAUCCGCGAUCG
>chain 2
CCCUCACGCCAGGGGUGCCCCGUUUCGGGGCUGAGAAGGGGG-CCCCCAGACCCUUUGAACCUGAACUGGGUAAUACCAGCGGAGGGAAGCGGAUUGUUCGACGG
>chain 3
AAAAGAUGCUAGGGGAGUCCGCUG-GGGCGCUGAGAGUGCGGUCCGC-UGACCCUUUGAACUUGAUCAGGGUCAUACCAGCGGAAGGAAGCUAAACCGUUUUAUA
>chain 4
GCCUUCCGCUGGGGGAGCCCAACAGGUGGGCUGAGAGGGCGCUGCGCC-GACCGUUGA-AACUG-UCCGGCUCGUUCCGGCGUAGGGAAGCGGGCCUCAUGUUUA
>chain 5
AACAAAAGGUAGGGGUGCUGUUAGCGAACGCUGAGAUC----------AUACCCUCU-AACCUGAUUUGGGUAAUGCCAACGAAGGGAACCUAAGGACAUACAUU
>chain 6
U-AA-AUGCCAGGGGAGCUCGUCGAACGAGCUGAGAGGGCGUUGCGCC-GACUCUGAU-ACCUGAUCUGGAUAAUGCCUGCGGAGGAAUGCACUUACGAUCCAUA
>chain 7
GAUAGUAUUCGCGGGUGCUUGUAUGACGAGCUGAGAAA----------ACUCCGCGAGCACCUGAUUUGGAUAAUGCCAACGUAGGGAGAGUAUUCUUUUGGCUU
>chain 8
AGUAAUCCCUGGGGGUGUCCUUUCUCGGGACUGAGAAUAGGACUCCU-UGACCCC--GAACCUGAUGCGGUUCAGACCGCCGUAGGGAUGGGAGCCAGGUACGGA
>chain 9
AUCACCCGCUAGGGGUGCUGCCCGGGGCGGCUGAGAUC----------AAACCCUUAGAACCUGAUCCGGAUAAUACCGGCGUAGGGAAGCGGCUUUCUACGGGG
>chain 10
UAAGUUCGCUCGGGGAGUCUCCGGUGUUGGCUGAGAGGGAGACGAUCC-GACCCGUUGAACCUGAUCCGGGUCAUACCGGCGCAGGGAAGCCCGAAGAGUUUUAU
```

Fig. 3 Example of RNA sequences formatted in FASTA format

- p.1 Remove the sequences having some tokens not included in the chosen alphabet.
- p.2 If needed, compute the importance weights for the sequences in the MSA.
- p.3 Apply a pseudocount to compute the MSA statistics.

Their precise implementation is described in the following.

Computing the Importance Weights The sequence weights are computed to mitigate as much as possible the systematic biases in the data, such as correlations due to the phylogeny or over-representation of some regions of the sequence space because of a sequencing bias. Given an MSA of M sequences, to compute the importance weight of each sequence $\boldsymbol{a}^{(m)}$, $m = 1, \ldots, M$, we consider $N^{(m)}$ as the number of sequences in the dataset having sequence identity from $\boldsymbol{a}^{(m)}$ greater or equal to $0.8 \cdot L$ (this threshold can be tuned by the user). Then, the importance weight of $\boldsymbol{a}^{(m)}$ will be

$$w^{(m)} = \frac{1}{N^{(m)}} . \tag{7}$$

This reweighting allows us to give less importance to sequences found in very densely populated regions of the sequence space, while enhancing the importance of isolated sequences.

Alternatively, the user can decide to provide an external file containing the weights, or to ask the program to assign uniform weights across the sequences in the MSA.

Pseudocount and Reweighted Statistics DCA models are trained to reproduce the one- and two-site frequencies of the empirical data. To compute these, we introduce in the computation of the empirical statistics a small parameter α, called pseudocount, that allows us to deal with unobserved (pairs of) symbols in one (or two) column(s) of the MSA. The one- and two-site frequencies are given by

$$f_i(a) = (1-\alpha) f_i^{\text{data}}(a) + \frac{\alpha}{q}, \tag{8}$$

$$f_{ij}(a,b) = (1-\alpha) f_{ij}^{\text{data}}(a,b) + \frac{\alpha}{q^2}, \tag{9}$$

where $f_i^{\text{data}}(a)$ and $f_{ij}^{\text{data}}(a,b)$ are computed from the MSA as in Eq. 3.

3 Implementation

All the software implementations that we propose (Python, Julia, and C++) offer the same interface from the terminal through the `adabmDCA` command. The complete list of training options can be listed through the command (Table 2):

```
$ adabmDCA train -h
```

The standard command for starting the training of a DCA model is

```
$ adabmDCA train -m <model> -d <fasta_file> -o <output_folder> -l <label>
```

where:

- `<model>` $\in$ {bmDCA, eaDCA, edDCA} selects the training routine. By default, the fully connected bmDCA algorithm is used. edDCA can follow two different routines: Either it decimates a pretrained bmDCA model, or it first trains a bmDCA model and then decimates it. The corresponding commands are shown below (*see* Subheading 3.2).
- `<fasta_file>` is the FASTA file, with the complete path, containing the training MSA.
- `<output_folder>` is the path to a (existing or not) folder where to store the output files.

Table 2
Command line arguments for training bmDCA, eaDCA, or edDCA

Command	Default value	Description
`-d, --data`	N/A	Filename of the dataset to be used for training the model
`-o, --output`	DCA_model	Path to the folder where to save the model
`-m, --model`	bmDCA	Type of model to be trained. Possible options are `bmDCA`, `eaDCA`, and `edDCA`
`-w, --weights`	None	Path to the file containing the weights of the sequences. If `None`, the weights are computed automatically
`--clustering_seqid`	0.8	Sequence identity threshold to be used for computing the sequence weights
`--no_reweighting`	N/A	If this flag is used, the routine assigns uniform weights to the sequences
`-p, --path_params`	None	Path to the file containing the model's parameters. Required for restoring the training
`-c, --path_chains`	None	Path to the FASTA file containing the model's chains. Required for restoring the training
`-l, --label`	None	A label to identify different algorithm runs. It prefixes the output files with this label
`--alphabet`	protein	Type of encoding for the sequences. Choose among `protein`, `rna`, `dna` or a user-defined string of tokens
`--lr`	0.01	Learning rate
`--nsweeps`	10	Number of sweeps for each gradient estimation
`--sampler`	gibbs	Sampling method to be used. Possible options are `gibbs` and `metropolis`
`--nchains`	10000	Number of Markov chains to run in parallel
`--target`	0.95	Pearson correlation coefficient on the two-site statistics to be reached
`--nepochs`	50000	Maximum number of epochs allowed
`--pseudocount`	None	Pseudocount for the single- and two-site statistics. Acts as a regularization. If `None`, it is set to $1/M_{\mathrm{eff}}$
`--seed`	0	Random seed
`--nthreads`*	1	Number of threads used in the Julia multithreads version
`--device`*	cuda	Device to be used between cuda (GPU) and CPU. Used in the Python version
`--dtype`*	float32	Data type to be used between float32 and float64. Used in the Python version

(continued)

Table 2
(continued)

Command	Default value	Description
eaDCA options		
`--gsteps`	10	Number of gradient updates to be performed on a given graph
`--factivate`	0.001	Fraction of inactive couplings to try to activate at each graph update
edDCA options		
`--gsteps`	10	The number of gradient updates applied at each step of the graph convergence process
`--density`	0.02	Target density to be reached
`--drate`	0.01	Fraction of remaining couplings to be pruned at each decimation step

* These arguments are specific to the code implementation

- `<label>` is an optional argument. If provided, it will label the output files. This is helpful when running the algorithm multiple times in the same output folder.

Once started, the training will continue until the Pearson correlation coefficient between the two-point connected correlations of the model and the empirical ones obtained from the data reaches a modifiable target value (set by default at `target = 0.95`). In Fig. 4, we show in a log-log plot the evolution of (one minus) the Pearson correlation coefficient as a function of the training time. The long-time evolution of the training is well approximated by a straight line, meaning that the Pearson growth follows a power law. As such, one should keep in mind that the training procedure converges very rapidly at the beginning (say, it reaches Pearson values of ~ 0.9 in about 100 iterations in Fig. 4), while reaching higher values of the Pearson requires comparatively more algorithmic time. Therefore, if needed, it is possible to fit a coarse model in a very short amount of time by just lowering the `target` value somewhere around 0.9.

Output Files By default the training algorithm outputs three kinds of text files:

- `<label>_params.dat`: *File containing the parameters of the model saved in this format (Fig. 5):*
  - *Lines starting with* `J` *represent entries of the coupling matrix, followed by the two interacting positions in the sequence and the two amino acids or nucleotides involved.*

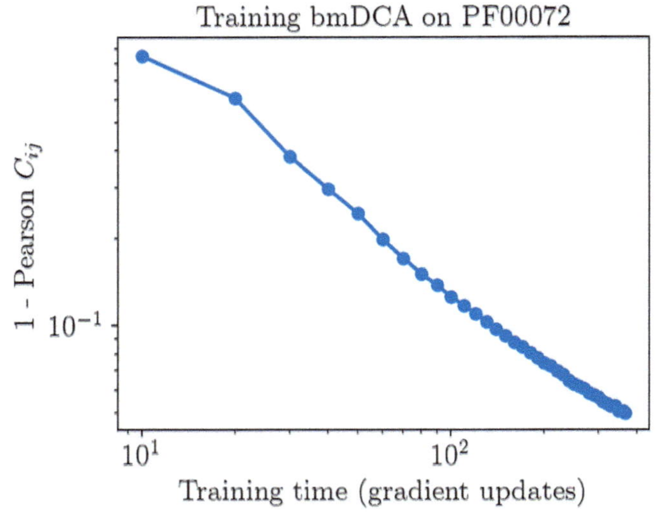

Fig. 4 One minus the Pearson correlation coefficient between the data and chains correlation matrices as a function of the training time for the protein family PF00072. The curve is well approximated by a power law decay

```
J 5 14 A A 0.0002308107
J 5 14 A U -5.8257352e-5
J 5 14 A C -0.00010025737
J 5 14 A G 0.001547538
J 5 14 A - -0.0016467333
J 5 14 U A -0.00046549534
h 12 U -2.2951539
h 12 C -0.9893654
h 12 G 4.0110064
h 12 - -0.44967586
h 13 A 0.415377
h 13 U -1.7539572
```

Fig. 5 Format for saving the coupling matrix *J* and biases *h* for an RNA domain

  – *Lines starting with* `h` *represent the bias, followed by a number and a letter indicating the position and the amino acid or nucleotide subject to the bias.*

  Note that inactive, i.e., zero, couplings are not included in the file.

- <label>_chains.fasta: *FASTA file containing the sequences corresponding to the last state of the Markov chains used during the learning.*
- <label>_adabmDCA.log: *.log file containing information collected during the training procedure.*

During the training, the output files containing the parameters and the chains are overwritten, and the log file is updated, every 50 gradient updates (epochs) for `bmDCA`, and every 10 graph updates for `eaDCA` and `edDCA`.

Restore an Interrupted Training It is possible to start the training by initializing the parameters of the model and the chains at a given checkpoint. To do so, two arguments specifying the path of the parameters and the chains are needed:

```
$ adabmDCA train [...] -p <file_params> -c <file_chains>
```

Importance Weights It is possible to provide the algorithm with a precomputed list of importance weights to be assigned to the sequences (*see* Subheading 2.3) by giving the path to the text file to the argument `-w`. If this argument is not provided, the algorithm will automatically compute the weights using Eq. 7, and it will store them into the folder <output_folder> as <label>_weights.dat. The default sequence identity threshold used for computing the weights is set to 0.8, but this value can be modified using the argument `--clustering_seqid`. It is also possible to ask the routine to use uniform weights for all sequences by means of the flag `--no_reweighting`.

Choosing the Alphabet By default, the algorithm will assume that the input MSA belongs to a protein family, and it will use the preset alphabet defined in Table 1 (by default: `--alphabet protein`). If the input data comes from RNA or DNA sequences, it has to be specified by passing respectively `rna` or `dna` to the `--alphabet` argument. There is also the possibility of passing a user-defined alphabet, provided that all the tokens match with those that are found in the input MSA. This can be useful if one wants to use a different order than the default one for the tokens, or in the eventuality that one wants to handle additional symbols present in the alignment. This is done using `--alphabet ABCD-` if, for example, the alphabet contains the symbols `A, B, C, D, -`.

3.1 eaDCA

To train an eaDCA model, we just have to specify `--model eaDCA`. Two more hyperparameters can be changed:

- `--factivate`: The fraction of inactive couplings that are selected for the activation at each update of the graph. By default, it is set to 0.001.
- `--gsteps`: The number of parameter updates to be performed on the given graph. By default, it is set to 10.

For this routine, the number of sweeps for updating the chains can be typically reduced to 5, since only a fraction of all the possible couplings have to be updated at each iteration.

3.2 edDCA

To launch a decimation with default hyperparameters, use the command:

```
$ adabmDCA train –m edDCA –d <fasta_file> –p <file_params> –c <file_chains>
```

where <file_params> and <file_chains> are, respectively, the file names of the parameters and the chains (including the path) of a previously trained bmDCA model. The edDCA can perform two routines as described above. In the first routine, it uses a pretrained bmDCA model and its associated chains provided through the parameters <file_params> and <file_chains>. The routine makes sure everything has converged before starting the decimation of couplings. It repeats this process for up to 10,000 iterations if needed. If these parameters are not supplied, the second routine initializes the model and chains randomly, trains the bmDCA model to meet convergence criteria, and then starts the decimation process as described. Some important parameters that can be changed are:

- `--gsteps`: The number of parameter updates to be performed at each step of the convergence process on the given graph. By default, it is set to 10.
- `--drate`: Fraction of active couplings to be pruned at each graph update. By default, it is set to 0.01.
- `--density`: Density of the graph that has to be reached after the decimation. The density is defined as the ratio between the number of active couplings and the number of couplings of the fully connected graph. By default, it is set to 0.02.
- `--target`: The Pearson correlation coefficient to be reached to assume the model to have converged. By default, it is set to 0.95.

3.3 How to Choose the Hyperparameters?

The default values for the hyperparameters are chosen to be a good compromise between having a relatively short training time and a good quality of the learned model for most of the *typical* input MSA, where for *typical* we mean a clean MSA with only a few gaps for each sequence and a not too structured dataset (not too clustered in subfamilies) that could create some ergodicity problems during the training. It may happen, though, that for some datasets some adjustments of the hyperparameters are needed to get a properly trained model. The most important ones are:

Learning Rate By default, the learning rate is set to 0.01, which is a reasonable value in most cases. If the resampling of the model is bad (very long thermalization time or mode collapse), one may try to decrease the learning rate through the argument `--lr` to some smaller value (e.g., 0.005 or 0.001).

Number of Markov Chains By default, the number of Markov chains is set to 10000, which works well in most cases. Using fewer chains reduces the memory required to train the model, but it may also lead to a longer algorithm convergence time. To change the number of chains, we can use the argument `--nchains`.

Number of Monte Carlo Steps The argument `--nsweeps` defines the number of Monte Carlo chain updates (sweeps) between one gradient update and the following. A single *sweep* is obtained once we propose a mutation for all the residues of the sequence. By default, this parameter is set to 10, which is a good choice for easily tractable MSAs. The higher this number is chosen, the better the quality of the training will be, because in this way we allow the Markov chains to decorrelate more to the previous configuration. However, this parameter heavily impacts the training time, so we recommend choosing it in the interval 10–50.

Regularization Another parameter that can be adjusted if the model does not resample correctly is the pseudocount, α, which can be changed using the key `--pseudocount`. The pseudocount is a regularization term that introduces a flat prior on the frequency profiles, modifying the frequencies as in Eqs. 8 and 9. If $\alpha = 1$, we impose an equal probability to all the amino acids to be found in the residue at position i, while if $\alpha = 0$, we just use the empirical frequencies of the data. By default, the pseudocount is set as the inverse of the effective number of sequences, $1/M_{\mathrm{eff}}$, where

$$M_{\mathrm{eff}} = \sum_{m=1}^{M} w^{(m)}. \tag{10}$$

4 Applications

4.1 Generate Sequences

Once we have a trained model, we can use it to generate new sequences. This can be done using the command:

```
$ adabmDCA sample –p <path_params> –d <fasta_file> –o <output_folder> ––ngen <num_gen>
```

where `output_folder` is the directory where to save the data and `num_gen` is the number of sequences to be generated. All the possible options for the routine are listed in Table 3. The routine will first compute the mixing time (t^{mix}) of the model by running a simulation starting from the sequences of the input MSA. After that, it will randomly initialize `num_gen` Markov chains and run for `nmix` $\times$ t^{mix} sweeps to ensure that the model equilibrates. It will save in the output directory a FASTA file containing the sequences sampled with the model and a text file containing the records used

Table 3
Command line arguments for sampling from a DCA model

Command	Default value	Description
`-p, --path_params`	N/A	Path to the file containing the parameters of the DCA model to sample from
`-d, --data`	N/A	Filename of the dataset MSA
`-o, --output`	N/A	Path to the folder where to save the output
`--ngen`	None	Number of samples to generate
`-l, --label`	None	A label to identify different algorithm runs. It prefixes the output files with this label
`-w, --weights`	None	Path to the file containing the weights of the sequences. If `None`, the weights are computed automatically
`--clustering_seqid`	0.8	Sequence identity threshold to be used for computing the sequence weights
`--no_reweighting`	N/A	If this flag is used, the routine assigns uniform weights to the sequences
`--nmeasure`	10000	Number of data sequences to use for computing the mixing time. The value min(`nmeasure`, len(data)) is taken
`--nmix`	2	Number of mixing times used to generate "ngen" sequences starting from random
`--max_nsweeps`	10000	Maximum number of sweeps allowed
`--alphabet`	protein	Type of encoding for the sequences. Choose among `protein`, `rna`, `dna`, or a user-defined string of tokens
`--sampler`	gibbs	Sampling method to be used. Possible options are `gibbs` and `metropolis`
`--beta`	1.0	Inverse temperature to be used for the sampling
`--pseudocount`	None	Pseudocount for the single- and two-site statistics. Acts as a regularization. If `None`, it is set to $1/M_{\text{eff}}$
`--device*`	cuda	Device to be used between cuda (GPU) and CPU. Used in the Python version
`--dtype*`	float32	Data type to be used between float32 and float64. Used in the Python version

to determine the convergence of the algorithm. Figure 6 right shows the comparison between the entries of the covariance matrices obtained from the data and from the generated sequences. The Pearson correlation coefficient is the same used as a target for the training and the slope is close to 1, meaning that the model is able to correctly recover the two-site statistics of the data MSA.

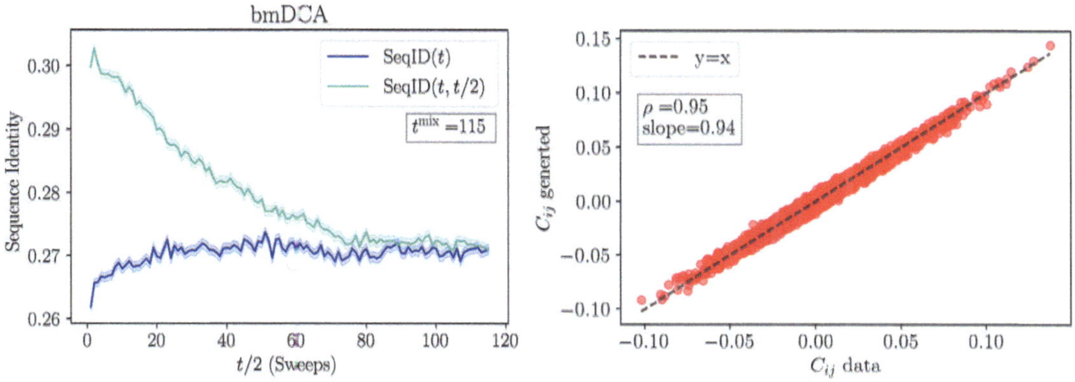

Fig. 6 Analysis of a bmDCA model. **Left**: Measuring the mixing time of the model using 10^4 chains. The curves represent the average overlap among randomly initialized samples (dark blue) and the one among the same sequences between times t and $t/2$ (light blue). Shaded areas represent the error of the mean. When the two curves merge, we can assume that the chains at time t forgot the memory of the chains at time $t/2$. This point gives us an estimate of the mixing time of the model, t^{mix}. Notice that the times start from 1, so the starting conditions are not shown. **Right**: Scatter plot of the entries of the covariance matrix of the data versus that of the generated samples

Convergence Criterion To determine the convergence of the Monte Carlo simulation, the following strategy is used. We extract `nmeasure` $= N$ sequences from the data MSA according to their statistical weight, and we make a copy of them, which are used to initialize a set of N chains. We then compare two sets of chains. The first set represents the chains simulated up to time t, and we denote them as $\boldsymbol{A}(t) = \{\boldsymbol{a}_1(t), \ldots, \boldsymbol{a}_N(t)\}$, while the sequences of the second set are the chains simulated until time $t/2$, and we call them $\boldsymbol{A}(t/2) = \{\boldsymbol{a}_1(t/2), \ldots, \boldsymbol{a}_N(t/2)\}$. With some abuse of notation, we define the *intrachain correlation* and *autocorrelation* as

$$\mathrm{SeqID}(t) = \frac{1}{N} \times \sum_{i=1}^{N} \mathrm{SeqID}(\boldsymbol{a}_i(t), \boldsymbol{a}_{\sigma(i)}(t)), \quad \mathrm{SeqID}(t, t/2) = \frac{1}{N} \times \sum_{i=1}^{N} \mathrm{SeqID}(\boldsymbol{a}_i(t), \boldsymbol{a}_i(t/2)), \tag{11}$$

where $\sigma(i)$ is a random permutation of the index i and

$$\mathrm{SeqID}(\boldsymbol{a}, \boldsymbol{b}) = \frac{1}{L} \sum_{i=1}^{L} \delta_{a_i, b_i} \in [0, 1] \tag{12}$$

is the normalized sequence identity (or overlap) between the sequences $\boldsymbol{a}$ and $\boldsymbol{b}$. By construction, at the initialization, we have $\mathrm{SeqID}(t, t/2) = 1$ and $\mathrm{SeqID}(t)$ somewhat close to the average sequence identity of the MSA. The convergence is obtained when chains are *mixed*, meaning that the system has completely forgotten the initial configuration. This requirement is satisfied when the statistics of a set of independent chains is the same as the one between the initialization and the evolved chains, meaning $\mathrm{SeqID}(t) \cong \mathrm{SeqID}(t, t/2)$. The point at which the two curves merge is called *mixing time*, and we denote it as t^{mix}. After reaching the mixing time of the model, the algorithm will initialize `ngen` chains at random. It will run a sampling for other `nmix` $\times$ t^{mix} steps to guarantee complete thermalization, with `nmix=2` by default. Together with the generated sequences, the script will output a text file containing the records of $\mathrm{SeqID}(t)$ and $\mathrm{SeqID}(t, t/2)$ and their standard deviations (Fig. 6 left).

4.2 Contact Prediction

One of the principal applications of the DCA models has been that of predicting a tertiary structure of a protein or RNA domain. In particular, with each pair of sites *i* and *j* in the MSA, `adabmDCA 2.0` computes a contact score that quantifies how likely the two associated positions in the chains are in contact in the three-dimensional structure. Formally, it corresponds to the average-product-corrected (APC) Frobenius norms of the coupling matrices [10], i.e.,

$$F_{i,j}^{\mathrm{APC}} = F_{i,j} - \frac{\sum_k F_{i,k} \sum_k F_{k,j}}{\sum_{kl} F_{k,l}}, \qquad F_{i,j} = \sqrt{\sum_{a,\, b \neq '-'} J_{i,j}(a,b)^2}\,. \tag{13}$$

Note that the coupling parameters are usually transformed in a zero-sum gauge before computing the scores, and the gap symbol should be neglected while computing the sum in Eq. 13 [13]. The scores for all site pairs are provided in the output folder in a separate file called `<label>_frobenius.txt`. The first two columns indicate the site indices, and the third one contains the associated APC Frobenius norm. The command for computing the matrix of Frobenius norms is

```
$ adabmDCA contacts -p <file_params> -o <output_folder>
```

The possible options are listed in Table 4.

4.3 Scoring a Sequence Set

At convergence, users can score a set of input sequences according to a trained DCA model by using the command line

```
$ adabmDCA energies -d <fasta_file> -p <file_params> -o <output_folder>
```

`adabmDCA 2.0` will produce a new FASTA file in the output folder <output_folder> where the input sequences in <fasta_file> have an additional field in the name that accounts for

Table 4
Command line arguments for computing the Frobenius contact matrix of a DCA model

Command	Default value	Description
`-p, --path_params`	N/A	Path to the file containing the parameters of the DCA model
`-o, --output`	N/A	Path to the folder where to save the output
`-l, --label`	None	If provided, adds a label to the output files inside the output folder
`--alphabet`	protein	Type of encoding for the sequences. Choose among `protein`, `rna`, `dna`, or a user-defined string of tokens
`--device`*	cuda	Device to be used between cuda (GPU) and CPU. Used in the Python version
`--dtype`*	float32	Data type to be used between float32 and float64. Used in the Python version

Table 5
Command line arguments for computing the DCA energies of an MSA

Command	Default value	Description
`-d, --data`	N/A	Filename of the input MSA
`-p, --path_params`	N/A	Path to the file containing the parameters of the DCA model
`-o, --output`	N/A	Path to the folder where to save the output
`--alphabet`	protein	Type of encoding for the sequences. Choose among `protein`, `rna`, `dna`, or a user-defined string of tokens
`--device`*	cuda	Device to be used between cuda (GPU) and CPU. Used in the Python version
`--dtype`*	float32	Data type to be used between float32 and float64. Used in the Python version

the DCA energy function computed according to the model in <file_params>. Note that low energies correspond to good sequences. *See* Table 5 for the available options.

4.4 Single-Mutant Library

Another possible application exploits the sequence-fitness score computable according to the energy function E in Eq. 1. This routine provides a single-mutant library for a given wild type to possibly guide deep mutational scanning (DMS) experiments. In particular, `adabmDCA 2.0` allows one to predict the fitness reduction (increase) in terms of $\Delta E = E(mutant) - E(wildtype)$ for

Table 6
Command line arguments for generating the deep mutational scan of a provided wild-type sequence

Command	Default value	Description
`-d, --data`	N/A	Filename of the input MSA containing the wild type. If multiple sequences are present, the first one is used
`-p, --path_params`	N/A	Path to the file containing the parameters of the DCA model
`-o, --output`	N/A	Path to the folder where to save the output
`--alphabet`	protein	Type of encoding for the sequences. Choose among `protein`, `rna`, `dna`, or a user-defined string of tokens
`--device`*	cuda	Device to be used between cuda (GPU) and CPU. Used in the Python version
`--dtype`*	float32	Data type to be used between float32 and float64. Used in the Python version

positive (negative) value of ΔE, respectively [14]. To produce a FASTA file containing all weighted single mutants, one has to run

```
$ adabmDCA DMS –d <WT> –p <file_params> –o <output_folder>
```

where `<WT>` is the name of the FASTA file containing the wild-type sequence, `<file_params>` is a model file in a format compatible with `adabmDCA 2.0` output, and `<output_folder>` corresponds to the folder that will contain the library file. The sequences in the output FASTA file are named after the introduced mutation and the corresponding ΔE ; for instance, `>G27A | DCAscore: -0.6` denotes that position `27` has been changed from `G` to `A` and $\Delta E = -0.6$. The routine arguments are listed in Table 6.

5 Source Code

The code and the documentation for the `adabmDCA 2.0` package can be found at https://github.com/spqb/adabmDCA. We also prepared a Colab notebook with a simple tutorial for training and sampling a `bmDCA` model.

6 Notes

6.1 Commands

- Train a bmDCA model with default arguments:

```
$ adabmDCA train –d <fasta_file> –o <output_folder>
```

- Restore the training of a bmDCA model:

```
$ adabmDCA train –d <fasta_file> –o <output_folder> –p <file_params> –c <file_chains>
```

- Train an eaDCA model with default arguments:

```
$ adabmDCA train –m eaDCA –d <fasta_file> –o <output_folder> ––nsweeps 5
```

- Restore the training of an eaDCA model:

```
$ adabmDCA train –m eaDCA –d <fasta_file> –o <output_folder> –p <file_params> –c <file_chains>
```

- Decimate a bmDCA model at 2% of density:

```
$ adabmDCA train –m edDCA –d <fasta_file> –p <file_params> –c <file_chains>
```

- Train and decimate a bmDCA model at 2% of density:

```
$ adabmDCA train –m edDCA –d <fasta_file>
```

- Sample from a previously trained DCA model:

```
$ adabmDCA sample –p <file_params> –d <fasta_file> –o <output_folder> ––ngen <num_gen>
```

- Scoring a sequence set:

```
$ adabmDCA energies –d <fasta_file> –p <file_params> –o <output_folder>
```

- Generating a single-mutant library starting from a wild type:

```
$ adabmDCA DMS –d <WT> –p <file_params> –o <output_folder>
```

- Computing the matrix of Frobenius norms for the contact prediction:

```
$ adabmDCA contacts –p <file_params> –o <output_folder>
```

References

1. Cocco S, Feinauer C, Figliuzzi M, Monasson R, Weigt M (2018) Inverse statistical physics of protein sequences: a key issues review. Rep Progr Phys 81(3):032601.
2. Russ WP, Figliuzzi M, Stocker C, Barrat-Charlaix P, Socolich M, Kast P, Hilvert D, Monasson R, Cocco S, Weigt M (2020) An evolution-based model for designing chorismate mutase enzymes. Science 369(6502): 440–445
3. Morcos F, Pagnani A, Lunt B, Bertolino A, Marks DS, Sander C, Zecchina R, Onuchic JN, Hwa T, Weigt M (2011) Direct-coupling analysis of residue coevolution captures native contacts across many protein families. Proc Natl Acad Sci 108(49):E1293–E1301
4. Figliuzzi M, Jacquier H, Schug A, Tenaillon O, Weigt M (2016) Coevolutionary landscape inference and the context-dependence of mutations in beta-lactamase TEM-1. Mol Biol Evol 33(1):268–280
5. Cuturello F, Tiana G, Bussi G (2020) Assessing the accuracy of direct-coupling analysis for RNA contact prediction. RNA 26(5):637–647
6. Muntoni AP, Pagnani A, Weigt M, Zamponi F (2021) adabmDCA: adaptive Boltzmann machine learning for biological sequences. BMC Bioinform 22(1):528
7. Figliuzzi M, Barrat-Charlaix P, Weigt M (2018) How pairwise coevolutionary models capture the collective residue variability in proteins? Mol Biol Evol 35(4):1018–1027
8. Calvanese F, Lambert CN, Nghe P, Zamponi F, Weigt M (2024) Towards parsimonious generative modeling of RNA families. Nucleic Acids Res 52(10):5465–5477
9. Barrat-Charlaix P, Muntoni AP, Shimagaki K, Weigt M, Zamponi F (2021) Sparse generative modeling via parameter reduction of Boltzmann machines: application to protein-sequence families. Phys Rev E 104(2):024407
10. Ekeberg M, Lövkvist C, Lan Y, Weigt M, Aurell E (2013) Improved contact prediction in proteins: using pseudolikelihoods to infer Potts models. Phys Rev E: Stat Nonlinear Soft Matter Phys 87(1):012707
11. Decelle A, Furtlehner C, Seoane B (2021) Equilibrium and non-equilibrium regimes in the learning of restricted Boltzmann machines. In: Advances in neural information processing systems, vol. 34. Curran Associates, pp 5345–5359
12. Pearson WR, Lipman DJ (1988) Improved tools for biological sequence comparison. Proc Natl Acad Sci 85(8):2444–2448
13. Feinauer C, Skwark MJ, Pagnani A, Aurell E (2014) Improving contact prediction along three dimensions. PLoS Comput Biol 10(10): e1003847
14. Hopf TA, Ingraham JB, Poelwijk FJ, Schärfe CP, Springer M, Sander C, Marks DS Mutation effects predicted from sequence co-variation. Nat Biotechnol 35(2):128–135

Chapter 7

Unlocking Protein Evolution Insights: Efficient and Interpretable Mutational Effect Predictions with GEMME

A. Carbone, E. Laine, and G. Lombardi

Abstract

The Global Epistatic Model for predicting Mutational Effects (GEMME) is a computational method that reconstructs protein mutational landscapes using sequence data alone. In this chapter, we delve into the broader biological questions that GEMME can address beyond mere landscape reconstruction. We provide several examples to guide users in maximizing the utility of GEMME. Additionally, we discuss recent advancements that enhance GEMME's predictive accuracy and extend its capabilities by integrating sequence data with structural information and allele frequency data, when available. These innovations enable GEMME to answer new biological questions and offer deeper insights into protein evolution and function.

Key words Evolution, Sequence, Protein, Mutation, Mutational effect

1 Introduction

The systematic and accurate assessment of the impact of mutations on protein functioning is of critical importance in biology, bioengineering, and medicine. The Global Epistatic Model for predicting Mutational Effects (GEMME) is a computational tool that reconstructs these landscapes using only protein sequence data across species, significantly contributing to recent advancements in this field [1].

GEMME is an unsupervised and rapid method that predicts the effects of mutations by explicitly modeling the evolutionary history of natural sequences. It employs a few biologically meaningful and interpretable parameters and globally considers epistasis by evaluating the whole sequence context when assessing a mutation's impact. Given an input alignment, GEMME can generate a protein's full mutational landscape by assessing the impact of all possible single amino acid substitutions in just a few minutes. It

Shahid M. Khan and Florencio Pazos (eds.), *Protein Evolution: Methods and Protocols*, Methods in Molecular Biology, vol. 2979, https://doi.org/10.1007/978-1-0716-4828-5_7,

performs on par with high-capacity deep-learning-based architectures [2]. It is applicable to both single-site mutations and combinations of mutations. It is freely available for the community through a standalone package and a webserver at http://www.lcqb.upmc.fr/GEMME.

GEMME proved useful for discovering functionally important residues in proteins and distinguishing them from residues essential to protein thermodynamic stability [3–5]. It helped to decipher the molecular mechanisms underlying various diseases [6, 7] and to estimate mutation-induced binding affinity changes in protein-protein interactions [8]. Moreover, using GEMME scores as surrogates for experimental measurements allowed for boosting protein language model (pLM) accuracy and speed [9] with the resulting predictor, VespaG, combining the best of both worlds, the robustness and transparency of GEMME with pLM universal protein representation space.

This contribution explores broader biological questions that GEMME can address beyond landscape reconstruction by providing four examples to help users maximize GEMME's utility. They are presented with a fine-tuned version of GEMME, named iGEMME, with optimized parameters enhancing its performance [10] (*see* **Note 1**).

We also highlight recent advancements, ESCOTT and PRESCOTT [10], which improve GEMME's predictive accuracy and extend its capabilities. ESCOTT integrates sequence data with structural information, while PRESCOTT incorporates both structural information and allele frequency data when available. These innovations enable GEMME to answer new biological questions and provide deeper insights into protein evolution and function.

1.1 Biological Questions GEMME Can Help to Address

Multiple questions arise directly from mutational landscape reconstructions:

Do mutational landscapes provide insights into which regions along the protein sequence are more sensitive to mutations than others?

When structural data is available, do mutational landscapes indicate which three-dimensional regions of the protein are more sensitive to mutations?

Can GEMME be used to discover and annotate sites in protein sequences with distinguished functional and structural roles?

Can GEMME identify key regions in intrinsically disordered proteins that might suggest interaction sites?

More broadly, what are the mechanisms underlying disease mutations at a proteome-wide scale?

Is there a correlation between regions predicted to be sensitive by GEMME and regions where tools reconstructing the structure, like AlphaFold [11], ESMFold [12], and RosettaFold [13], provide confident predictions?

Are regions that AlphaFold, ESMFold, RosettaFold predict with uncertainty discovered to be sensitive to mutations by GEMME? In other words, can GEMME provide complementary information about a protein structure that structure predictors do not reveal?

How can we visualize sensitive regions within a protein sequence? and on a protein structure, whenever available?

Can GEMME model go beyond one mutation? Is GEMME model useful to predict the effect of two or more mutations?

Based on the analysis of four distinct proteins, we will guide the user in utilizing GEMME to answer to these questions: The thioredoxin TRX is a widespread protein family with a highly conserved structure across the tree of life [14–16]; the homeobox protein ARX is a disordered protein with no structural prediction available from AlphaFold [17, 18]; the chromodomain of the heterochromatine protein 1 HP1, a major component of heterochromatin, directly binds to H3K9me [19–21]; the nuclease NucB is an enzyme used to explore generated sequences for designing optimized proteins in various environments [22, 23].

2 Methods

2.1 A Guide to Generating Mutational Landscapes

GEMME/iGEMME takes a protein sequence as a reference and generates a mutational landscape for that sequence. We provide here a brief description of the necessary input, expected output, and execution protocol for running GEMME/iGEMME. Additionally, practical examples are included to guide users through various analyses that can be performed with the software. The recent development of ESCOTT, which integrates structural information into GEMME/iGEMME, prompts us to outline the ways to use this new generation software. This section will be followed by several examples to practice on the analyses you can run with GEMME/iGEMME and ESCOTT integrating it with other available software tools.

Input Requirements To use GEMME/iGEMME, the following input files are required:

- *A multiple sequence alignment (MSA) in FASTA format (mandatory). The query protein must be the first sequence in the MSA and should not contain any gap, ensuring that the alignment length matches the query sequence length* L.

- *A TXT file with a list of mutations to analyze (optional). Each mutation should be listed on a different line in the format* <wild-type aa><position><mutated aa> *(e.g., A26P), where positions are 1-indexed. Multiple-point mutations should be separated by a colon character (e.g., C13T:A37L). The format is case sensitive, and whitespace should be avoided. If this file is not provided, the software will generate the full mutational landscape by default.*
- *A structure file in PDB format (only for ESCOTT). If no PDB file is provided, ESCOTT will default with iGEMME computation. Ensure that the sequence in the PDB file precisely matches the query sequence, with no extra residues or chains.*

Input Preparation A quick and reliable way to obtain both the input MSA and a PDB file for a query protein is by using ColabFold (https://colab.research.google.com/github/sokrypton/ColabFold/blob/main/AlphaFold2.ipynb) or using the AlphaFold server (https://alphafoldserver.com/). *See* **Notes 2** and *3*. The obtained alignments (a3m file) can be converted into the correct input format with the following commands, relying on the hhsuite package (https://github.com/soedinglab/hh-suite) [24]:

```
# Convert a3m -> fasta format, with hhsuite reformat.pl script
reformat.pl a3m fas ali.a3m ali.fasta -r

# Reduce length of headers in fasta file (the program might throw error otherwise)
awk 'BEGIN{FS="[ |]"}{if(NF>1) {printf(">%s\n", $1)}else{print $0}}' ali.fasta >
    ali_processed.fasta
```

A more detailed description of inputs formats is found in the scripts repository.

Program Execution GEMME and iGEMME/ESCOTT can be installed locally or accessed through dedicated webservers (*see* **Note 4** for related links). To run the software locally, we recommend using Docker images. Once they are downloaded from Docker Hub, the basic syntax for the execution of the programs is the following:

```
IMAGE=elodielaine/gemme:gemme  # for GEMME
IMAGE=tekpinar/prescott-docker:v1.6.0  # for ESCOTT/PRESCOTT
cd inputs_directory  # all inputs have to be in the same directory

# Run docker container
docker run -ti --rm --mount type=bind,source=$PWD,target=/example $IMAGE

# Move to mounted directory and execute the software
cd /example
python2.7 $GEMME_PATH/gemme.py ali.fasta -f ali.fasta <-m mutations.txt>  # for GEMME
escott ali.fasta <-p pdbfile> <-m mutations.txt>  # for ESCOTT
```

The full list of input options is found in the documentation.

Python Scripts We provide Python script to facilitate inputs preparation, Docker container execution, and outputs visualization, available at http://gitlab.lcqb.upmc.fr/lombardi/gemme-utils:

– `prepare_alignment.py` *(requirements: None): Convert FASTA-like alignment file (A2M, A3M, FASTA) into GEMME/ESCOTT input format, by removing gaps from query sequence, shortening headers and converting to FASTA. The output is saved as* `{alignment_filename}_processed.fasta`. *Usage and example:*

```
prepare_alignment.py [-h] alignment_file
python prepare_alignment.py test_data/trxf2.a3m
```

– `prepare_pdb.py` (requirements: BioPython package): Remove all entities but the query sequence from PDB input file for ESCOTT execution. The output is saved as `{pdb_filename}_processed.fasta`. Usage and example:

```
prepare_pdb.py [-h] pdb_file alignment_file
python prepare_pdb.py test_data/blat-af2.pdb test_data/aliBLAT.fasta
```

– `docker_run.py` (requirements: Docker): Run GEMME/ESCOTT Docker container (corresponding Docker image is downloaded if not present) and generate output files. It is also possible to run predictions for different queries in parallel. Usage and example:

```
docker_run.py [-h] [-n NUM_PROCESSES] software alignment_files [alignment_files ...] ...
python docker_run.py escott test_data/aliBLAT.fasta -n 1 --pdbfile test_data/blat-af2.pdb
    \
-m test_data/BLAT_ECOLX_Stiffler_2015_experimental.dat
```

– `plot_on_structure.py` (requirements: Pandas package, PyMOL): Load protein structure in new PyMOL session and plot average per-residue predicted score. Red color is assigned to positions more sensible to mutations, while blue to positions with no mutational effect. Usage and example:

```
plot_on_structure.py [-h] [--save] gemme_output_file pdb_file
python plot_on_structure.py test_data/BLAT_normPred_evolCombi_test.txt test_data/blat-af2.
    pdb
```

Outputs There are several output files that are provided:

- normPred_evolCombi.txt. *This is the main output of the software. It can be a* $L \times 20$ *matrix describing the full predicted mutational landscape through scores associated to each mutation, or a 2-column dataframe, listing mutations and corresponding scores, if the optional TXT file is provided. A graphical representation of the 2D matrix is generated and saved in* normPred_evolCombi.<png, jpg> *file.*
- normPred_evolInd.txt *and* normPred_evolEpi.txt *(optional). Intermediate predictions that contribute to GEMME/iGEMME and ESCOTT outputs, derived from site-independent frequencies, evolutionary and structural (for ESCOTT) fit, respectively, with the same format as* normPred_evolCombi.txt *and corresponding images. In ESCOTT, these outputs are not returned if* verbose=False *argument is set.*
- normPred_evolCombi_TransposedRanksorted.csv *(only for ESCOTT). A transposed and ranksorted version of* normPred_evolCombi.txt*, where scores vary from 0 to 1, and higher values are assigned to more impactful mutations.*

Output matrices can be loaded as DataFrames with the following Python function:

```python
import pandas as pd

def load_scores(prediction_file: str) -> pd.DataFrame:
    """Load the scores from the prediction file."""
    if prediction_file.endswith(".csv"):
        scores = pd.read_csv(prediction_file, index_col=0)
    elif prediction_file.endswith(".txt"):
        scores = pd.read_csv(prediction_file, sep=r'\s+', index_col=0).transpose()
    scores.fillna(0, inplace=True)  # set NaN values to 0
    return scores
```

Comprehensive documentation is available for both the software and online servers, detailing installation, execution, and a complete list of parameters, inputs, and outputs (*see* **Note 4**). GEMME/iGEMME ensures high computational efficiency and rapid execution (*see* **Note 5**). The integration of structural data with ESCOTT and allele frequency data with PRESCOTT further enhances the accuracy of mutational effect predictions (*see* **Note 6**). Additionally, GEMME, iGEMME, ESCOTT, and PRESCOTT adhere to the Atlas of Variant Effects (AVE) Alliance Guidelines (*see* **Note 7**).

2.2 GEMME/iGEMME Reconstructs Mutational Landscapes of Proteins: Input and Output

In this first example, we show the reconstruction of the iGEMME mutational landscape matrix for the thioredoxin protein TRX of type f from *Chlamidomonas reinhardtii* (Fig. 1a). GEMME/iGEMME predicts the mutational landscape of the reference protein sequence by analyzing evolutionary constraints encoded in a MSA (Fig. 1b). The reference sequence must be gap-free, meaning any MSA columns containing gaps in the reference should be

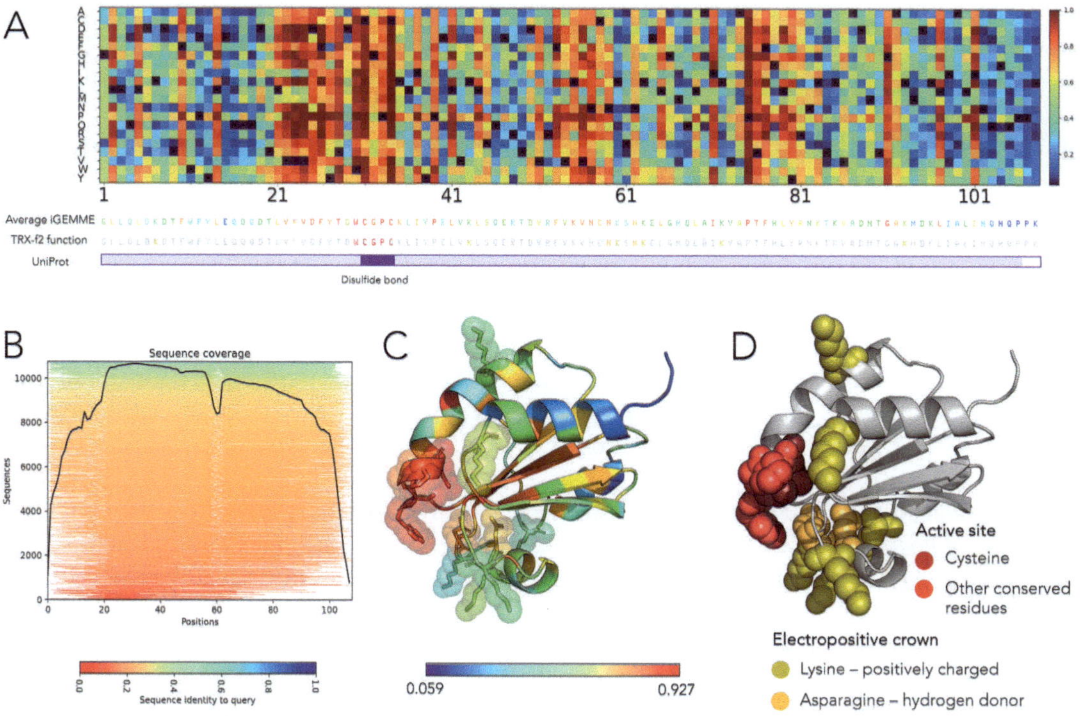

Fig. 1 iGEMME analysis of the thioredoxin, type f, TRX-f2. (**a**) iGEMME matrix of TRX-f2. iGEMME scores are ranksorted from high effect (red) to no effect (blue) based on the color bar on the right. Residues associated to the 108 TRX-f2 sequence positions are colored with the average score of the columns in the iGEMME matrix. Residues known to be functionally important for the TRX-f2 protein are reported with colors as in the legend in C, and all others are in gray. Uniprot (ID A0A2K3DSC9) domain annotation with the TRX domain highlighted in light purple and the disulfide bond motif in dark purple. (**b**) ColabFold profile of sequence coverage (black line) for the MSA used by iGEMME to predict TRX mutational effects. The sequence identity to the query is described by the colors. (**c**) TRX-f2 structure (PDB 6i1c) reports average iGEMME scores. Same colors scale as in A. (**d**) TRX-f2 structure reports residues known to play an important biological role for TRX-f2: the extended disulfide bond motif (WCGPC) in red tones, positively charged crown surrounding the active site, comprised of lysines and asparagines, in orange tones. All other residues are left gray. The two cysteines in the WCGPC motif, colored in dark red, are positioned on the back of the red motif on the left-hand side of the structure and are barely visible

removed, maintaining a fixed length L. The tool outputs an $L \times 20$ matrix, where each position is assigned a score across the 20 amino acids, capturing the predicted effects of single-site mutations. The iGEMME scores, corresponding to single mutations relative to the reference sequence, are visualized using a color scale: Red indicates highly sensitive mutations, whereas blue signifies neutral mutations with no mutational effect.

Several positions in the matrix exhibit red tones for most or all mutations at that position, indicating highly sensitive sites. These sites are often grouped in consecutive regions and typically correspond to structured areas with potential functional roles. They could also be located on small loops or be indicators of the formation of transient structures. In contrast, positions displaying blue tones indicate areas with no mutational effect, likely in parts of the protein that remain unstructured throughout its lifetime, but not only, as illustrated by the blue helix in Fig. 1c.

GEMME/iGEMME scores averaged per column reflect residues' sensitivity to mutations. Hence, a useful way to analyze GEMME/iGEMME results and capture positional relevance in the global context is to map these average values on an experimental or predicted three-dimensional (3D) structure for the reference protein sequence. This approach allows for identifying 3D regions sensitive to mutations. *See* Fig. 1a for an example of average scores associated with residues in the sequence, and Fig. 1c for the plot of these averages on the structure. Sensitive residues are highlighted in red in both panels.

2.3 Recognition and Discovery of Positions and Regions Highly Sensitive to Mutations

Analyzing the average sensitivity of positions in a sequence or structure with GEMME/iGEMME provides biological insights into 2D/3D regions that are more mutation sensitive than others. For instance, in Fig. 1, the known disulfite bond motif at positions 102–105 (CGPC) is identified by iGEMME as highly important (red) for all thioredoxins. Specifically, iGEMME identifies the extended motif WCGPCK as highly significant (red residues in Fig. 1d), supporting existing observations that additional residues are involved in an extended active-site three-dimensional motif controlling the reactivity of the thioredoxin fold [25]. Key residues are located on the beta sheets in front of the motif, which iGEMME identifies as highly sensitive regions.

For the specific TRX-f2 protein, it has been observed [26] that the disulfite bond motif is surrounded by a crown of lysines (Lys7, Lys43, Lys60, Lys63, Lys72, Lys93; olive green residues in Fig. 1d) and asparagines (Asn59, Asn62; orange residues in Fig. 1d). Most of these residues are identified as important (score > 0.4) by iGEMME. Compare Fig. 1d with Fig. 1c, where scores are represented by residue colors, ranging from red to green.

GEMME/iGEMME can also help to discover new regions of the protein that have not yet been identified with biological significance. Several regions in the TRX-f2 sequence are highlighted by iGEMME as important (red regions in Fig. 1c), but these regions have not yet been mentioned in the literature. In conclusion, by inspecting the global matrix of mutational effects, iGEMME can help discover important functional sites that have not been previously annotated in databases like UniProt [27] and InterPro [28] and guide the design of mutational experiments to explore these sites further.

2.4 Finding Functional Motifs in Intrinsically Disordered Regions or Regions with Few Homologs

Intrinsically disordered proteins or disordered regions within proteins are expected to interact with other molecules, including structured proteins [29, 30]. During complex formation, some transient structures may form. The precise location of these transient motifs, which help in binding with partners, remains an open question. GEMME/iGEMME can likely contribute to this area of study. These ephemeral regions are thought to be encoded in sequences that meet specific physico-chemical and geometrical conditions, which should be detectable by evolutionary sequence constraints.

As illustrated in Fig. 1, relevant sensitive sites can be highlighted using iGEMME matrix, and new suggested motifs can be further annotated to understand protein function. This is particularly important when no structural model is available for the protein, as shown for the protein ARX, which is associated with mental retardation and epilepsy [31], in Fig. 2. Annotations from Uniprot indicate that ARX exhibits a high degree of disorder primarily concentrated in the initial segment of the protein (1–255 in Fig. 2a), while the folded portion encompasses the homeobox domain (328–390) and the OAR region (530–543). iGEMME identifies distinct signals of sensitivity within the disordered regions, indicating potential structure formation. Specifically, these signals correspond to residues 1–21, 50–81, 157–181 in the initial segment, and also between the homeobox and OAR domains (391–529). These regions are highly disordered in the AlphaFold model (Fig. 2c). The prediction provided by ESMFold shows a more structured pattern (Fig. 2d) than AlphaFold, but with comparable confidence for both predictors (Fig. 2e). Thus, GEMME/iGEMME can serve as a valuable complementary tool to state-of-the-art protein structure predictors.

A recent study used ARX to illustrate the ability of the protein language model ESM1b to detect signals independently of MSAs, allowing it to transfer information from well-covered regions in proteins to poorly covered ones [32]. However, we found that GEMME/iGEMME can recover similar signals directly and solely from an input MSA, even if the latter does not uniformly cover the sequence and includes regions with very reduced coverage or highly divergent sequences (Fig. 2b). This finding is further supported by the comparison of iGEMME results with the zero-shot scores computed with the pLM ESM2 (650M parameters) [12], which are derived from log-odds ratios of predicted amino acids probabilities for each position (Fig. 2f). Both iGEMME and ESM2 managed to pick up signals in disordered regions or regions with few homologs, and the identified sensitive regions overlap to some extent. Note that ESM2 identifies several regions (green boxes in Fig. 2g) that iGEMME does not mark as significant. Interestingly, these regions consist of stretches of identical amino acids, which

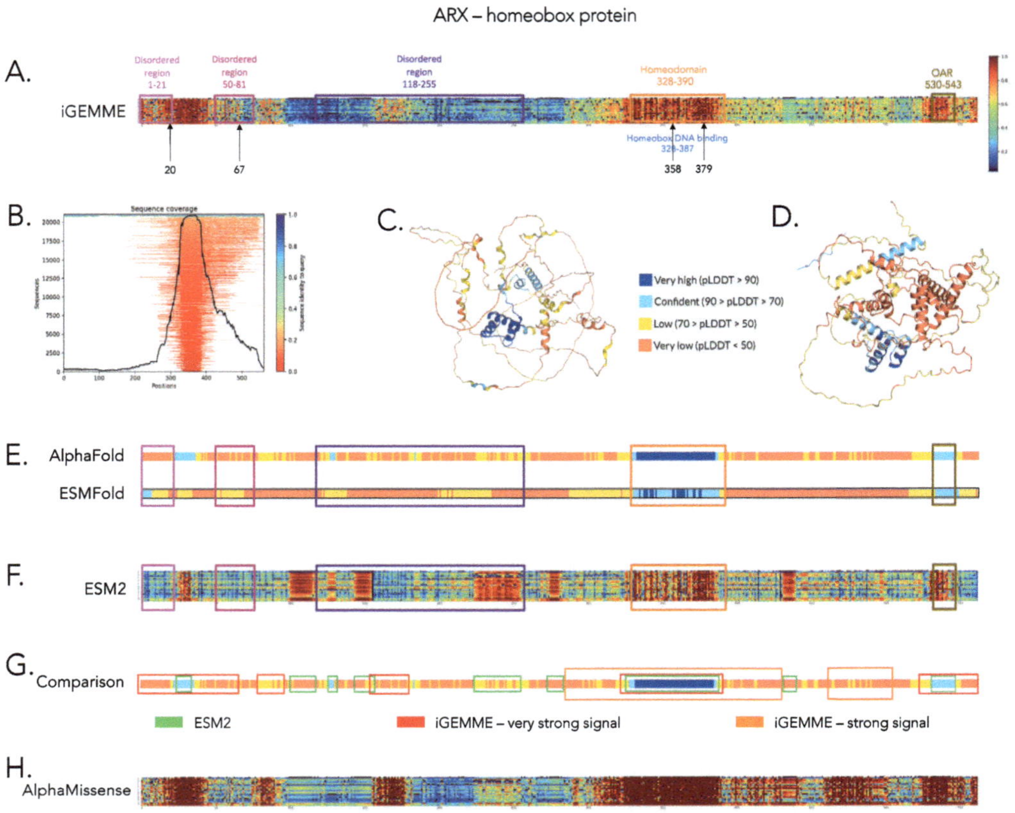

Fig. 2 The human homeobox protein ARX analyzed with iGEMME and other methods. (**a**) iGEMME matrix is annotated with domain information coming from UniProt (www.uniprot.org/uniprotkb/Q96QS3/entry). Residues 20 and 67 are annotated as phosphoserines. Positions 358 and 379 in the homeobox domain are known to abolish sequence-specific DNA binding, to reduce transcriptional repression of LMO1 and SHOX2, and abolish transcriptional activation of the enzyme KDM5C. iGEMME scores are ranksorted. (**b**) ColabFold profile of sequence coverage (black line) for the MSA used by iGEMME to predict ARX mutational effects. The sequence identity to the query is described by the colors. (**c**) AlphaFold2 3D model, downloaded from UniProt, see B. (**d**) ESMFold 3D model, computed locally. (**e**) Positional confidence of the AlphaFold2 and ESMFold predicted structures. For each residue along the sequence, confidence is measured by pLDDT scores grouped into four main classes: high (blue), medium (light blue), low (yellow), and very low (orange). AlphaFold prediction is retrieved from UniProt (www.uniprot.org/uniprotkb/Q96QS3/entry). ESMFold prediction has been computed locally. Domain annotation is reported from A. (**f**) ESM2 zero-shot predictions of ARX mutational effects. ESM2 scores are ranksorted. (**g**) The AlphaFold2 confidence bar in B is aligned with regions in iGEMME (red boxes for very high scores and orange boxes for high scores—see red and orange colors in A) and in ESM2 (green boxes; see red colors in E). (**h**) AlphaMissense prediction of ARX mutational effects. Compare with panel A

ESM2 may have detected as relevant due to its contextual sequence analysis.

Hence, this example illustrates the ability of GEMME/iGEMME to detect signals even in regions where the available sequence data is limited (Fig. 2b). These results are consistent with our previous studies emphasizing the robustness of GEMME with respect to low MSA variability [1, 10, 33] and its high performance on viral proteins [1, 2]. This is a significant finding because, given the speed of GEMME/iGEMME and the rapid generation of MSAs with ColabFold [34], the advantage of using protein language models to predict mutational effects becomes less clear.

We also specifically want to highlight that GEMME's performance can be directly compared with more data-intensive tools like AlphaMissense [35], which relies on the AlphaFold model and incorporates extensive additional information including structural data and allele frequencies. This comparison is evident when examining Fig. 2g (red and orange regions) alongside Fig. 2h (red regions), demonstrating that iGEMME, which uses only an MSA, can achieve similar results despite the more limited data input.

2.5 Identification of Residues Essential for Protein Functioning and Not Folding

Contrasting GEMME/iGEMME evolutionary-based analysis with complementary information about thermodynamic stability can reveal functionally important residues not essential for folding. We illustrate this concept with the chromodomain of HP1 (Fig. 3) for which folding stability measurements of 90% of all possible single mutants are available [3]. Computed from an input MSA containing over 10,000 diverse sequences (Fig. 3a), GEMME mutational landscape (Fig. 3b) uncovered four residues, namely F2, Y26, D30, and E34, whose evolutionary constraints suggest a functional role, while they are unimportant for thermodynamic stability (Fig. 3c, orange dots). These residues are essential for the binding of methylated histone H3 (Fig. 3d, left panel), either by establishing a direct contact or by forming a cage into which the methylammonium group inserts [36]. In addition, GEMME identified six residues highly sensitive to mutations, namely I7, Y19, V21, W23, W33, F50, that are also crucial for the domain's thermodynamic stability (Fig. 3c, red dots). These residues are all hydrophobic and buried or part of well-defined secondary structures (Fig. 3d, right panel). Overall, GEMME evolutionary scores are consistent with physical constraints, and none of the residues predicted as tolerant to mutations are structurally critical (Fig. 3c). A more systematic comparison of GEMME scores and mutation-induced folding stability changes can be found in [3].

2.6 Prediction of Multiple Mutations

GEMME/iGEMME is designed to predict multiple mutations by independently evaluating each mutation. While this model does not fully capture the interdependence of residue positions within a sequence, iGEMME's performance remains competitive with many other zero-shot predictors, exclusively based on sequences. To demonstrate this, we analyzed one of the most extensive genotype-phenotype enzyme activity landscape to date. This dataset comprises 55,759 diverse variants of the nuclease enzyme NucB (github.com/google-deepmind/nuclease_design) [37]. The

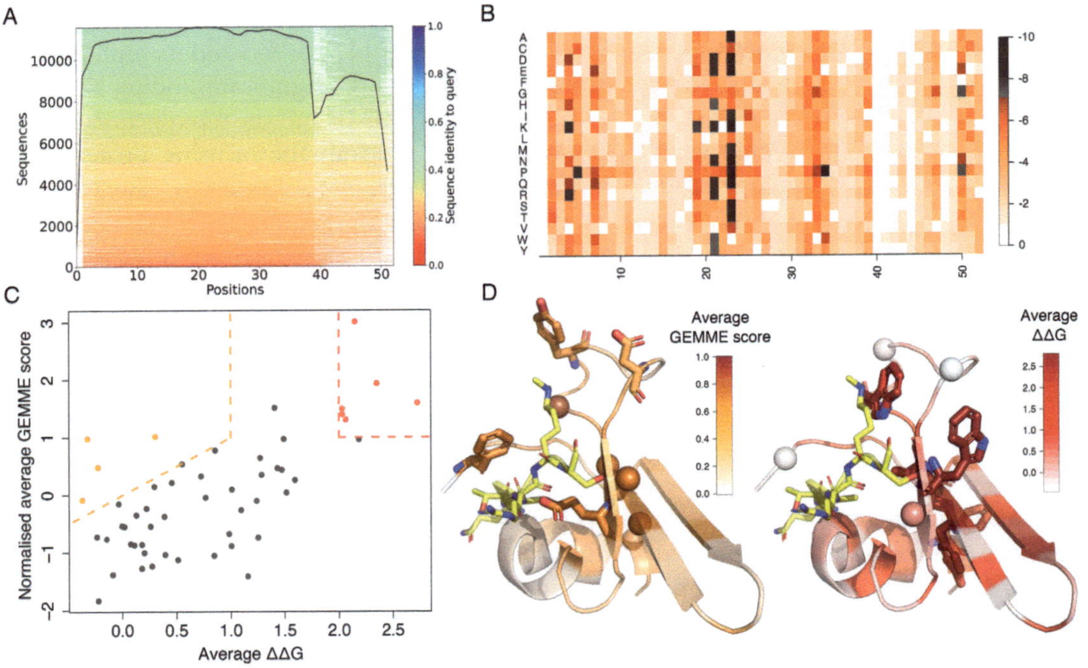

Fig. 3 Comparison between evolutionary constraints and thermodynamic structural stability in the chromodomain of HP1. (**a**) Alignment generated with ColabFold and given as input to GEMME. The sequence of the PDB entry 2M2L (residues 12 to 63) is used as query. (**b**) Single-site mutational landscape computed by GEMME. (**c**) Normalized average GEMME scores in function of average experimental $\Delta\Delta G$ values taken from [3]. Each dot represents a residue/position. Normalization consists in subtracting the mean and dividing by the standard deviation. Residues above the orange dotted line are sensitive to mutations according to GEMME/iGEMME but unimportant for stability; we refer to them as "functional" sites. Residues enclosed by the red dotted lines are sensitive to mutations from both evolutionary and structural perspectives. (**d**) Mapping of average GEMME scores (left, in orange tones) and average $\Delta\Delta G$ (right, in red tones) on the 3D structure 2M2L. The ligand, positioned by superimposing the structure 1KNA, is shown in yellow. On the left panel, functional residues are in sticks, and structurally critical residues detected as sensitive by GEMME are in spheres. Reciprocally, on the right panel

dataset has been used to optimize NucB for functioning in novel chemical environments, a key objective in synthetic biology. The goal is to show that models trained solely on evolutionary data, without any experimental data, can design functional variants at a significantly higher rate than traditional approaches to initial library generation.

The dataset comprises three rounds of mutational campaigns (G2, G3, G4), in which mutations were selected through a variety of techniques including direct evolution (DE), hit recombination (HR), and machine-learning models fit on evolutionary data and mutational data from previous generations (ML). G2, G3, and G4 campaigns were all built upon the same initial library (G1), obtained by applying error-prone PCR (epPCR) to a single-site saturation mutagenesis library of the wild-type sequence from

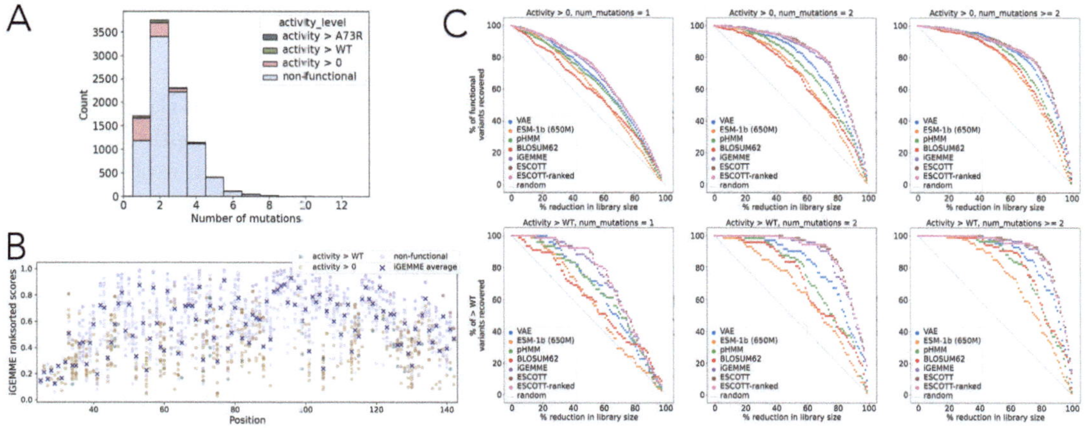

Fig. 4 Comparative zero-shot analysis of the nuclease activity, NucB, with different tools predicting multiple mutations. (**a**) Activity level distribution for the dataset of approximately 9400 variants in G1 [37] colored by estimated activity levels. G1 includes a single-site saturation mutagenesis library as well as variants generated by error-prone PCR. On average, G1 variants have 2.5 mutations, while functional variants have, on average, 1.6 mutations (from Figure 5f or Supplementary Figure 28 in [37]). (**b**) iGEMME ranked sorted analysis of the 2088 single-point mutations present in the dataset. This analysis includes the approximately 1700 single-point mutations in G1 described in A. The prediction concerns three phenotypic classes of variants: with activity $> WT$ (green), with activity > 0 (brown), nonfunctional (light purple). For each position in the sequence, iGEMME averages are indicated by a cross. Low iGEMME scores indicate no mutational effect, hence functional activity, while high iGEMME scores indicate no functional activity. (**c**) VAE, ESM-1b, pHMM, BLOSUM62, iGEMME, ESCOTT (raw scores), and ESCOTT (ranked) are compared on two tasks, for sequences with 1, 2 or more mutations: the percentage of functional variants recovered (top); the percentage of sequences that after mutation improve their activity with respect to the WT (bottom). The dataset used for this analysis is a subset of G1, previously used and described in [37]. See Supplementary Figure 6 in [37]

B. licheniformis. Among more than 211K initial mutations generated by epPCR, employed for the DE campaign, only 9.4K were included in G1, corresponding to those for which catalytic activity was measured (Fig. 4a). Based on their activity, mutated sequences were divided into four classes, namely *nonfunctional*, *activity* > 0, *activity* $>$ *WT*, and *activity* $>$ *A73R*, the latter being the top-performing variant from the initial epPCR library, with 8-fold higher activity than the WT. iGEMME prediction of 2088 single-point mutations in this dataset is reported in Fig. 4b.

Figure 4c illustrates the superior performance of iGEMME compared to VAE [37], ESM-1b [38], pHMM [39], BLOSUM62 [40]. This comparison is based on two different evaluations: the percentage of functional variants recovered (Fig. 4c, top) and the percentage of sequences that after mutation improve their activity with respect to the WT (Fig. 4c, bottom). Notably, ESCOTT, which incorporates structural information into iGEMME, further enhances performance [10]. The analysis was conducted on a subset of the G1 variants: 874 have an activity higher than 0, and 103 have an activity higher than WT (Fig. 4a).

3 Notes: Good Practice and Limitations

1. **Optimization of GEMME parameters in iGEMME.** The default parameters in iGEMME have been optimized compared to the original version of GEMME. The modified parameters concern the calculation of evolutionary conservation values by the JET algorithm [41]. Namely, the maximum number of sequences processed has been increased from 20,000 to 40,000, and the parameter limiting the number of sequences due to computational load (CPU load or max_load parameter) has been raised from 500,000 to 8,000,000. Additionally, the maximum heap size has been expanded from 1024 MB to 8192 MB. These adjustments enable the handling of large proteins with MSA files containing thousands of sequences [10]. While users can tune these parameters based on their specific systems, the recommended values are generally suitable for most cases [10]. In addition, JET includes a Gibbs sampling-like procedure to subsample the input sequences. This stochasticity can generate different results from one run to another. To mitigate this effect, JET offers the possibility to run its algorithm several times (iJET) and retain the maximum conservation values over all runs. Our experiments with the tool indicate that setting the number of iterations to 7 leads to stable results. We thus recommend this value to users.
2. **Generation of the input alignments.** While GEMME initial design used two sets of aligned sequences as input [1], we advise users to give only one input MSA to the tool. This simplification improves interpretability without compromising on predictive accuracy. Moreover, as shown in [33] on a comprehensive benchmark, GEMME achieves similarly competitive performance with MSAs generated either by the highly efficient MMseqs2-based strategy implemented in ColabFold [34] or by highly sensitive profile Hidden Markov Model (HMM) workflows [42]. Users can generate an input MSA with the profile HMM-based method JackHMMer [42] against UniRef100 [43] via GEMME webserver, available at http://www.lcqb.upmc.fr/GEMME. The webserver also facilitates the import of alignments precomputed by ColabFold, simplifying accessibility to both strategies.

   The iGEMME results reported here were produced starting from MSAs computed by ColabFold, based on the uniref30_2103 and colabfold_envdb_202108 databases of sequences. iGEMME was used through the webserver at http://prescott.lcqb.upmc.fr [10] without giving any structural model as input.

   In the general case, we recommend users to run GEMME/iGEMME on MSAs obtained from ColabFold with the most recent versions of the Uniref30 and environmental databases, particularly if they wish to perform high-throughput predictions. When the number of sequences retrieved is lower than 200, we recommend to remove ColabFold default filter [33].

3. **What might go wrong on GEMME/iGEMME.** Some common issues include having a very long reference sequence ($>$ 3500 residues), poor representation of the reference sequence throughout the MSA, positions with many gaps, insufficient sequences in the MSA, or overly long sequence names. If you encounter problems accessing results from the GEMME/iGEMME servers or need to perform large-scale analysis on multiple proteins, we recommend downloading the Docker image, which includes all necessary dependencies. Most of these issues can be resolved; for example, the number of sequences in the MSA can be adjusted to exceed 40,000. Please note that processing MSAs with a large number of sequences or very long reference sequences may take more time.

4. **Software availability.** GEMME is available online at http://www.lcqb.upmc.fr/GEMME/ under the MIT Free Software License. It can be accessed as a standalone package, a Docker image (https://hub.docker.com/r/elodielaine/gemme), or through a webserver. All GEMME parameters can be modified depending on users' needs through configuration files directly in the Docker image. The webserver default parameters are set so as to ensure rapid execution—the number of iterations and of sequences for conservation calculation can be modified in the submission form. In addition, we provide public access to GEMME predictions of single mutational landscapes for the entire human proteome (20,586 proteins from the Swiss-Prot reviewed human proteome available in UniProt, as of August 2023) on Dryad [33].

   iGEMME is accessible online at http://prescott.lcqb.upmc.fr, by providing to the ESCOTT webserver a sequence alignment without structure. ESCOTT and PRESCOTT are available at http://prescott.lcqb.upmc.fr, together with a database of ESCOTT precomputed prediction for more than 19,000 human proteins, accessible for visualization. iGEMME and ESCOTT are a part of the PRESCOTT software; a Docker image of PRESCOTT is available at https://hub.docker.com/repository/docker/tekpinar/PRESCOTT-docker/general, and the source code is provided at http://gitlab.lcqb.upmc.fr/tekpinar/PRESCOTT under the CC BY-NC-SA 4.0 International license.

   Python scripts for inputs preparation, Docker execution, and outputs visualization are available at http://gitlab.lcqb.upmc.fr/lombardi/gemme-utils.

5. **Computational time.** GEMME/iGEMME is very fast since it does not rely on any training due to learning required by the methodological approach. The computational evaluation for GEMME is found in [1] and [9] and for iGEMME in [10].
6. **Additional inputs in subsequent developments.** Users can enhance predictions by utilizing additional information available through ESCOTT and PRESCOTT [10]. ESCOTT takes as additional input a 3D experimental or predicted structure, and PRESCOTT takes allele frequency data [10]. ESCOTT and PRESCOTT follow the AVE Alliance guidelines for variant effect predictors [44].
7. **Compliance with the Atlas of Variant Effects (AVE) Alliance guidelines.** GEMME/iGEMME follows several guidelines from the AVE Alliance for variant effect predictors [44]. Specifically, GEMME/iGEMME codebase is publicly accessible, clearly documented, and open source. GEMME/iGEMME algorithm is transparent, making it possible to rationalize the predictions with respect to the input sequence data. Moreover, since GEMME/iGEMME is not based on learning, it avoids circularity issues. We made GEMME/iGEMME predictions Findable, Accessible, Interoperable, and Reusable (FAIR) by the community. They include human proteome-wide scores and also scores computed on ProteinGym [2] to facilitate independent benchmarking and analysis. GEMME/iGEMME scores can be easily scaled between 0 and 1 for facilitating comparison with other variant effect predictors. Finally, GEMME/iGEMME is population-free, *i.e.*, it does not use any human population data, thus avoiding pervasive ancestry bias [45].

Competing Interests
The authors declare that there are no financial nor nonfinancial competing interests.

Funding
This work was supported by grants from the Institut Universitaire de France (IUF) (AC, EL), and the Sorbonne Center for Artificial Intelligence (SCAI) PhD fellowship at Sorbonne Université (GL).

References

1. Laine E, Karami Y, Carbone A (2019) GEMME: a simple and fast global epistatic model predicting mutational effects. Mol Biol Evol 36(11):2604–2619
2. Notin P, Kollasch A, Ritter D, Van Niekerk L, Paul S, Spinner H et al (2024) ProteinGym: large-scale benchmarks for protein fitness prediction and design. In Advances in neural information processing systems, vol. 36
3. Tsuboyama K, Dauparas J, Chen J, Laine E, Mohseni Behbahani Y, Weinstein JJ et al (2023) Mega-scale experimental analysis of

protein folding stability in biology and design. Nature. https://doi.org/10.1038/s41586-023-06328-6

4. Cagiada M, Bottaro S, Lindemose S, Schenstrøm SM, Stein A, Hartmann-Petersen R et al (2023) Discovering functionally important sites in proteins. Nat Commun 14(1): 4175. https://doi.org/10.1038/s41467-023-39909-0
5. Gersing S, Cagiada M, Gebbia M, Gjesing AP, Coté AG, Seesankar G et al (2023) A comprehensive map of human glucokinase variant activity. Genome Biol 24(1):1–23
6. Tiemann JKS, Zschach H, Lindorff-Larsen K, Stein A (2023) Interpreting the molecular mechanisms of disease variants in human transmembrane proteins. Biophys J 122(11): 2176–2191
7. Abildgaard AB, Nielsen SV, Bernstein I, Stein A, Lindorff-Larsen K, Hartmann-Petersen R (2023) Lynch syndrome, molecular mechanisms and variant classification. Br J Cancer 128(5):726–734
8. Mohseni Behbahani Y, Laine E, Carbone A (2023) Deep Local Analysis deconstructs protein–protein interfaces and accurately estimates binding affinity changes upon mutation. Bioinformatics 39(Supplement_1):i544–i552
9. Marquet C, Schlensok J, Abakarova M, Rost B, Laine E (2024) VespaG: expert-guided protein Language Models enable accurate and blazingly fast fitness prediction. bioRxiv p 2024–04
10. Tekpinar M, David L, Henry T, Carbone A (2024) PRESCOTT: a population aware, epistatic and structural model accurately predicts missense effect. medRxiv p 2024–02
11. Jumper J, Evans R, Pritzel A, Green T, Figurnov M, Ronneberger O et al (2021) Highly accurate protein structure prediction with AlphaFold. Nature 596(7873):583–589
12. Lin Z, Akin H, Rao R, Hie B, Zhu Z, Lu W et al (2023) Evolutionary-scale prediction of atomic-level protein structure with a language model. Science 379(6637):1123–1130
13. Baek M, DiMaio F, Anishchenko I, Dauparas J, Ovchinnikov S, Lee GR et al (2021) Accurate prediction of protein structures and interactions using a three-track neural network. Science 373(6557):871–876
14. Lu J, Holmgren A (2014) The thioredoxin antioxidant system. Free Radical Biol Med 66: 75–87
15. Mustacich D, Powis G (2000) Thioredoxin reductase. Biochem J 346(1):1–8
16. Schürmann P, Jacquot JP (2000) Plant thioredoxin systems revisited. Annu Rev Plant Biol 51(1):371–400
17. Gécz J, Cloosterman D, Partington M (2006) ARX: a gene for all seasons. Curr Opin Genet Dev 16(3):308–316
18. Shoubridge C, Fullston T, Gécz J (2010) ARX spectrum disorders: making inroads into the molecular pathology. Hum Mutation 31(8): 889–900
19. Zeng W, Ball Jr. AR, Yokomori K (2010) HP1: heterochromatin binding proteins working the genome. Epigenetics 5(4):287–292
20. Grewal SI (2023) The molecular basis of heterochromatin assembly and epigenetic inheritance. Mol Cell 83(11):1767–1785
21. Lomberk G, Wallrath L, Urrutia R (2006) The heterochromatin protein 1 family. Genome Biol 7:1–8
22. Van Sinderen D, Kiewiet R, Venema G (1995) Differential expression of two closely related deoxyribonuclease genes, nucA and nucB, in Bacillus subtilis. Mol Microbiol 15(2): 213–223
23. Baslé A, Hewitt L, Koh A, Lamb HK, Thompson P, Burgess JG et al (2018) Crystal structure of NucB, a biofilm-degrading endonuclease. Nucleic Acids Res 46(1):473–484
24. Steinegger M, Meier M, Mirdita M, Vöhringer H, Haunsberger SJ, Söding J (2019) HH-suite3 for fast remote homology detection and deep protein annotation. BMC Bioinform 20:1–15
25. Mavridou DA, Saridakis E, Kritsiligkou P, Mozley EC, Ferguson SJ, Redfield C (2014) An extended active-site motif controls the reactivity of the thioredoxin fold. J Biol Chem 289(12):8681–8696
26. Lemaire SD, Tedesco D, Crozet P, Michelet L, Fermani S, Zaffagnini M et al (2018) Crystal structure of chloroplastic thioredoxin f2 from Chlamydomonas reinhardtii reveals distinct surface properties. Antioxidants 7(12):171
27. The-UniProt-Consortium (2023) UniProt: the universal protein knowledgebase in 2023. Nucleic Acids Res 51(D1):D523–D531
28. Paysan-Lafosse T, Blum M, Chuguransky S, Grego T, Pinto BL, Salazar GA et al (2023) InterPro in 2022. Nucleic Acids Res 51(D1): D418–D427
29. Holehouse AS, Kragelund BB (2024) The molecular basis for cellular function of intrinsically disordered protein regions. Nat Rev Mol Cell Biol 25(3):187–211
30. Liu C, Wu K, Choi H, Han H, Zhang X, Watson JL et al (2024) Diffusing protein binders to intrinsically disordered proteins. bioRxiv p 2024–07
31. Strømme P, Mangelsdorf ME, Scheffer IE, Gécz J (2002) Infantile spasms, dystonia, and

other X-linked phenotypes caused by mutations in Aristaless related homeobox gene, ARX. Brain Dev 24(5):266–268

32. Brandes N, Goldman G, Wang CH, Ye CJ, Ntranos V (2023) Genome-wide prediction of disease variant effects with a deep protein language model. Nat Genet 55(9):1512–1522
33. Abakarova M, Marquet C, Rera M, Rost B, Laine E (2023) Alignment-based protein mutational landscape prediction: doing more with less. Genome Biol Evol 15(11):evad201
34. Mirdita M, Schütze K, Moriwaki Y, Heo L, Ovchinnikov S, Steinegger M (2022) ColabFold: making protein folding accessible to all. Nat Methods 19(6):679–682
35. Cheng J, Novati G, Pan J, Bycroft C, Žemgulyte A, Applebaum T et al (2023) Accurate proteome-wide missense variant effect prediction with AlphaMissense. Science 381(6664): eadg7492
36. Jacobs SA, Khorasanizadeh S (2002) Structure of HP1 chromodomain bound to a lysine 9-methylated histone H3 tail. Science 295(5562):2080–2083
37. Thomas N, Belanger D, Xu C, Lee H, Hirano K, Iwai K et al (2024) Engineering highly active and diverse nuclease enzymes by combining machine learning and ultra-high-throughput screening. bioRxiv p 2024–03
38. Rives A, Meier J, Sercu T, Goyal S, Lin Z, Liu J et al (2021) Biological structure and function emerge from scaling unsupervised learning to 250 million protein sequences. Proc Natl Acad Sci 118(15):e2016239118
39. Finn RD, Clements J, Eddy SR (2011) HMMER web server: interactive sequence similarity searching. Nucleic Acids Res 39 (suppl_2):W29–W37
40. Henikoff S, Henikoff JG (1992) Amino acid substitution matrices from protein blocks. Proc Natl Acad Sci 89(22):10915–10919
41. Engelen S, Trojan LA, Sacquin-Mora S, Lavery R, Carbone A (2009) Joint Evolutionary Trees: a large-scale method to predict protein interfaces based on sequence sampling. PLoS Comput Biol 5(1):e1000267. https://journals.plos.org/ploscompbiol/article?id=10.1371/journal.pcbi.1000267
42. Eddy SR (2011) Accelerated profile HMM searches. PLoS Comput Biol 7(10):e1002195
43. Suzek BE, Wang Y, Huang H, McGarvey PB, Wu CH, Consortium U (2015) UniRef clusters: a comprehensive and scalable alternative for improving sequence similarity searches. Bioinformatics 31(6):926–932
44. Livesey BJ, Badonyi M, Dias M, Frazer J, Kumar S, Lindorff-Larsen K et al (2024) Guidelines for releasing a variant effect predictor. ArXiv
45. Pathak AK, Bora N, Badonyi M, Livesey BJ, Consortium S, Ngeow J et al (2024) Pervasive ancestry bias in variant effect predictors. bioRxiv p 2024–05

Part II

3D Structure Analysis: Design and Dynamics

Chapter 8

Using Total Network Dissimilarity to Analyze Protein Structures Across Homologs

Vasam Manjveekar Prabantu, Vasundhara Gadiyaram, and Saraswathi Vishveshwara

Abstract

A novel methodology for comparing protein structural networks (PSNs) of homologous proteins is presented here, leveraging graph spectral analysis and quantitative comparison techniques. The primary objective of this approach is to systematically explore the similarities and differences between homologous protein structures, thereby elucidating the intricate relationships between protein sequence, structure, and function. By employing a combination of network similarity score, total network dissimilarity, and Fiedler vector analysis, this method provides a robust framework for identifying subtle patterns and relationships in protein structure evolution. The insights gained from this approach will contribute significantly to our understanding of protein biology.

Key words Protein structure networks, Graph spectral analysis, Network connectivity, Total network dissimilarity, Wnt proteins

1 Introduction

The comparison of homologous proteins has revolutionized our understanding of protein structure biology. Homologous proteins, sharing a common evolutionary origin, provide a unique framework for elucidating the intricate relationships between protein sequence, structure, and function [1–4]. By contrasting homologous proteins, researchers can identify conserved residues, motifs, and domains that are crucial for protein function, stability, and interactions [5, 6].

The significance of homologous protein comparison extends beyond the realm of basic research, with far-reaching implications for our understanding of protein-related diseases, the development of novel therapeutics, and the design of enzymes with improved catalytic properties [7]. Furthermore, the integration of homologous protein comparison with other structural bioinformatics

Shahid M. Khan and Florencio Pazos (eds.), *Protein Evolution: Methods and Protocols*, Methods in Molecular Biology, vol. 2979, https://doi.org/10.1007/978-1-0716-4828-5_8,

approaches has greatly enhanced our ability to predict protein structures, model protein-ligand interactions, and simulate protein dynamics [8–10].

Traditional methods for comparing protein structures rely heavily on geometric measures, such as the root-mean-square deviation (RMSD) of topologically equivalent residues. While RMSD provides a quantitative measure of structural similarity, it has limitations while dealing with proteins that have undergone large conformational changes and in the comparison of remote homologs. Moreover, RMSD calculations can become increasingly complex and computationally challenging when considering atomic-level deviations that arise from the inclusion of side chains in the comparisons. This is because side chains can exhibit significant flexibility and variability, making it difficult to accurately assess their contributions to overall structural similarity. Other existing methods, such as the TM-score and GDT-score, offer alternative approaches for evaluating structural similarity yet have their limitations.

Recent advances in network science and graph theory have led to the development of novel methods for comparing protein structures. In this approach, the topology of the protein structure is represented as a network, where nodes correspond to amino acids and edges represent interactions between them. Analyzing the topological features of these networks, one can identify subtle patterns and relationships that are not apparent through traditional geometric measures [11–13]. Earlier work has demonstrated the application of spectral and graph methods in comparing networks, highlighting the potential of these approaches for protein structure comparison [14–18].

This representation of proteins, called as protein structural network (PSN), offers a powerful framework for comparing homologous proteins and has the capacity to identify conserved connectivity motifs that may even be essential for protein function. However, a caveat is that the structures of homologous proteins have to be aligned and the information from nodes that are not aligned is lost. Using the information of the entire structural connectivity in the comparison of PSNs from homologous proteins yields valuable insights into the evolution of protein structure and function.

This work outlines the protocol to a methodology of protein structural network comparison of homologous proteins, with a focus on the use of graph spectral analysis and quantitative comparison techniques. The aim of the method is to explore the similarities and differences between homologous protein structures [19]. This approach captures the connectivity that is not readily apparent through traditional geometric measures, such as RMSD. Unlike traditional methods, which are limited in their ability to compare proteins that have undergone substantial conformational changes or are distant homologs, PSNs offer a more nuanced understanding of protein structure and hence their function. It will also help to gain a deeper understanding of the complex relationships between protein sequence, structure, and function and to provide new insights into the biology of proteins, ultimately aiding in the development of novel therapeutic strategies and protein engineering applications.

2 Materials and Methods

2.1 Protein Structural Network (PSN)

The residue–residue interconnectivity that captures the topology of a protein structure is modeled as a protein structural network (PSN). The residues in the protein structure are nodes, and the interactions between interacting residues are manifested as weighted or binary edges. An all-atom model of the PSN is used, which is a weighted nondirected graph. The structure of a protein is represented using the Cα atoms as the nodes and an edge between them is drawn based on whether they are interacting. The interactions are defined by atom contacts. Any pair of nonadjacent residues whose atoms come within a distance of 4.5 Å make atom contacts. In real time, the maximum number of atom contacts is determined by iterating through all pairs of atoms between all pairs of nonadjacent residues across a dataset of protein structures that are to be analyzed. The ratio of the number of atom contacts made between a pair of residues and the maximum number of atom contacts possible between the given pair of amino acids calls for the weight of the interaction between the residues. This information stored in an adjacency/Kirchhoff matrix which makes the structural network. The structural network can also be represented in the graphical molecular representation where the three-dimensional (3D) structure of the protein is drawn with the Cα atoms of interacting residues connected by edges whose weight is as defined above (*see* **Note 3**).

2.2 PSN Comparison Using a Graph Spectral Approach

The graph spectral method involves using the structural network for obtaining their eigenvalues and eigenvectors. These parameters help in the analysis of the clustering of residues in the network. For instance, the eigenvector corresponding to the second smallest eigenvalue (nonzero) is called the fielder vector which is used to cluster the constituents of the network modularly. The process of determining the clusters using fielder vectors has been shown in the case study. The adjacency matrices of a pair of equivalent structures can be compared quantitatively using the network similarity score (NSS) approach.

However, in the case of homologs, the structures may not be equivalent requiring to filter the constituent residues that can be compared. The topologically equivalent residues between the pair of structures are determined. TM-align tool is used to determine

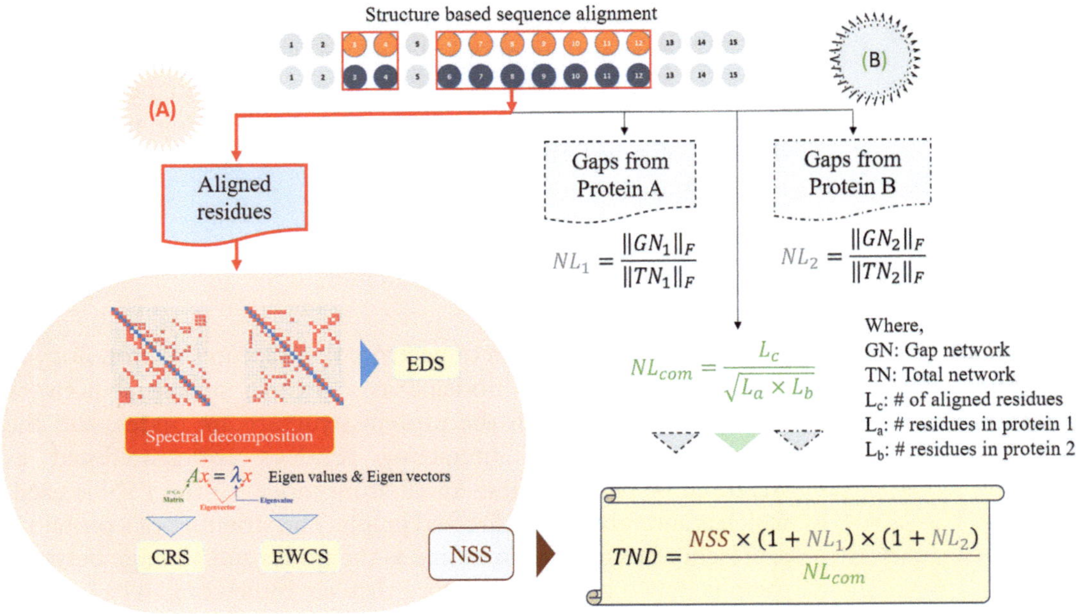

Fig. 1 Schematic overview for the computation of Total Network Dissimilarity (TND). The process begins with structure-based sequence alignments to identify topologically equivalent residues, followed by two key steps: (1) calculating the overall similarity in conserved connectivity between structures as shown in part A and (2) normalizing with unaligned network regions as shown in part B

the topologically equivalent residues [20]. The equivalent residues are aligned residues, and the ones that do not align to any equivalent residue are a nonaligned residue. A gap node of nonzero value is inserted corresponding to the nonaligned residues having no interactions with any other residue in the corresponding protein structures. Here, an extended method of comparison that includes differences in the gap nodes is presented as the total network dissimilarity (TND). Figure 1 shows a schematic of how the two metrics are used with different inputs and at different stages of the comparison. When the structures superpose well and have full alignment, the NSS can be used directly, and in the later scenario, the unaligned regions can be incorporated to arrive at the TND.

2.2.1 Network Similarity Score (NSS)

The network similarity score was developed for the comparison of equivalent networks. This method has been found to be immense use in the analysis and comparison of protein structures. An understanding of the significance of this methodology is detailed in earlier works [14]. This score is determined using components which capture differences at various levels: (1) edge difference score (EDS) quantifies edge weight differences between the networks, (2) eigenvalue weighted cosine score (EWCS) quantifies the differences in local node clustering, and (3) correspondence score (CRS) capturing global level changes (*see* **Note 1**).

Let network A and network B be two different networks that are being compared, which are equivalent: Let A and B be the adjacency matrices of network A and network B of size n (number of nodes is n), M is the difference of the two adjacency matrices A and B, and $||M||_F$ be the Frobenius norm of the difference matrix M, then edge difference score (EDS) is given by

$$\text{EDS} = \frac{||M||_F}{\sqrt{(\sum A \times \sum B)}} \tag{1}$$

The eigenvectors of the normalized Laplacian matrices of networks A and B are aligned by considering the maximum of cosine values between all eigenvectors of A and all eigenvectors of B. The cosine between the vectors is given by

$$\text{cosine}\ (\theta_{i,j}) = \frac{\left(\text{Evec}_i^A.\text{Evec}_j^B\right)}{\left\|\text{Evec}_i^A\right\|\left\|\text{Evec}_j^B\right\|}\quad i,j \in N, 1 \leq i,j \leq n \tag{2}$$

where $Evec_i^A$ is the ith eigenvector of network A and $Evec_j^B$is the ith eigenvector of network B, which is aligned with $Evec_i^A$, $Evec_i^A$, and $Evec_j^B$are the eigenvalues of $Evec_i^A$ and $Evec_j^B$. The difference at local clustering level (within a module of nodes) is calculated by EWCS, which is the weighted sum of deviations of cosine values from 1 (which is optimal in the case of perfect match of the vectors), the weights considered by dominance of the associated eigenvalues.

$$\text{EWCS} = \frac{\sum \text{cosine}\ (\theta_{i,j})^2 |1 - \text{Eval}_A||1 - \text{Eval}_B|}{\sum |1 - \text{Eval}_A||1 - \text{Eval}_B|} \tag{3}$$

The global difference which is the change in the node clusters is captured by the shift in the alignment of the eigenvectors. This is captured by CRS, which is Spearman's correlation between the indices of aligned eigenvectors.

$$\text{CRS} = 1 - \frac{6\sum (\text{Index } Evec_A - \text{Index } Evec_B)^2}{n(n^2 - 1)} \tag{4}$$

where n is the size of these networks and Index $Evec_A$ and Index $Evec_B$ are the indices of n pairs of aligned eigenvectors of networks A and B with maximum cosine values. Then, NSS for comparing the pair of networks is shown as

$$\text{NSS} = \sqrt{\text{EDS}^2 + \text{EWCS}^2 + (1 - \text{CRS})^2} \tag{5}$$

It should be noted that lower NSS reflects higher similarity between two networks. Exactly, identical networks will have an NSS of 1 and NSS of $\sqrt{3}$ between networks that have no match [14] (*see* **Note 5**).

2.2.2 Total Network Dissimilarity (TND) (see Note 4)

The total network dissimilarity is an extension of the NSS that can be computed between networks that are not equivalent. This process involves using the NSS computed between the equivalent nodes of a network and then accounting for the gaps in the networks. In the context of homologous proteins, suppose a protein with a sequence of length L_a is being compared with its homolog of length L_b. Using structure-based sequence alignment, L_c is calculated, that is, number of residues aligned between the homologous sequences. Hence, the length of corresponding aligned residues is L_c, and all the remaining residues are nonaligned having gaps in the corresponding sequence. The adjacency matrices of the PSNs are generated as TN_1 and TN_2, respectively. The remaining network composed of only edges that connect to the nonaligned residues in the sequence is generated as GN_1 and GN_2, respectively. The following three gap parameters are computed (*see* **Note 2**):

$$NL_1 = \frac{\|GN_1\|_F}{\|TN_1\|_F}, NL_2 = \frac{\|GN_2\|_F}{\|TN_2\|_F}, NL_{\text{com}} = \frac{L_c}{\sqrt{L_a \times L_b}} \tag{6}$$

The total network dissimilarity (TND) is obtained using the computed gap parameters as shown below (*see* **Note 6**):

$$\text{TND} = \frac{\text{NSS} \times (1 + NL_1) \times (1 + NL_2)}{NL_{\text{com}}} \tag{7}$$

The TND score depends on the residue alignment, and the gap parameters (NL_1, NL_2, and NL_{com}), normalizing the NSS score across homologous proteins [19] (*see* **Note 7**). A flowchart illustrating the stepwise calculation of the total network dissimilarity is presented in Fig. 2. To demonstrate the application of this method, we provide a case study that highlights the utility of TND scores in comparing homologous proteins.

Previous work by our group employed fielder vectors to investigate changes in clustering patterns, using the example of β2 adrenergic receptor bound to an agonist and an antagonist [21]. The focus of studying this membrane protein was to look at the connectivity between the extracellular domain and the intracellular domain, examining how ligand interactions altered network properties and connectivity and did a more detailed depiction of the global change of the network in terms of the edge difference, clustering changes, and also their Fiedler vector components. Here, we extend this concept to analyze the connectivity differences between functionally active and inactive homologous proteins, providing new insights into the relationship between protein structure, function, and evolution.

2.2.3 Data Availability (see Note 8)

The program for computing the NSS is available in the GraSp-PSN (Graph Spectral analysis of Protein Structure Networks) Web server which can be accessed at the website: pople.mbu.iisc.ac.in. The NSS module must be used to compute the network similarity

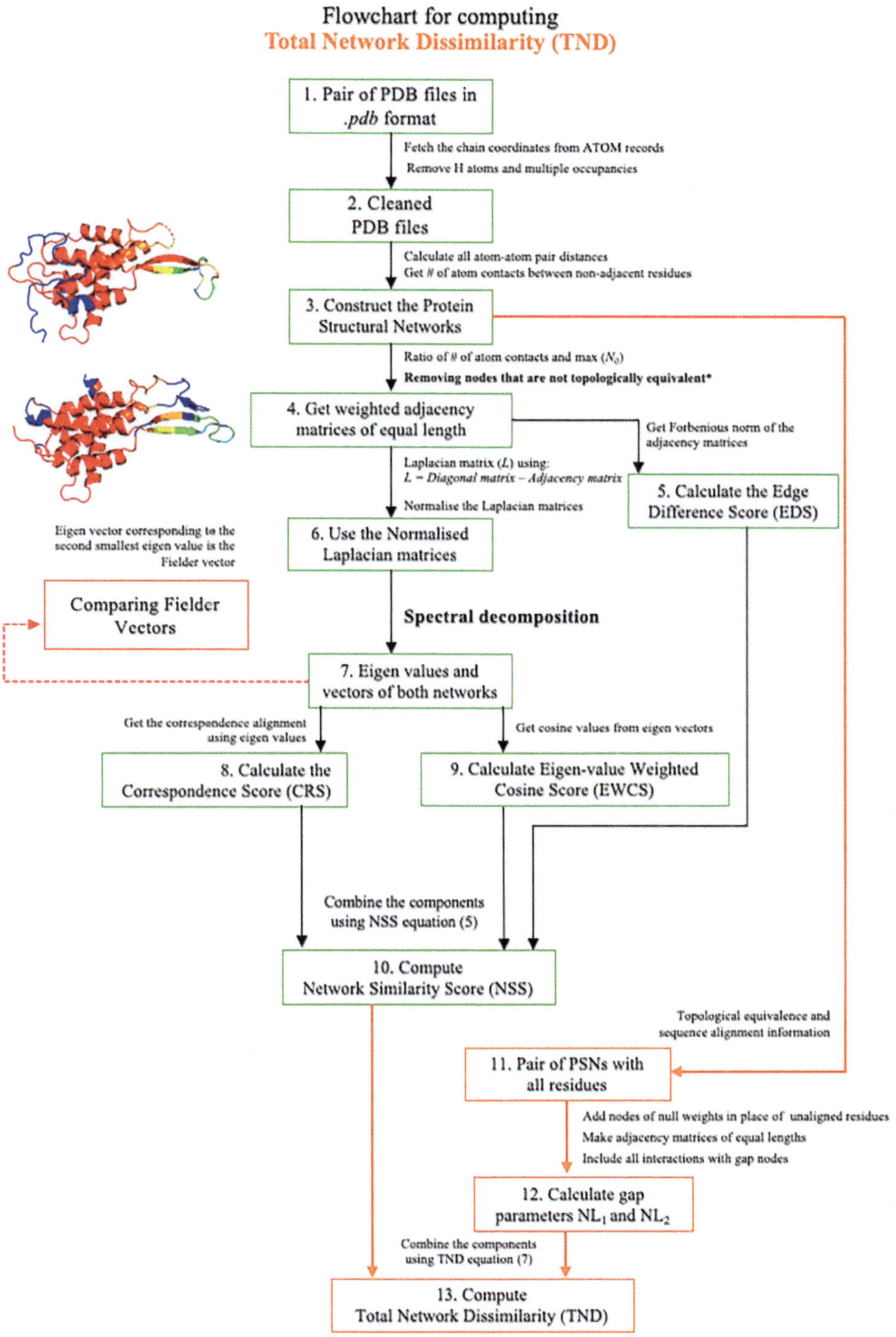

Fig. 2 Flowchart summarizing the steps involved in computing the TND between a pair of homologous protein structures. The input required to compute the TND is a pair of structures. First, the PSN is stored mathematically, and then, several operations are performed to obtain the features of the network. The obtained features are quantified into a standardized metric of dissimilarity

score between two input coordinate files. These programs can be accessed in accordance with the terms and conditions specified on the website. Also, the python-based code for computing the total network dissimilarity (TND) is available on GitHub using the following link: http://github.com/manjveekar/tnd

3 Demonstration with Wnt Protein

As a case study, we demonstrate the application of the total network dissimilarity (TND) score in comparing homologous protein structures using two available crystal structures of the Wnt protein. Wnt proteins comprise a family of secreted signaling molecules that play pivotal roles in development, tissue homeostasis, and disease pathology. The name "Wnt" is an abbreviation of "Wingless" and "Int-1" (Integration-1), reflecting the discovery of the first Wnt protein in Drosophila (Wingless) and the identification of Int-1 as a proto-oncogene in mice.

The signaling processes downstream of Wnt proteins are complex and multifaceted, involving several key pathways such as the Wnt/β-catenin pathway (canonical Wnt pathway) that regulates gene transcription and cell fate decisions, planar cell polarity (PCP) pathway, and the calcium pathway [22, 23]. Wnt proteins also play critical roles in embryonic development and adult, regulating several key factors including those related to disease pathogenesis especially cancer and neurodegenerative disorders. Dysregulation of this system has significant implications for human disease, thus implicating that studying the structure and molecular mechanisms of this protein family is important for understanding signaling mechanisms, disease pathogenesis, and therapeutic interventions [24–28].

A pair of Wnt protein structures was selected for this case study from the Protein Data Bank (PDB). The crystal structure of XWnt8 in complex with the frizzled eight domains (RCSB PDB ID: 4F0A) was obtained from the PDB, with a resolution of 3.25 Å and featuring the conserved Wnt domain, in .pdb format [29]. Using PyMOL, a cleaned PDB file was generated, encompassing only the N-terminal domain (residues 36–221) of the *Xenopus laevis* Wnt protein. This structure was compared to its paralog from *Drosophila melanogaster*, WntD N-terminal protein (RCSB PDB ID: 4KRR), which was similarly cleaned to include residues 33–205 [30]. Notably, the latter structure has a slightly better resolution of 2.5 Å and lacks bound interacting proteins.

The two structures represent distinct functional states of Wnt proteins, with the Xenopus Wnt structure (XWnt8 bound to frizzled receptor domain) mimicking the "on" state of the canonical Wnt/β-catenin pathway, and Drosophila Wnt structure (WntD)

representing the "off" state. The two structures exhibit good superposition using TM-align, with a root-mean-square deviation (RMSD) of 2.68 Å. Figure 3 shows the original structures before cleaning and a superposed cartoon of the trimmed structures.

The alignment of the structures results in the information of the topological equivalences; there are 159 aligned residues (L_c). First, construct the PSN of the pair of proteins by computing the interaction strength between all pair of nonadjacent residues (i and j) in the protein. The interaction strength is the ratio of actual number of atom contacts (n_{ij}) and the maximum number of atom contacts (N_{ij}) between the amino acid pair [19]. The resultant is an adjacency matrix which is used to compute the normalized Laplacian matrix (L*). The spectral decomposition of the L* reveals the spectra of the proteins being compared using which NSS is computed. The NSS obtained by comparing the Wnt domains of 4F0A and 4KRR is 0.269. Since the structures superposed well, an EDS of 0.025 and correspondence of 0.955 (1-CRS) were observed. However, the networks were in fact affected by the change in clustering from the regions that did not superpose well, which contributed to an EWCS of 0.264, collectively resulting in the high NSS.

To identify the regions of the protein that resulted in the altered network, the Fielder vectors of the two proteins are analyzed [21]. The comparison of the Fielder vector (the eigenvector corresponding to the second smallest eigenvalue) provides insight into the regions of variable clustering. In the case of the Xenopus Wnt, this was the third eigenvalue of 0.02 and the second eigenvalue of 0.027 in the Drosophila Wnt spectra. The fielder vectors are aligned to determine the sites of variable clustering. It was observed that the unaligned loop regions of the protein structures significantly affected the local clustering that is captured by EWCS. Figure 4 illustrates the clusters obtained from the Fiedler vectors in a cartoon representation of the structures, showing the variable clustering that contribute to the network differences. Notably, the cystine residues that may interact with the cystine-rich frizzled receptor domain (bound in Xenopus PDB) are found in the first cluster that is involved in protein binding. A mixture of aliphatic and polar residues is observed in the second cluster, and the third cluster contains several hydrophobic residues.

An extension of this is to now include all residues irrespective of the topological equivalence. Those residues that are unaligned in one of the proteins have null nodes/dummy residues in the corresponding position of the other matrices. A pair of total networks (TN_1 and TN_2), containing interactions between all residues including the dummy nodes and gap networks (GN_1 and GN_2), with only interactions of the gap nodes, is derived. Using all the computed parameters and weights, the total network dissimilarity between the pair of structures is computed. An in-house python

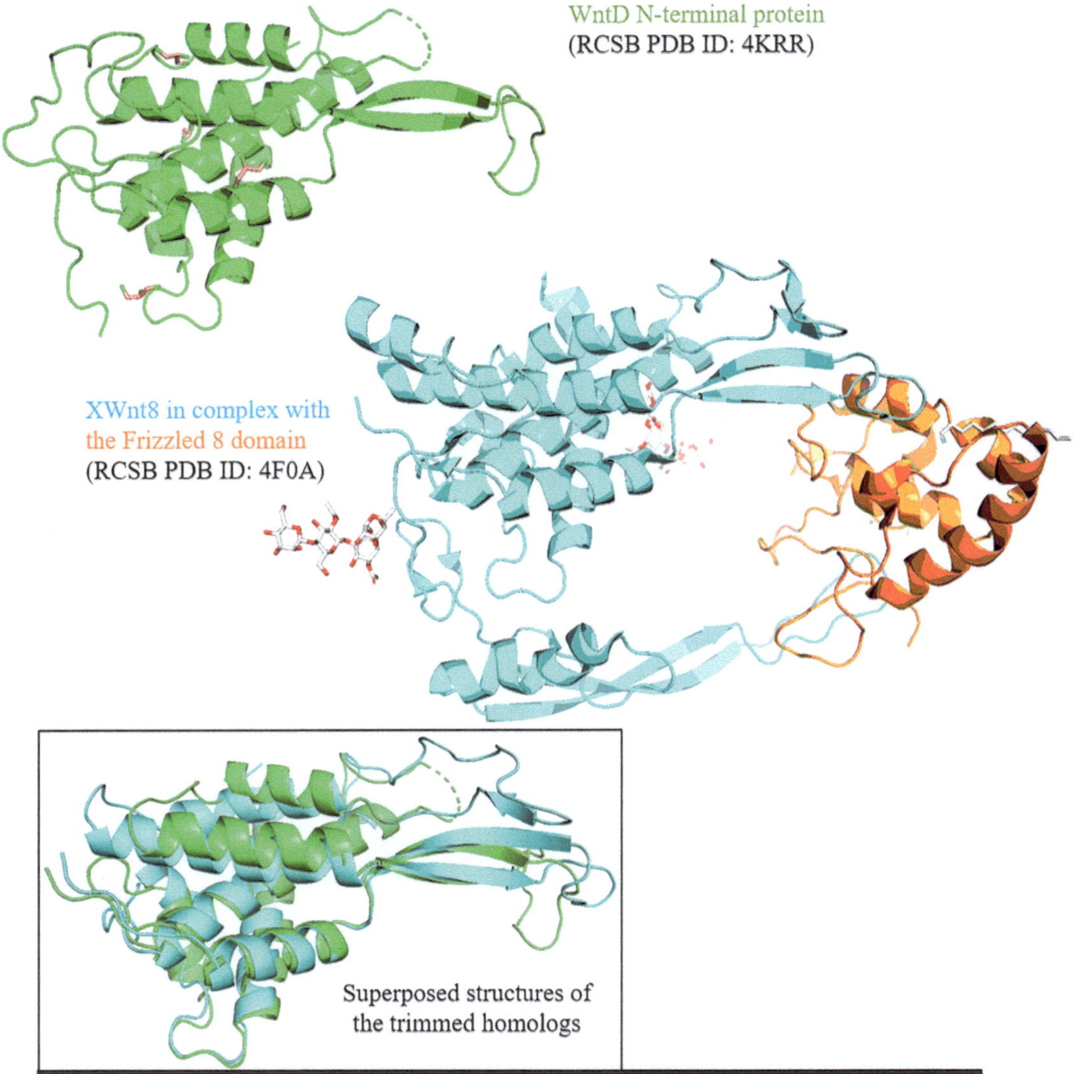


```
Aligned length=  159, RMSD=   2.68, Seq_ID=n_identical/n_aligned= 0.252
TM-score= 0.79016 (if normalized by length of Chain_1)
TM-score= 0.71913 (if normalized by length of Chain_2)
(You should use TM-score normalized by length of the reference protein)

(":" denotes aligned residue pairs of d < 5.0 A, "." denotes other aligned residues)
P-LSWEDITGKGLKQALDSCQQSFQWQRWNCPSQDFVQKNSKPE-ENSPNREDVYVAAISMAAIVHTLTKDCANGVIAGCGC-T
. .:::::::::::::::::::::::::::::  :::::::::. ...:::::::::::::::::::::::::::::::::: .
AYLTYSASVAVGAQNGIEECKYQFAWERWNCP--ESTLQLATHNGLRSATRETSFVHAISSAGVMYTLTRNCSMGDFDNCGCDD


--------------PCAHEPTKALEQYEKHFGS-G------SGAIGHNRRVVGALLQRSLEQECRCKQPGAVQG-ECQ--EEEC
              ::::::::::::::::::. .      ::::::::::::::::::::::::.:.       .::  ::::
SRNGRIGGRGWVWGGCSDNAEFGERISKLFVDGLETGQDARALMNLHNNEAGRLAVKETMKRTCKCHG------ISGSCSIQTC

VAVLKPFEAIAQDLLQMYDDAIQLEG
::::::::::::::::::::::::::
WLQLAEFRDIGNHLKIKHDQALKLEM
```

Fig. 3 Structural comparison of a pair of homologous Wnt protein structures. The original crystal structures of XWnt8 bound to frizzled receptor domain (RCSB PDB ID: 4F0A) and WntD (RCSB PDB ID: 4KRR) before cleaning are depicted. In the panel below is the superposed cartoon representation of the trimmed N-terminal domains after cleaning and structural alignment

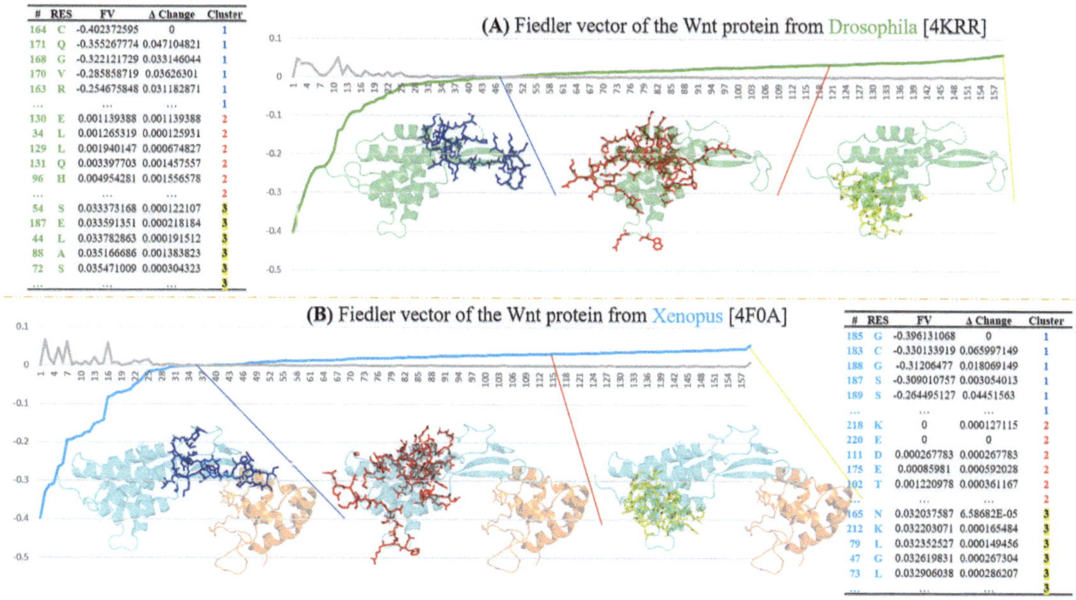

#	RES	FV	Δ Change	Cluster
164	C	-0.402372595	0	1
171	Q	-0.355267774	0.047104821	1
168	G	-0.322121729	0.033146044	1
170	V	-0.285858719	0.03626301	1
163	R	-0.254675848	0.031182871	1
...		...	...	1
130	E	0.001139388	0.001139388	2
34	L	0.001265319	0.000125931	2
129	L	0.001940147	0.000674827	2
131	Q	0.003397703	0.001457557	2
96	H	0.004954281	0.001556578	2
...		...	...	2
54	S	0.033373168	0.000122107	3
187	E	0.033591351	0.000218184	3
44	L	0.033782863	0.000191512	3
88	A	0.035166686	0.001383823	3
72	S	0.035471009	0.000304323	3
...		...	...	3

#	RES	FV	Δ Change	Cluster
185	G	-0.396131068	0	1
183	C	-0.330133919	0.065997149	1
188	G	-0.31206477	0.018069149	1
187	S	-0.309010757	0.003054013	1
189	S	-0.264495127	0.04451563	1
...		...	...	1
218	K	0	0.000127115	2
220	E	0	0	2
111	D	0.000267783	0.000267783	2
175	E	0.00085981	0.000592028	2
102	T	0.001220978	0.000361167	2
...		...	...	2
165	N	0.032037587	6.58682E-05	3
212	K	0.032203071	0.000165484	3
79	L	0.032352527	0.000149456	3
47	G	0.032619831	0.000267304	3
73	L	0.032906038	0.000286207	3
...		...	...	3

Fig. 4 Fiedler vector analysis revealing changes in clustering patterns between homologous Wnt proteins. The bound (Xenopus) and unbound (Drosophila) forms of Wnt proteins are shown, along with their Fiedler vector components and slope changes. Additionally, the components are arranged in ascending order, the top five members of each cluster are reported, and the residue number and the amino acid type have been depicted alongside the cluster images in the table. (**a**) The unbound structure of the Wnt protein from Drosophila is shown, and there are three main clusters that have been identified. Each comprises 48, 73, and 38 residues, respectively. (**b**) The Wnt protein from Xenopus which is bound to the cystine-rich frizzled receptor domain is shown, and three clusters have been identified in this protein based on their Fiedler vector components, consisting of 36, 79, and 44 residues, respectively. The altered residues between the corresponding clusters contribute to the change in network connectivity

code was used to perform the calculations of TND, which is also available on GitHub. A total of 27 and 8 gap sites were identified in the respective protein structures, which yielded a TND of 0.325.

By computing the network similarity score and normalizing using the total network dissimilarity (TND), a quantifiable and interpretable measure of dissimilarity between PSNs is obtained. Analyzing the changes in clustering from Fiedler vectors informs about the alteration of the connectivity patterns, due to subtle alterations in side-chain interactions which is an added advantage of using the graph spectra [14, 17, 31]. The TND metric enables the classification and analysis of variability in network connectivity, providing valuable insights into the relationship between protein structure and function. A metric such as this can be employed to construct phylogenetic trees based on protein structure comparisons, ultimately revealing evolutionary relationships between proteins [19]. By doing so, it will contribute significantly to our understanding of protein functional evolution, shedding light on the intricate mechanisms that have shaped protein biology over time.

4 Notes

1. Unlike traditional protein structure comparison methods that capture structural deviations, the network method discussed in this work captures changes in global and local clustering of residues along with information of structural alterations. These features can be used to understand the conformational diversity of proteins, study the effect of perturbations, and validate the protein models.

2. To compute TND parameters, adjacency matrices of the complete networks (TN_1 and TN_2), including nonequivalent and gap nodes, are constructed. The corresponding gap networks (GN_1 and GN_2) are derived by excluding interactions involving the equivalent nodes from these matrices.

3. A protein structure is represented as a protein structural network (PSN) by transforming it into an adjacency matrix, which serves as a mathematical framework for analysis. To compare a pair of PSNs, structurally equivalent residues are selected as nodes. The NSS method operates on this adjacency matrix. TM-align has been used to identify equivalent residues.

4. Despite low sequence identity in remote homologs, structural conservation often ensures enough equivalent residues for meaningful comparisons. However, when the number of equivalent residues drops below a critical threshold, the reliability of the comparison declines. In such cases, the TND may yield an inefficient or less informative comparison [19].

5. An NSS score of 0 indicates that the compared networks are identical. The maximum possible NSS score is 1.732, which occurs when EDS = 1, EWCS = 1, and CRS = 0, though this scenario only arises when there are no equivalent nodes between the networks. TND, on the other hand, is unbounded and can range from 0 to any positive number.

6. When there is no dissimilarity in the aligned regions, the total dissimilarity is zero, resulting in both NSS and TND values equal to 0.

7. While NSS shows a correlation with sequence identity, like RMSD, TND does not necessarily share such a relationship, despite having a similar range to RMSD.

8. A module for pairwise structural comparison using NSS is currently available on the GraSp-PSN Web server and supports equivalent networks. An algorithm for TND-based comparison of homologous structures, designed to be more practical and

user-friendly, will be added to the server soon, along with application programming interface (API) tools for batch processing.

Acknowledgments

VMP thanks support from the laboratory of Prof N Srinivasan. VG was supported by CSIR-RA fellowship. SV is an Honorary Scientist of National Academy of Sciences (NASI, Allahabad, India).

References

1. Dayhoff M, Schwartz R (2025), B. O.-A. of protein sequence, and undefined 1978, "22 a model of evolutionary change in proteins," *profs.scienze.univr.it*, Accessed 01 Feb 2025. [Online]. Available: http://profs.scienze.univr.it/~liptak/ALBioinfo/2011_2012/files/pam1.pdf
2. Chothia C, Lesk AM (1986) The relation between the divergence of sequence and structure in proteins. EMBO J 5(4):823–826
3. Sander C, Schneider R (1991) Database of homology-derived protein structures and the structural meaning of sequence alignment. Proteins Struct Funct Bioinf 9(1):56–68. https://doi.org/10.1002/PROT.340090107
4. Hegyi H, Gerstein M (1999) The relationship between protein structure and function: a comprehensive survey with application to the yeast genome. J Mol Biol 288(1):147–164. https://doi.org/10.1006/JMBI.1999.2661
5. Orengo C, Thornton J, Holm L, Sander C (2006) Protein folds and motifs: representation, comparison and classification. In: International tables for crystallography, pp 575–578. https://doi.org/10.1107/97809553602060000714
6. Bordin N, Sillitoe I, Lees JG, Orengo C (2021) Tracing evolution through protein structures: nature captured in a few thousand folds. Front Mol Biosci 8:408. https://doi.org/10.3389/FMOLB.2021.668184/BIBTEX
7. Blundell TL, Jhoti H, Abell C (2002) High-throughput crystallography for lead discovery in drug design. Nat Rev Drug Discov 1(1):45–54. https://doi.org/10.1038/nrd706
8. Miller MD, Phillips GN (2021) Moving beyond static snapshots: protein dynamics and the protein data Bank. J Biol Chem 296:100749. https://doi.org/10.1016/J.JBC.2021.100749
9. Eisenmesser EZ et al (2005) Intrinsic dynamics of an enzyme underlies catalysis. Nature 438(7064):117–121. https://doi.org/10.1038/nature04105
10. Knoverek CR, Amarasinghe GK, Bowman GR (2019) Advanced methods for accessing protein shape-shifting present new therapeutic opportunities. Trends Biochem Sci 44(4):351–364. https://doi.org/10.1016/J.TIBS.2018.11.007
11. Bhattacharyya M, Ghosh S, Vishveshwara S (2015) Protein structure and function: looking through the network of side-chain interactions. Curr Protein Pept Sci 17(1):4–25. https://doi.org/10.2174/1389203716666150923105727
12. Bhattacharyya M, Bhat CR, Vishveshwara S (2013) An automated approach to network features of protein structure ensembles. Protein Sci 22(10):1399–1416. https://doi.org/10.1002/PRO.2333
13. Schieber TA, Carpi L, Díaz-Guilera A, Pardalos PM, Masoller C, Ravetti MG (2017) Quantification of network structural dissimilarities. Nat Commun 8. https://doi.org/10.1038/ncomms13928
14. Gadiyaram V, Ghosh S, Vishveshwara S (2017) A graph spectral-based scoring scheme for network comparison. J Complex Netw 5(2):219–244. https://doi.org/10.1093/comnet/cnw016
15. Ghosh S, Gadiyaram V, Vishveshwara S (2017) Validation of protein structure models using network similarity score. Proteins Struct Funct Bioinf 85(9):1759–1776. https://doi.org/10.1002/prot.25332
16. Gadiyaram V, Vishveshwara S, Vishveshwara S (2019) From quantum chemistry to networks in biology: a graph spectral approach to protein structure analyses. J Chem Inf Model 59(5):1715–1727. https://doi.org/10.1021/ACS.JCIM.9B00002

17. Prabantu VM, Gadiyaram V, Vishveshwara S, Srinivasan N (2022) Understanding structural variability in proteins using protein structural networks. Curr Res Struct Biol 4:134–145. https://doi.org/10.1016/J.CRSTBI.2022.04.002
18. Prabantu VM, Tandon H, Sandhya S, Sowdhamini R, Srinivasan N (2023) The alteration of structural network upon transient association between proteins studied using graph theory. Proteins 93. https://doi.org/10.1002/PROT.26606
19. Prabantu VM, Gadiyaram V, Vishveshwara S, Srinivasan N (2023) Comparison of structural networks across homologous proteins. Proteins 93. https://doi.org/10.1002/PROT.26650
20. Zhang Y, Skolnick J (2005) TM-align: a protein structure alignment algorithm based on the TM-score. Nucleic Acids Res 33(7): 2302–2309. https://doi.org/10.1093/nar/gki524
21. Gadiyaram V, Dighe A, Ghosh S, Vishveshwara S (2021) Network re-wiring during Allostery and protein-protein interactions: a graph spectral approach. Methods Mol Biol 2253:89–112. https://doi.org/10.1007/978-1-0716-1154-8_7
22. Moon RT (2005) Wnt/β-catenin pathway. Sci STKE 271:2005. https://doi.org/10.1126/STKE.2712005CM1
23. Miller JR, Hocking AM, Brown JD, Moon RT (2000) Mechanism and function of signal transduction by the Wnt/β-catenin and Wnt/Ca2+ pathways. Oncogene 18(55): 7860–7872. https://doi.org/10.1038/sj.onc.1203245
24. Logan CY, Nusse R (2004) The Wnt signaling pathway in development and disease. Annu Rev Cell Dev Biol 20(2004):781–810
25. Reya T, Clevers H (2005) Wnt signalling in stem cells and cancer. Nature 434(7035): 843–850. https://doi.org/10.1038/nature03319
26. Klaus A, Birchmeier W (2008) Wnt signalling and its impact on development and cancer. Nat Rev Cancer 8(5):387–398. https://doi.org/10.1038/nrc2389
27. Chien AJ, Conrad WH, Moon RT (2009) A Wnt survival guide: from flies to human disease. J Invest Dermatol 129(7):1614. https://doi.org/10.1038/JID.2008.445
28. MacDonald BT, Tamai K, He X (2009) Wnt/β-catenin signaling: components, mechanisms, and diseases. Dev Cell 17(1): 9–26. https://doi.org/10.1016/J.DEVCEL.2009.06.016
29. Janda CY, Waghray D, Levin AM, Thomas C, Garcia KC (2012) Structural basis of Wnt recognition by frizzled. Science 337(6090): 59–64. https://doi.org/10.1126/SCIENCE.1222879
30. Chu MLH, Ahn VE, Choi HJ, Daniels DL, Nusse R, Weis WI (2013) Structural studies of Wnts and identification of an LRP6 binding site. Structure 21(7):1235–1242. https://doi.org/10.1016/J.STR.2013.05.006
31. Prabantu VM, Naveenkumar N, Srinivasan N (2021) Influence of disease-causing mutations on protein structural networks. Front Mol Biosci 7:492. https://doi.org/10.3389/FMOLB.2020.620554/BIBTEX

Chapter 9

Allosteric Network Analysis Toolkit for Single-Domain Phosphoproteins

Maham Hamid, Safee Ullah Chaudhary, Alessandro Pandini, and Shahid M. Khan

Abstract

Single-domain protein superfamilies mediate essential signaling functions in both prokaryotes and eukaryotes. Signal transduction mediated by single-residue phosphorylation involves subtle secondary structure, side chain rotamer, and hydrogen-bond fluctuations for transitions between active and inactive conformational states, rather than large fold variations. Its study requires molecular dynamics simulations for adequate sampling of the conformational landscape. Discrimination between functional states and subfamilies is achieved by subsequent network analysis designed to couple variations at the local residue or fragment levels to functional collective motions. We describe an integrated protocol for quantitative measurement of local fluctuations and their coupling to functional motions with phosphorylation of the CheY bacterial chemotaxis signal protein as a case study.

Key words Signal phosphorelays, Covalent modification, Response regulators, Molecular dynamics, Mutual information, Structural alphabet

1 Introduction

Phosphorelays are a dominant intracellular signaling strategy in both eukaryotes and prokaryotes. Covalent modification (phosphorylation) at a single residue triggers changes in distant target binding sites. In eukaryotes, the phosphorylated residue is typically a serine, threonine, or tyrosine. In bacteria, aspartate and histidines are the residues that are modified. This chapter outlines a generic protocol for the preparation of the parameter files, the preproduction and production molecular dynamics (MD) runs, and a network analysis toolkit for subsequent elucidation of various aspects of phosphorylation-mediated allosteric communication.

We illustrate key features of this approach with a study of native and phosphorylated states of the *Escherichia coli* chemotaxis protein, CheY—the representative member of the bacterial response

Shahid M. Khan and Florencio Pazos (eds.), *Protein Evolution: Methods and Protocols*, Methods in Molecular Biology, vol. 2979, https://doi.org/10.1007/978-1-0716-4828-5_9,

regulator (RR) superfamily. This family is one of the largest known with functionally diverse subfamilies characterized by a single αβα sandwich. The RR fold is closely related to the eukaryotic Ras superfamily [1]. Phosphorylation of a single CheY aspartate residue (D57) affects binding interactions at a distant surface. This surface binds multiple targets that trigger distinct allosteric pathways to phosphorylate D57, facilitate D57 dephosphorylation, or couple the aspartyl phosphate to bacterial flagellar motor occupancy. It has been difficult to deconvolve these pathways within CheY; let alone identify distinct subfamily functions from RR sequence diversity alone. There is extensive crystal structure data of native and mutated CheY proteins that have given insight into the mechanism, but the sampling of the conformational landscape by crystal structures alone is too sparse to adequately sample the conformational landscape—a common issue for signal protein families. Molecular simulations by us [2–4] and others [5, 6] have therefore exploited the extensive structural data for enteric CheY RRs to advance the knowledge of mechanism.

In this chapter, we study molecular dynamic (MD) trajectories obtained from two CheY crystal structures, an inactive structure and an active structure phosphorylated in silico after the removal of the phosphate analog, beryllium fluoride (BeF3). We report that phosphorylation triggers the compaction of the hydrophobic core as revealed by analysis of the physically connected hydrophobic network and the dynamically coupled information theoretic network. We localize the transition to the reduction in conformational flexibility of a single α-helix. Finally, we measure how coupled changes in the global fold and the local orientation of an aromatic side chain, a classic indicator of activation, are triggered by the phosphorylation switch. Our comparison of the unphosphorylated and phosphorylated CheY states illustrates what can be learnt about allosteric communication from network architectures, flexibility analysis, and local–global correlations.

Structural data are not available for the large fraction of the RR superfamily. We hope that the availability of atomic AlphaFold Foldseek models [7] for the structural elucidation of large superfamilies will permit extensive use of this protocol for the allosteric signal pathways that underlie the functional diversity mediated by single-domain phosphoproteins.

2 Materials

1. Computer Workstations

   The simulations and analyses can be done on a graphics processing unit (GPU)-supported personal computer (PC), but performance is enhanced on dedicated nodes allocated within a cluster. The Biomedical Informatics and Engineering Research Laboratory (BIRL) cluster has four compute nodes. The

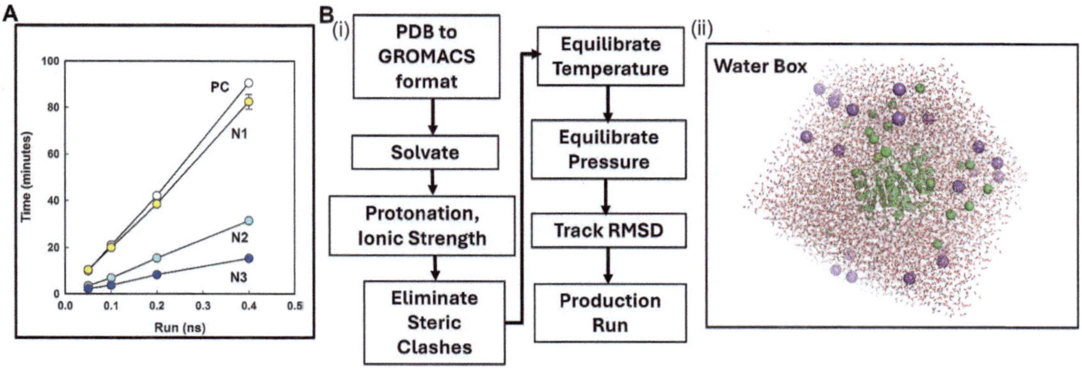

Fig. 1 (**A**) Benchmarks. Performance of a PC (white) versus BIRL compute nodes equipped with commercially available processors (N1 (yellow)), (N2 (cyan)), (N3 (blue)). (**B**) MD simulation. (*i*) Workflow for structure, topology, and parameter file preparation. (*ii*) Simulation Box. CheY protein (cartoon representation), Na ions (blue spheres), and Cl ions (green spheres). Water molecules are shown as small tripods (oxygen (red), hydrogen (white)). The typical run times on Biowulf, by comparison, were 0.15ns/s.

performance of three of these nodes is compared against a PC (Intel® Core ™ i7-8700, NVIDIA 1650x GPU) in Fig. 1a. Node 1 (N1) has 4 × Intel Xeon Processor E7-4830 v3, NVIDIA Tesla K40m GPU. Node 2 (N2) has 2 × Intel Xeon CPU ES-2678 v3, NVIDIA Tesla K40C GPU. Node 3 (N3) has 4 × Intel Xeon Processor E7-8880 v3 2.30 GHz. The nodes are connected via the Ray open-source framework (https://docs.ray.io/) for distributed jobs when needed.

2. Software

   The Groningen Machine for Chemical Simulations (GROMACS) v.2022 MD package was utilized for the simulations (<100 ns) and analyses detailed in this chapter (*see* **Note 1**). The long (1-µs) simulations run on the National Institute of Health (NIH)_ high-performance Biowulf supercomputer used 2 compute nodes (4xNVIDIA, A100 GPUS (80 GB VRAM, 6921 cores, 432 tensor cores)). Bridge2 (https://github.com/maltesie/bridge2) was used for the construction of physical networks. AlloHub.Py (https://github.com/Fraternalilab/AlloHubPy) was used for the construction of information-theoretic networks. These packages and the associated C and Python libraries were downloaded and installed on the BIRL computers.

3. Internet

   Internet access to structure and sequence databases, download of software and libraries, and the use of online Web servers is an important limitation to throughput globally, particularly in the developing world [8]. We used the online distance matrix alignment (DALI) server (http:/ekhidna2.biocenter.helsinki.fi/dali/) to compare fold similarity as detailed below. The broadband connection speed at BIRL ranged from 200 to 600 Mb/s.

3 Methods

3.1 MD Simulations

1. Input Files.
   (a) The required inputs for the GROMACS MD simulation package [9, 10] are a three-dimensional (3D) PDB file, plus parameter and topology files. The atomic structure file, ideally <3.0 angstrom resolution, is first curated to add hydrogen atoms and residues missing from the electron density map. Modeller [11] is a comparative modeling suite that can be also used to add missing residues in incomplete structures from knowledge of the one-dimensional (1D) sequence.

   The CheY crystal structures in the Mg^{2+}-bound inactive (*Salmonella Typhimurium* (2CHE.pdb)) and active states (*E. coli* (1FQW)) are downloaded from the Protein Data Bank. The proteins differ at three residue positions (51FstYec, 54IstVec, and 78SstGec). The active CheY structure has beryllium fluoride (BeF_3) to mimic the phosphorylated state since aspartyl phosphate is labile [12].
   (b) Web servers with compatible force fields are available to generate topology parameters for phosphorylation of most residues. The BeF3 is removed, and CheY D57 is phosphorylated in silico using the online Vienna-PTM Web server [13]. The aspartyl phosphorylation is selected, followed by the dianionic phosphate option (*see* **Note 2**). While there are several force fields that all employ Langevin dynamics useful for protocol detailed here, the most straightforward option to circumvent compatibility issues is to retain the force field (gromos-54a7) used by Vienna-PTM for subsequent simulations.
   (c) The divalent metal cofactor, typically Mg^{2+}, is essential for coordination of the phosphorylated residue. The Mg^{2+} coordination site is highly conserved. The Mg^{2+} is added to the PDB file with initial coordinates that approximate the coordination geometry [14]. The coordination is optimized during the energy minimization. The correct parameter and topology files that recognize the Mg^{2+} and the PTM aspartyl phosphate must be used.
   (d) The curated PDB files are converted to GROMACS-compatible format. Use

```
gmx pdb2gmx -f *.pdb/gro -o *.pdb.gro -p topol.top
-i posre.itp,
```

   where "-f" and "-o" expect input and output files, respectively, while topology and positional restraint files are entered with the -p and -i options. These defaults will be adopted for all subsequent operations. The curated 2CHE and phosphorylated 1FQW will henceforth be denoted as (−) PO_4 and (+) PO_4, respectively.

(e) The structures are solvated within a water box. Use

gmx editconf -f complex.gro -o complex_newbox.gro -c -d 1.0 -bt cubic

to set box type (cubic is the common option) and dimensions (in nm).

Then, *gmx solvate -cp complex_newbox.gro -cs spc216.gro -o complex_solv.gro.*

The "*-cs*" option specifies the water model. The common choices are the three-point TIP3 and SOL models [15]. The box dimensions should be such that the protein is separated from the box boundaries by at least 5 angstroms on each side. Periodic boundaries ensure bulk effects for the solute. The topology tile can be specified at the stage if not done earlier with *gmx-pdb2gmx*

(f) Finally, the system (protein + solvation water box) is then ionized. Residue protonation states are set for the desired pH. The default options are 150 mM ionic strength (monovalent Na^+, K^+, Cl^- ions) and neutral pH (7.0).

Use *gmx grompp -f ions.mdp -complex_solv.gro -p topol.top -n complex.ndx -o ions.tpr*, where the "-n" option specifies the index file.

Then, *gmx genion -s ions.tpr -o complex_solv.gro -p topol.top -pname NA -nname CL - neutral -conc 0.15.*

The GROMACS preprocessor "*gmx grompp*" command expands the topology file to specify the atomic parameters (atom type and number).

2. Preproduction

Short simulations prepared the system for the production runs. Energy minimization using the steepest descent algorithm is followed by NVT (canonical ensemble) (100–500 ps) and NPT (isothermal-isobaric ensemble) (100–500 ps) equilibration (100–500 ps). Temperature control was set at 300 K and 0.1 ps coupling constant with the Berendsen algorithm. Multiple steps may be used for the temperature equilibration [16]. Isotropic pressure was set at 1 bar with the Parrinello-Rahman barostat [17]. Nonbonded interactions had a 1.0-nm Verlet cutoff, with the long-range interactions approximated by the particle-mesh Ewald (PME) method.

The energy minimization run processed by the "*gmx mdrun*" command corrects for steric clashes and optimizes bond and coordination geometries based on the topology file. Important presets are the step time (2 fs for backbone, 1 fs for water dynamics), the frame- to-frame saving frequency (5–10 ps), the simulation duration, the force field, and the distance threshold for the PME approximation (*see* **Note 3**).

Use `gmx grompp -f minim.mdp -c complex_solv.gro -p topol.top -o em.tpr`.

Then, `gmx mdrun -v -deffnm em`.

The tpr files are compressed binary files, in contrast to a pdb or gro files, and in addition to the structural coordinates and topology specify velocities, forces and other parameters are needed for the simulation.

(a) The system is equilibrated to ambient temperature and one-atmosphere pressure prior to the production run.

Use `gmx grompp -f nvt.mdp -c em.gro -r em.gro -p topol.top -o nvt.tpr`.

Then, `gmx grompp -f npt.mdp -c nvt.gro -r nvt.gro -t nvt.cpt -p topol.top -o npt.tpr`.

In each case, ***gmx mdrun*** should be used for short (1-ns) simulations once the topology file is updated. The "***nohup***" option can be used for workstations that experience frequent interruptions.

(b) The production run can now be started after the temperature and pressure equilibrated presets have been met, after the final update of the *.tpr file.

```
gmx grompp -f md.mdp -c npt.gro -t npt.cpt -p topol.top
-o rep1_md.tpr
```

where "***rep1***" denotes replicate-1. Multiple replicate runs, three or more, are recommended.

The subsequent `gmx mdrun` command creates a set of files that include the log file and checkpoint files to restart runs that terminate prematurely or require extended time to improve sampling. These steps are schematized in Fig. 1b.

The system is prepared for analysis after completion of the replicate runs. The protein is centered, rotationally and translationally aligned for each replicate (rep*) with the *gmx trjconv* command.

```
gmx trjconv -s rep*_md.tpr -f rep*_md.xtc -o rep*_md_noPBC.xtc
-pbc mol -ur compact – center
gmx trjconv -s rep*_md.tpr -f rep*_md_noPBC.xtc -o rep*_rot-
trans.xtc -fit rot+trans
```

The command can also be used to strip water and solvent ions for the trajectories if the protein dynamics are to be studied in isolation from the solute (i.e., Allohub.py), but not for the analyses of physical contacts for which water inclusion is important (i.e., Bridge2).

3.2 Network Analysis

1. Representation and Encoding

A generic network consists of nodes and edges. For proteins, the nodes are the residues, and the edges are the interactions between them. Major nodes may then be designated based on the prominence of their role in network connectivity (*see* **Note 4**). The network representation of the *E. coli* CheY protein is shown in Fig. 2a. The nodes (residues) have characteristic properties that include chemical type, mass, and solvent accessibility. The interactions (edges) may be grouped into two categories—direct physical

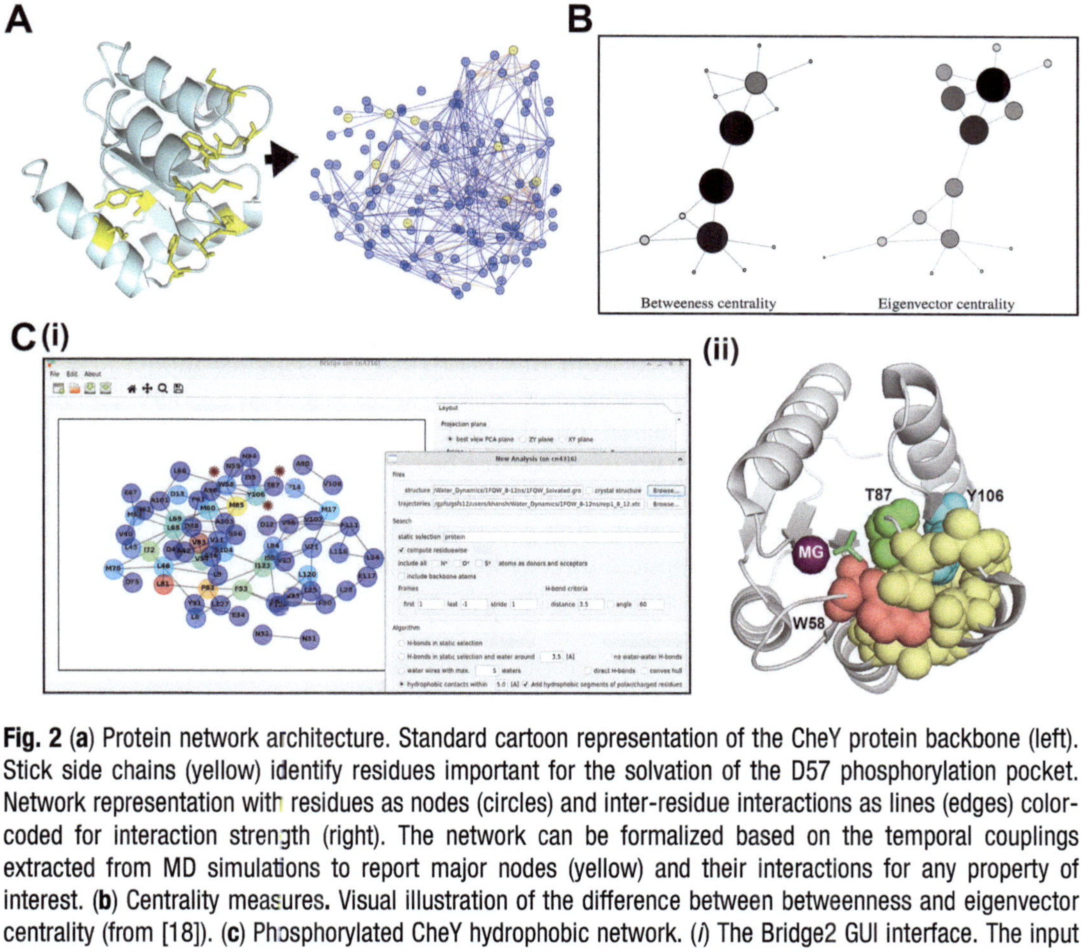

Fig. 2 (**a**) Protein network architecture. Standard cartoon representation of the CheY protein backbone (left). Stick side chains (yellow) identify residues important for the solvation of the D57 phosphorylation pocket. Network representation with residues as nodes (circles) and inter-residue interactions as lines (edges) color-coded for interaction strength (right). The network can be formalized based on the temporal couplings extracted from MD simulations to report major nodes (yellow) and their interactions for any property of interest. (**b**) Centrality measures. Visual illustration of the difference between betweenness and eigenvector centrality (from [18]). (**c**) Phosphorylated CheY hydrophobic network. (*i*) The Bridge2 GUI interface. The input menu window shown (bottom right) has options to load structures and trajectories. Once these are loaded, the analysis option is selected—hydrophobic bonds in the present case. The results/analysis window shows the network color-coded by residue betweenness centrality. (*ii*) Extended hydrophobic network. Three 4 ns replicates of the (−)PO_4 and (+)PO_4 states were compared. The contiguous hydrophobic patch (pale yellow) formed by phosphorylation, mapped onto 1FQW, connects W58 (salmon) adjacent to D57, with the switch T87 (green) and interfacial Y106 (cyan) in all (+)- PO_4, but not in any of the (−)-PO_4 replicates. The patch residues are L66, L69, M85, E89, A90, N94, I95, A98, A101, and A103. Mg^{2+} (purple sphere), BeF3 (green tripod). The α4-β5 segment is a major target

contact or dynamic correlations. There are multiple options within each category. The specific selection will depend on the research question under study.

2. Centrality

One of the fundamental properties of a network is its node centrality. Several measures of centrality have been proposed for different data domains. In this chapter, we will use "betweenness centrality" to study a physically connected network and "eigenvector centrality" to study information transfer via dynamic couplings. In a sense, all residues are serially connected as components of the linear polypeptide, but physical connectivity as used here refers to long-range spatial contacts measured in 3D Cartesian coordinates.

The betweenness centrality, Cbw, is

$$C_{\mathrm{bw}}^{n} = \sum_{n1=1-N,\, n2=1-n} \left(X_{n1n2}/(M_{n1n2})\right)$$

It measures how often a node, n, lies on the shortest path between a pair of nodes N1 and N2 [4].

A network can be represented by an adjacency matrix describing its nodes' connectivity. The eigenvector centrality can directly be computed from this matrix. In the dynamic model here presented, the adjacency matrix is a matrix of locally correlated motions, based on the normalized mutual information (nMI) between pairs of nodes (see below). In vector notation, the overall connectivity of a given node is reported by its eigenvector centrality, E_v. The contribution of a node to the network, as estimated by its E_v, is computed directly from the correlation adjacency matrix:

$$E_{\mathrm{v}}.\{M\}_{\mathrm{corr}} = E_{\mathrm{v}}.\lambda$$

where the $\{M\}_{\mathrm{corr}}$ is the correlation matrix. The λ is the eigenvalue.

The difference between these centrality measures is illustrated by the schematics in Fig. 2b.

3. Connectivity

We will use the Bridge2 algorithm for the construction of a physically connected network. Our choice is based on two key considerations. First, Bridge2 employs a k-d tree search that scales with n(log(n)), where n is the number of nodes. It is two orders of magnitude faster than the computation of all pairwise distances that scale as n^2 (see the original publication for further details). Second, Bridge2 has an interactive graphical user interface (GUI) that greatly facilitates the initial assessment of the constructed network. Figure. 2c.i shows a snapshot of the Bridge2 GUI. Bridge2 can be set up on a PC but has enhanced performance on a computer cluster (*see* **Note 5**).

Here, we use Bridge2 to compare the hydrophobic networks that connect the target binding surface to the D57 between the (−) PO_4 and (+)PO_4 states. Load the individual replicate trajectories together with the structure file in the INPUT menu. Select "hydrophobic contacts (<5 angstroms)" and "add hydrophobic segments of polar/charged residues." Visual inspection can be used to assess the Bridge2 OUTPUTS before download and further analysis. Comparison, based on all six replicates, reveals the hydrophobic core of the state becomes more compact in the (+)PO_4 state (*see* **Note 6**). The differential compaction of the core is shown in the map of the connected network onto the structure (*see* Fig. 2c.ii).

Additional options available in the Bridge2 GUI options menu are the occupancy, the degree centrality, and the computation of the shortest path between any two nodes. In Bridge2, as in many other software packages, shortest paths are computed with the Dijkstra algorithm [19].

4. Protein Flexibility

Proteins and other biopolymers are flexible, and the modulation of the flexibility underlies functional adaptations and mediates conformational changes. The common measure of flexibility reported in GROMACS is the root-mean-square fluctuations (RMSF) of the positions of C^{α} backbone atoms.

```
gmx rmsf -f *.xtc -s *.tpr -n *.ndx -o *.xvg
```

where the s option specifies the reference frame for the trajectory fit.

The *gmx rmsf* function reports backbone flexibility in Cartesian coordinates. The backbone is characterized by its secondary structure architecture. It is frequently desirable to extract mechanistically significant secondary structure fluctuations such as α-helix melting transitions. This is done most simply by readout of the residue dihedral angles [20] that can then inform on secondary structure transitions with the aid of the Ramachandran plot.

A more potent approach is the use of structural alphabets (SAs). The SAs are secondary structure motifs assigned to sequence fragments based on the propensity of local arrangement as reported by 3D protein structure databases. Here, we will illustrate the use of the M32K25 structural alphabet for four-residue fragments developed over a decade ago [21] within the Python package AlloHub.py [22]. The package together with its libraries and Jupyter notebooks was installed on server S-2 and executed with Python v.3.11.5. The trajectory is first encoded by assigning a letter from M32K25 SA for each four-residue fragment using a rolling window. The choice of the best letter for each fragment is done by local RMSD fit. Shannon entropy and the normalized MI for the dynamic couplings between fragments are then compared between conformations, here the (−)PO_4 and (+)PO_4 states.

The Shannon entropy, $\Delta S(X)$, of fragment X reports on the number and probability of its microstates within a trajectory. It is a measure of "conformational" flexibility.

$$\Delta S(X) = -\sum_{x} p(x) \log_2 p(x)$$

The eigenvector centrality profiles are derived from the nMI matrix. The mutual information, $I^n_{LL}(C_i; C_j)$, between two fragments (i) and (j) is

$$I^n_{LL}(C_i; C_j) = \left(\frac{I(C_i; C_j) - \varepsilon(C_i; C_j)}{H(C_i, C_j)} \right)$$

where i and j are the relevant residue column. $I(C_i; C_j)$ is the mutual information, $H(C_i; C_j)$ is the joint entropy, and $\varepsilon(C_i; C_j)$is the finite size error.

$$MI(X, \Upsilon) = \sum_{x, y} p(x, y) \, log \left(\frac{p(x, y)}{p(x) p(y)} \right)$$

where $P(X, \Upsilon)$ represents the joint probability distribution of the residue X and Υ, and $P(\mathrm{X})$, $P(\Upsilon)$ denote their marginal probabilities.

$$H(X, \Upsilon) = -\sum_{x, y} p(x, y) \log_2 p(x, y)$$

$$\epsilon(X, \Upsilon) = \frac{B_{X,\Upsilon} - B_X - B_\Upsilon + 1}{2N}$$

where N represents the sample size and B represents the probability of distinct states (see [19] for details).

AlloHub.Py does not have a GUI. The commands for the key steps are outlined below with 2CHE ((−)PO_4) as an example. First, initialize the SA handler

```
sa_2che= SAtraj.SAtraj(block_size=1000000, alphabet=SAtraj.
ALPHABETS{"M32K25"})
sa_2che.load_data("all.R1-2che.gsa.1f_str.out")
```

The large block size allows the encoded trajectory to be processed as a single block. Next, compute the MI with a spherical distance cutoff, typically 7 angstroms, so that only the strongest correlations are used. A distance matrix is created to employ the cutoff.

```
traj = md.load("2che.pdb")
ca_indices = traj.topology.select('name CA')

# Extract the coordinates of C-alpha atoms
ca_positions = traj.xyz(0)(ca_indices)
```

```
# Number of points
n_points = ca_positions.shape(0)
# Initialize an empty distance matrix
distance_matrix = np.zeros((n_points, n_points))
for i in range(n_points):
 for j in range(n_points):
# Compute Euclidean distance between points i and j
  distance_matrix(i, j) = np.sqrt(np.sum((ca_positions(i) -
ca_positions(j))**2))
# Convert to angstroms
 distance_matrix *= 10
```

The max_workers option controls the parallelization of the nMI computation. Finally, the eigenvector centrality is computed. Plots can be generated within AlloHub.Py or by the export of the ASCII text files.

```
mi_2che = sa_2che.compute_mis(max_workers=7) centrality_df =
SAgraph.compute_centrality()
```

Figure 3a reports the differences between the (−)PO_4 and (+) PO_4 states revealed by these measures.

AlloHub.Py incorporates additional features to the initial R script GSATools package [23]. The feature of the most interest for comparison of distinct states, as used here, is block partition of the trajectory with subsequent randomization that improves difference detection. Once randomization is complete, similarity matrices are generated by the overlap handler to record the differences within and between the replicate trajectories of the states (inactive = (−) PO_4, active = (+) PO_4) . "100_1" denotes block size =100, replicate = 1, etc.

Create the overlap handler to compute similarities between the trajectories.

```
overlap_100 = Overlap.Overlap((mi_inactive_100_1, mi_inac-
tive_100_2, mi_inactive_100_3,
mi_active_100_1, mi_active_100_2, mi_active_100_3), ev_list=
(0,1,2))
overlap_100.fill_overlap_matrix()
```

Overlap for block 100 randomized

```
Allohub_plots.plot_overlap(overlap_100.get_overlap_matrix(),
vmax=0.4,
action="show") similarity_matrix_100 = overlap_100.compute_-
similarities()
```

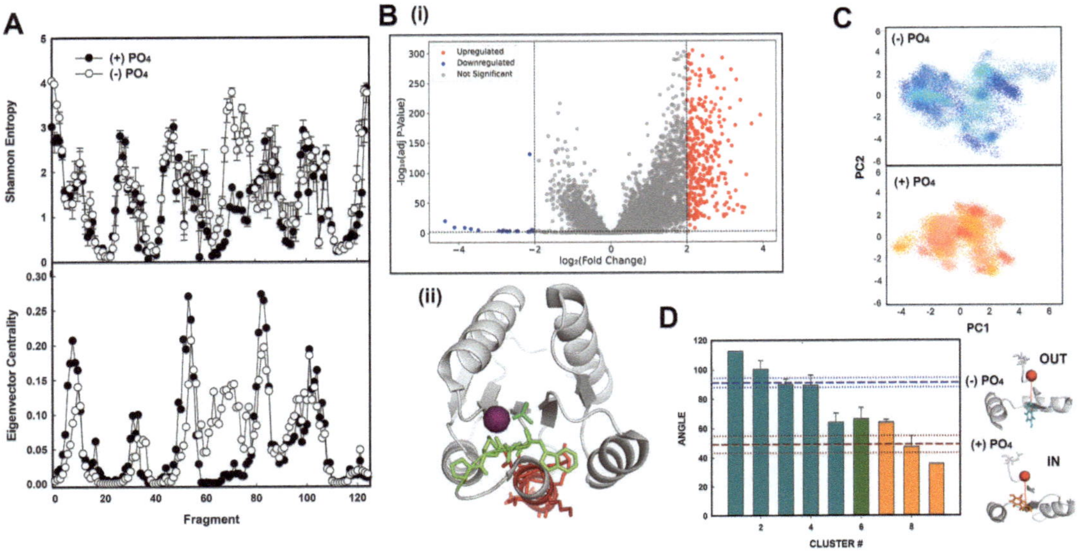

Fig. 3 (**A**) Properties of the M32K25-encoded network. Shannon entropy (top). Eigenvector centrality (bottom). (**B**). Phosphorylation-dependent network modulation. (*i*) Volcano plot reporting the number and statistical significance of the upregulated (red) and downregulated (blue) dynamic couplings upon the switch from the (−)PO_4 to the (+)PO_4 state. (*ii*) The central nodes (fragments), as identified by their frequency of occurrence in the differentially regulated couplings, mapped onto the BeF3-activated crystal structure (1FQW.pdb). The upregulated fragments (red side chains) map onto helix α3. The fragment adjacent to the D57 phosphorylation site occurs in both upregulated and downregulated couplings (green side chains). (**C**). Collective modes. The variations between replicates within each state are visualized in 2D plots of the principal PC1 and PC2 motions. (**D**) Global–local correlations. Two sequential cluster cycles for fold similarity reveal different fold subsets for (−)PO_4 and (+)PO_4, but also a shared fold. The orientation of the Y106 side chain, a classical marker for CheY activation, is coupled to fold conformation

The t-test is used to assess the significance of the differences.

```
tat, p_value = ttest_ind(within_conditions, between_condi-
tions, equal_var=False, alternative='greater')
```

Figure 3bi shows the volcano plot obtained when the (−)PO_4 and (+)PO_4 MI matrices were compared ($p > 0.01$). The adjusted p-values were calculated using the Benjamini–Hochberg procedure to control the false discovery rate. The plot reveals a greater number of upregulated versus downregulated couplings upon phosphorylation. The most frequent six fragments found in the differentially coupled pairs were then mapped onto the 1FQW structure used to generate the (+)PO_4 state (*see* Fig. 3bii). Five adjacent upregulated fragments (68, 69, 70, 71, and 73) map to helix α3. Fragment 58, adjacent to D57, occurs in both upregulated and downregulated couplings.

5. Local Versus Global Motions

Protein motions span a range of levels from side chain rotations and H-bond dynamics (picoseconds) to domain motions (microseconds). Once a protein network has been characterized, it is useful to compare its features to motions that occur on different size and time scales. Some GROMACS functions that extract this spatiotemporal information are illustrated below.

The overall conformational change in the protein over the trajectory time course is extracted with the *gmx rmsd* function. The RMSD (root mean square deviation) versus time plot can quickly assess conformational fluctuations to detect allosteric transitions between conformational states. Typically, functional motions entail collective modes that can be extracted with principal component analysis (PCA). The principal motions (PCs) are obtained by diagonalization of the C^a covariation matrix. To obtain the matrix for the inactive 2che ((−)PO_4 state replicates (rep.*)), use

```
gmx covar -f 2che_rep*.xtc -s md.tpr -v eigenvector -o eigenval.xvg -av average.pdb
```

Then,

```
gmx anaeig -f 2che_rep*.xtc -s npt.tpr -first 1 -last 2 -v -eig eigenval.xvg -2d 2dproj.xvg
```

The first two PCs, representing the most dominant collective motions, typically account for about 50% of the total motion. The PC1–PC2 plots for the two states are shown with the individual replicates distinguished by shade to visualize intra-state variation (*see* **Note 7**). Visual inspection of the plots reveals (a) a substantial overlap between replicates for each state and (b) the spread is greater for the (−)PO_4 versus the (+)PO_4 state (*see* Fig. 3c).

Next, the replicates for each state were concatenated. The concatenated trajectories were then analyzed with (*gmx cluster*). The function has different clustering options but the common one, used here, is the root-mean-square distance with a cutoff option to be adjusted according to the system. The -cl option lists the representative structures for each cluster.

```
gmx cluster -f combined_2che.xtc -s combined_2che.tpr -o cluster.xpm -cl clusters.pdb -cutoff 0.25.
```

The cluster representatives for both states were combined. The combined file was then used to relate the global fold to a local internal rotation of an aromatic side chain (Y106 for CheY) at the target binding surface that has been identified as a classic indicator of activation in the RR superfamily. We used the DALI GUI Web server to further group the combined state cluster representatives based on fold similarity. Use the "All against All" option in the DALI menu that can simultaneously compare up to 64 structures that can be simultaneously compared with this option. The resulting heat map revealed five (-)PO_4 clusters, three (+)PO_4 clusters, and one mixed cluster. The nine cluster representatives were combined into a trajectory file with *gmx trjconv* and the orientation of the Y106 side chain relative to the vector connecting the D57 and Y106 C^{α} atoms computed with `gmx gangle`.

```
gmx trjconv -f cluter_rep.pdb -o cluster_rep.xtc
gmx gangle -f cluster_rep.xtc -s cluster_rep.pdb -oav Y106_an-
gaver.xvg -g1 vector -g2 vector -group1 V1 -group2 V2
```

The Y106 rotamer angles coalesced around distinct IN and OUT orientations that mapped onto the (+)PO_4 and (−)PO_4 folds, respectively (*see* Fig. 3d). Thus, the phosphorylation-dependent switch in the CheY fold, albeit subtle, can be detected with sensitive sampling of the conformational landscape revealed by simulations. The switch involves both conformational selection and creation as argued by us [2] (*see* **Note 6**) and others. Furthermore, the global change in the fold is tightly coupled to the internal rotation of the Y106 side chain, with the caveat that the orientation varies over a limited range (+10°) for each state.

4 Notes

1. There are different MD packages. GROMACS, NAMD, and AMBER are the most widely used. They are compared in the chapter by Fatimi et al. The Windows PC performance (Fig. 1a) was benchmarked with NAMD. The GROMACS and NAMD runtime are comparable when benchmarked on BIRL N2. The trajectory files' output from these packages can be interconverted with the MDTraj [24] mdconvert operator. The online GROMACS tutorials [10] should be consulted for details of the protocol needed to set up the production runs.
2. The residue name for the Vienna-PTM entry of the dianionic aspartyl phosphate (D2P) in the PDB file needs to be renamed to ASP to be compatible with software for the Bridge2 and AlloHub.Py analyses. The PDB and parameter files are modified, in addition, by inclusion of the PO_4 atoms OZ1, OZ2, and OZ3. The polar D2P or ASP (aspartate) residues do not contribute to the hydrophobic networks presented in this chapter. The side chain differences between them should not

alter C^{α} backbone dynamics, but more data are needed to determine whether or not this is the case.

3. Textbook tutorials commonly state that the energy minimization can be concluded once temperature and pressure equilibration have been achieved. Such equilibration cannot be always extrapolated to fold stability, particularly in the case of large protein biopolymers and assemblies. Therefore, it is a good practice to check with *gmx rmsd* that the root-mean-square deviation (RMSD) of protein conformational fluctuations has reached a stationary value.
4. The basic network can be organized into communities with the Girvan–Newman algorithm [25]. A community in this higher-level network is a subset of nodes that are densely connected relative to other nodes. Community networks based on mutualinformation can detect the emergence of new dynamic sectors, as illustrated by our previous work that introduced this approach for CheY allostery [2].
5. The two-dimensional (2D) network display options in the Bridge2 GUI are limited. We have found that it is more useful to export the network node and edge lists as ASCII text or Python pkl files to present the networks using the dedicated network GUIs available in Cytoscape. In addition, memory limitations can become prohibitive for PCs and smaller workstations when the inputs include trajectory files in addition to the structure file.
6. This result extends our previous study that reported a similar effect for the phosphomimic CheY 13DK106YW residue substitution when compared against apo CheY, supported by time-resolved X-ray footprinting experiments of solvent accessibility [2]. A combined computational and experimental study of other CheY residue substitutions highlighted the role of the hydrophobic network around α4-β5 on CheY dephosphorylation kinetics [5].
7. The importance of replicates cannot be overestimated. The ergodic condition that the time average for any one simulation equals the ensemble average cannot be met due to the finite simulation time. The simulation could get stuck in local minima given the complex conformational landscapes for soft-matter polymers such as proteins. Three or more replicates are needed based on the system size and simulation time [26]. PCA can help assess convergence between replicates with the GROMACS *gmx anaeig* options "-over" and "-inpr" that calculate the subspace overlap and the inner product matrix between eigenvectors respectively between replicates (see [2]).

Acknowledgments

M.H. was supported by the SBA School of Science and Engineering PhD graduate program and the Pakistan Higher Education Commission (HEC) R&D Grants 20-3629/NRPU/R&D/HEC/14/585, RF-NCBC-015 (to S.U.C.). The sub-microsecond simulations and most analyses were performed on the LUMS-BIRL cluster supported by the above and other grants (https://birl.lums.edu.pk) (to S.U.C). The microsecond simulations utilized the NIH HPC Biowulf computational cluster (http://hpc.nih.gov) supported by the NIH, National Institute for Neurological Disorders and Stroke (NINDS) intramural program (S.K). Initial studies to develop this protocol were funded by the Royal Society Collaborative Exchange grant U1175.70592 (to S.K).

References

1. Goitre L, Trapani E, Trabalzini L, Retta SF (2014) The Ras superfamily of small GTPases: the unlocked secrets. Methods Mol Biol 1120: 1–18
2. Wheatley P, Gupta S, Chen Y, Petzold CJ, Ralston CR, Blair DF et al (2020) Allosteric priming of E. Coli CheY by the flagellar motor protein FliM. Biophys J 119:1108–1122
3. Hamid M, Jabeen I, Chaudhary SU, Khan S (2025) Conservation of the H-bond network in bacterial response regulators. BioRxiv/ 2025/674700
4. Hamid M, Khalid MF, Chaudhary SU, Khan S (2022) The solvation of the E. coli CheY phosphorylation site mapped by XFMS. Int J Mol Sci 23:12771
5. Foster CA, Silversmith RE, Immormino RM, Vass LR, Kennedy EN, Pazy Y et al (2021) Role of position K+4 in the phosphorylation and dephosphorylation reaction kinetics of the CheY response regulator. Biochemistry 60: 2130–2151
6. Foster CA, West AH (2017) Use of restrained molecular dynamics to predict the conformations of phosphorylated receiver domains in two-component signaling systems. Proteins 85:155–176
7. van Kempen M, Kim SS, Tumescheit C, Mirdita M, Lee J, Gilchrist CLM et al (2024) Fast and accurate protein structure search with Foldseek. Nat Biotechnol 42:243–246
8. Broom D (2023) These are the places in the world where internet access is still an issue—and why. World Economic Forum, Cologny, p 9
9. Kohnke B, Kutzner C, Grubmuller H (2020) A GPU-accelerated fast multipole method for GROMACS: performance and accuracy. J Chem Theory Comput 16:6938–6949
10. Lemkul JA (2024) Introductory tutorials for simulating protein dynamics with GROMACS. J Phys Chem B 128:9418–9435
11. Webb B, Sali A (2017) Protein structure modeling with MODELLER. Methods Mol Biol 1654:39–54
12. Petsko GA (2000) Chemistry and biology. Proc Natl Acad Sci USA 97:538–540
13. Margreitter C, Reif MM, Oostenbrink C (2017) Update on phosphate and charged post-translationally modified amino acid parameters in the GROMOS force field. J Comput Chem 38:714–720
14. Lee SY, Cho HS, Pelton JG, Yan D, Berry EA, Wemmer DE (2001) Crystal structure of activated CheY. Comparison with other activated receiver domains. J Biol Chem 276:16425–16431
15. Kadaoluwa Pathirannahalage SP, Meftahi N, Elbourne A, Weiss ACG, McConville CF, Padua A et al (2021) Systematic comparison of the structural and dynamic properties of commonly used water models for molecular dynamics simulations. J Chem Inf Model 61: 4521–4536
16. Motta S, Minici C, Corrada D, Bonati L, Pandini A (2018) Ligand-induced perturbation of the HIF-2alpha:ARNT dimer dynamics. PLoS Comput Biol 14:e1006021
17. Bussi G, Donadio D, Parrinello M (2007) Canonical sampling through velocity rescaling. J Chem Phys 126:014101

18. Rodrigues FA (2019) Network centrality: an introduction. arXiv. https://doi.org/10.48550/arXiv.1901.07901

19. Pandini A, Fornili A, Fraternali F, Kleinjung J (2012) Detection of allosteric signal transmission by information-theoretic analysis of protein dynamics. FASEB J 26:868–881

20. McClendon CL, Hua L, Barreiro A, Jacobson MP (2012) Comparing conformational ensembles using the Kullback-Leibler divergence expansion. J Chem Theory Comput 8:2115–2126

21. Pandini A, Fornili A, Kleinjung J (2010) Structural alphabets derived from attractors in conformational space. BMC Bioinformatics 11:97

22. Carmona OG, Kleinjung J, Anastasiou D, Oostenbrink C, Fraternali F (2025) AllohubPy: detecting allosteric signals through an information-theoretic approach. J Mol Biol 437:168969

23. Pandini A, Fornili A, Fraternali F, Kleinjung J (2013) GSATools: analysis of allosteric communication and functional local motions using a structural alphabet. Bioinformatics 29: 2053–2055

24. McGibbon RT, Beauchamp KA, Harrigan MP, Klein C, Swails JM, Hernandez CX et al (2015) MDTraj: a modern open library for the analysis of molecular dynamics trajectories. Biophys J 109:1528–1532

25. Newman ME (2006) Modularity and community structure in networks. Proc Natl Acad Sci USA 103:8577–8582

26. Brown AM, Lemkul JA (2025) Robustness in biomolecular simulations: addressing challenges in data generation, analysis, and curation. Cell Rep Phys Sci 6(5):102566

Chapter 10

Computational Identification of Potential Shape-Shifting Proteins from Structures

Francisco M. Pérez-Canales, Enrique Alanís-Domínguez, and Ana M. Rojas

Abstract

The "one structure–one function" paradigm is central to most computational methods for predicting protein function. However, a subset of proteins known as shape-shifters can adopt multiple well-defined structures to perform distinct biological functions. These proteins, along with intrinsically disordered proteins (IDPs) that fold upon binding, challenge conventional views of protein functionality. Despite their biological significance, identifying fold-switching proteins remains difficult due to limited structural data and the limitations of current prediction methods.

In this study, we present a computational protocol to identify potential shape-shifting proteins within structural databases. By integrating structural and sequence data, this approach enables the screening of the Protein Data Bank (PDB) for proteins exhibiting significant conformational variability.

Key words Shape-shifting proteins, Fold-switching, Structural alignment, Function

1 Introduction

The "one structure–one function" paradigm gained widespread acceptance following Anfinsen's 1961 work [1], which explained the native conformation of proteins. This led to the assumption that proteins achieve their physiological roles through a single, energetically stable fold. Although this concept has been challenged, it became foundational in computational methods for predicting protein function. The idea was further reinforced by Chothia and Lesk's 1986 findings [2] that protein structure is more conserved than sequence.

However, in the early 2000s, emerging evidence demonstrated that a single protein sequence can adopt multiple structures, each capable of performing distinct functions, especially in enzymes [3]. This discovery profoundly impacted the field of protein evolution, highlighting conformational diversity as a key factor in the

Shahid M. Khan and Florencio Pazos (eds.), *Protein Evolution: Methods and Protocols*, Methods in Molecular Biology, vol. 2979, https://doi.org/10.1007/978-1-0716-4828-5_10,

emergence of novel functions. Large-scale analyses of protein databases further supported this, revealing significant structural variability even among proteins with similar sequences [4, 5]. Such variability underscores the critical role of shape-shifting proteins in functional evolution.

These proteins mainly include intrinsically disordered proteins (IDPs) (reviewed by [6]) and fold-switching or "metamorphic" proteins [7]. IDPs transition into ordered states upon binding, while fold-switching proteins undergo substantial structural changes—such as shifts from β-sheet to α-helix or vice versa—in specific regions. It is estimated that about 4% of proteins in the Protein Data Bank (PDB) exhibit these fold-switching behaviors [8] and are often discovered serendipitously. Examples include regulator of fatty acid synthesis H (RfaH) [9, 10], lymphotactin (XLC1) [11], and chloride intracellular channel 1 (CLIC1) [12], where fold transitions are triggered by factors such as pH, temperature, salt concentrations, binding partners, redox states, or oligomerization.

Both fold-switching proteins and IDPs can perform multiple functions. For instance, lymphotactin (XLC1) can adopt two distinct folded states: One binds to a chemokine receptor, while the other interacts with glycosaminoglycans, with transitions regulated by salt and temperature [11]. Additionally, fold-switching can be engineered through mutations, as demonstrated by synthetic GA/GB proteins [13], suggesting that such folds may evolve naturally—though not all mutations result in new folds.

Given the limited structural data relative to sequence data, prediction methods focusing on sequence information have been developed to identify fold-switching properties. These methods often target regions with low secondary structure prediction scores and rely on energetic estimations [8, 14]. However, predicting fold-switching proteins by sequence alone remains challenging due to limitations in current tools; therefore, this will not be the focus of this work.

The protocol we present involves a multistep computational approach that integrates structural and sequence data to identify proteins with potential fold-switching capabilities. It consists of four main steps: (1) data collection, (2) sequence analysis, (3) structural analysis, and (4) literature search and interpretation. This study emphasizes using current standard methods to screen protein structural databases and identify potential fold-switching proteins using available structural data.

2 Materials

This approach requires several key resources, including in-house computational tools, access to a high-performance computing facility (HPC) for intensive computations, ample storage capacity, Internet access to download the PDB database, and programming skills for data processing. A standard desktop workstation with 24 cores, approximately 25 GB of RAM, and over 220 GB of available storage should suffice for most tasks. Additionally, around 24 h of computing time is required.

1. Sequence clustering/alignment tool: Cluster Database at High Identity with Tolerance (CD-HIT) [15] (available at https://github.com/weizhongli/cdhit/blob/master/doc/cdhit-user-guide.wiki).
2. Multiple sequence alignment program: Multiple Alignment using Fast Fourier Transform (MAFFT) [16] (available at https://mafft.cbrc.jp/alignment/software/).
3. Secondary structure assignment: Dictionary of Secondary Structure in Proteins (DSSP) [17] (accessible via https://github.com/MDAnalysis/mdanalysis/pull/4304).
4. Structure alignment method: Combinatorial Extension (CE) [18] (available as jCE at https://www.rcsb.org/docs/tools/pairwise-structure-alignment).
5. Three-dimensional (3D) visualization tool: PyMOL (available at http://www.pymol.org/pymol).

For alternative and suitable software tools, see the information contained in the Notes section.

3 Methods

3.1 Data Collection

3.1.1 Collect Proteins with Structures from the PDB Database

As protocols for accessing and retrieving data evolve over time, the best practice is to obtain information directly from repositories. In particular, instructions for programmatically downloading files are available at https://www.rcsb.org/docs/programmatic-access/file-download-services. Ideally, structures to search for metamorphism should have:

- Resolution <2.5 Å.
- Sequence length >100 amino acids.
- Reported in the last 10 years.
- Exploring methods beyond X-ray crystallography, such as nuclear magnetic resonance (NMR) or cryogenic electron microscopy (cryo-EM), may increase the likelihood of finding shape-shifting proteins.

3.1.2 Mapping Structures to Sequences and Extracting Sequences

Obtain UniProtKB IDs through application programming interfaces (APIs) from UniProtKB or PDB. A useful resource for this task is https://www.ebi.ac.uk/pdbe/api/sifts.html, which facilitates retrieving amino acid sequences of the mapped proteins.

3.2 Sequence Analyses

1. Create a small file with the sequences of mapped proteins

   The sequences should be in FASTA format (*see* **Note 1**), with one sequence listed after the other.

2. Cluster the amino acid sequences

   Use CD-HIT to cluster sequences at different identity levels (95–100%), maintaining an alignment coverage of >0.95 (*see* **Note 2**). This step will group similar sequences and create a dictionary of sequence clusters.

3. Multiple sequence alignment (MSA)

   Perform multiple sequence alignments using MAFFT [16] (*see* **Note 3**). Visualize the MSA results to check for errors and remove fragments, if necessary. Jalview is a recommended tool for visualizing MSA [19] (available at https://www.jalview.org/).

4. Secondary structure element assignment and alignment recoding

   Use DSSP [17] (updated at https://github.com/PDB-REDO/dssp or available at https://www3.cmbi.umcn.nl/xssp/) to assign secondary structure elements to each structure.

   Recoded MSA: Replace each amino acid in the alignment with its corresponding secondary structure assignment (*see* **Note 4**). Visualize the recoded alignment and focus on regions showing consistent transitions between β-sheets and α-helices. Concentrate on regions longer than ten residues which are not located at N- or C-terminal regions.

5. Identify protein domains

   From https://www.ebi.ac.uk/interpro/, collect domain information for each protein sequence. Regions with fold-switching behavior are often found in independently foldable domains.

3.3 Structural Analyses of Clusters

1. Pairwise structural alignment.

   For each cluster of amino acid sequences, conduct pairwise structural alignments using rigid-body algorithms that preserve sequence order, such as the Combinatorial Extension (CE) algorithm (*see* **Note 5**). The CE algorithm is also integrated into PyMOL and can be accessed through the "cealign" command for efficient structural comparison.

   - For sequences with 100% identity and a root mean square deviation (RMSD) <1 Å, retain the chain with the highest resolution and fewest unresolved regions. If multiple chains have the same resolution, keep the longest.

- For sequences with 99–100% identity but RMSD >3 Å, extract the structures as "potential" shape-shifters. Focus on structures with the highest RMSD values at the highest sequence identity.

2. Visualize structural alignments

   Align structures pairwise, from the same protein cluster, and visualize the structural alignment using PyMOL for accuracy.

3.4 Literature Search for Shape-Shifting Behavior

Fold-switching proteins can appear in various functional contexts. Therefore, it is crucial to perform a literature search to confirm different functional roles for identified proteins. PubMed (https://pubmed.ncbi.nlm.nih.gov/) is a comprehensive resource for finding such information.

Examples include RfaH that switches from transcription to translation [10], KaiB, which regulates the circadian clock via phosphorylation/dephosphorylation [20], and CLIC1 (*see* Fig. 1) that moves to the membrane to form an ion channel in response to oxidation states [12]. Sometimes proteins may only have one resolved structure but undergo conformational changes based on experimental evidence (e.g., Dishevelled undergoes conformational switching in response to β-catenin signaling, despite only one structure being resolved [21]).

4 Notes

1. The format of sequences may vary depending on the repository. For a comprehensive guide to sequence formats, refer to https://www.ebi.ac.uk/jdispatcher/docs/formats/.
2. Various tools can be used for clustering sequences. In addition to CD-HIT, basic local alignment search tool (BLAST) [22] is a suitable alternative for identifying similar sequences, especially when they are similar, and the data are small. For larger datasets, novel methods like MMseqs2 [23] (available at https://github.com/soedinglab/MMseqs2) can provide efficient clustering.
3. When aligning multiple sequences, other suitable tools include T-Coffee [24] or the MMseqs2 method (*see* **Note 2**). These methods can handle large and complex sequence alignments with high accuracy.
4. To accurately identify fold-switching regions, the aligned sequences should have 98–100% sequence identity. Differences at the N- or C-terminus due to insertions or deletions should

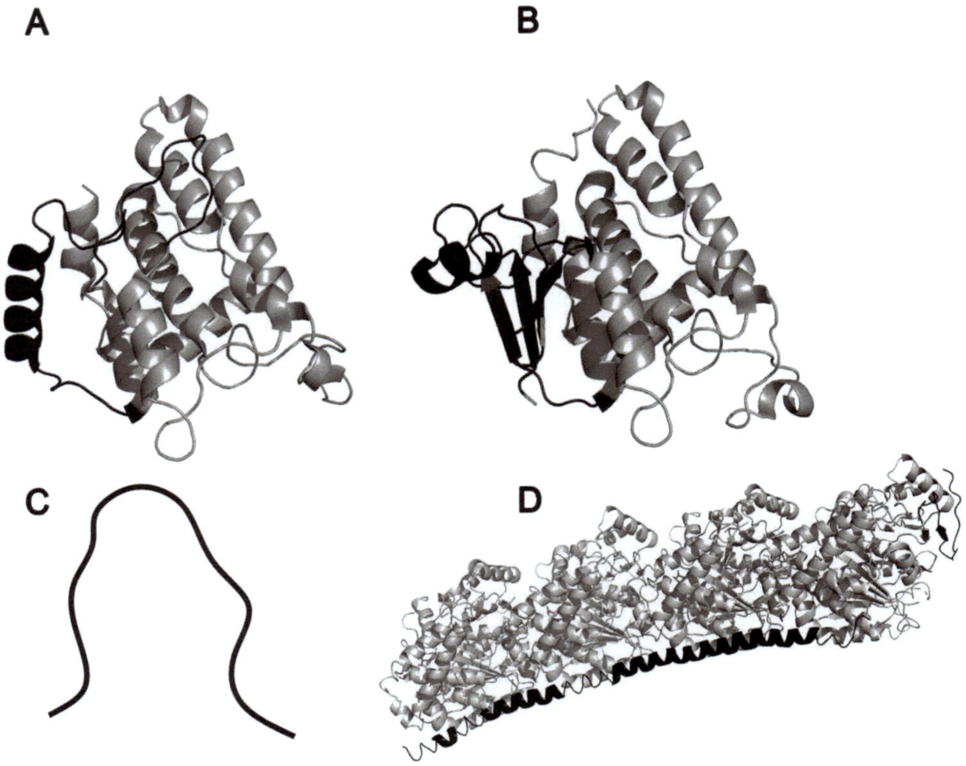

Fig. 1 Examples of shape-shifting proteins with regions of structural switching highlighted in black. (**a**) Chloride intracellular channel 1 (CLIC1) in its oxidized state (PDB ID: 1RK4) and (**b**) in its reduced state (PDB ID: 1K0M), illustrating a fold-switching event. (**c**) The intrinsically disordered protein stathmin in its unbound form (manually drawn) and (**d**) in its bound state in complex with tubulin (PDB ID: 1SA0). All structures were visualized using PyMOL

not be considered, as these are often nonessential for fold-switching behavior. Focus on regions where a given sequence corresponds to an α-helix in one structure and a β-strand in another. If variations are only in the length of the same secondary structure element (e.g., a longer or shorter helix or β-strand), these should not be considered fold-switching regions. If transitions occur from disordered to ordered regions, the protein may belong to the intrinsically disordered protein (IDP) category.

5. For structures with a resolution of 2.5 Å or better, the Cα RMSD due to experimental error is typically well below 1 Å [25], making this threshold appropriate for filtering out structures during further analysis. For the pairwise structural alignment, other alignment methods such as FATCAT [26] or TM-score [27] are also viable for identifying potential fold-switching regions. However, keep in mind that these methods aim to minimize local differences to achieve a global alignment, meaning RMSD values may not always indicate fold-switching

regions. Pay close attention to modified residues, unresolved regions, and the presence of additional molecules, as these factors might indicate different functional environments. Focus particularly on regions where transitions between α-helices and β-strands occur (*see* Fig. 1).

Acknowledgments

Ana M. Rojas acknowledges funding from the Spanish Ministry of Science and Innovation (MCIN/AEI/10.13039/501100011033), grant id: PID2021-127503OB-I00, and from the OSCARS project (European Commission's Horizon Europe Research and Innovation programme) under grant agreement No. 101129751. Enrique Alanís-Domínguez acknowledges funding from grants CEX2020-001088 and PRE2022-101986.

References

1. Anfinsen CB, Haber E, Sela M et al (1961) The kinetics of formation of native ribonuclease during oxidation of the reduced polypeptide chain. Proc Natl Acad Sci U S A 47:1309–1314
2. Chothia C, Lesk AM (1986) The relation between the divergence of sequence and structure in proteins. EMBO J 5:823–826
3. James LC, Tawfik DS (2003) Conformational diversity and protein evolution—a 60-year-old hypothesis revisited. Trends Biochem Sci 28: 361–368
4. Burra PV, Zhang Y, Godzik A et al (2009) Global distribution of conformational states derived from redundant models in the PDB points to non-uniqueness of the protein structure. PNAS 106:10505–10510
5. Kosloff M, Kolodny R (2008) Sequence-similar, structure-dissimilar protein pairs in the PDB. Proteins 71:891–902
6. Uversky VN, Kulkarni P (2021) Intrinsically disordered proteins: chronology of a discovery. Biophys Chem 279:106694
7. Murzin AG (2008) Metamorphic proteins. Science 320:1725–1726
8. Porter LL, Looger LL (2018) Extant fold-switching proteins are widespread. Proc Natl Acad Sci USA 115:5968–5973
9. Werner F, Grohmann D (2011) Evolution of multisubunit RNA polymerases in the three domains of life. Nat Rev Microbiol 9:85–98
10. Burmann BM, Knauer SH, Sevostyanova A et al (2012) An α helix to β barrel domain switch transforms the transcription factor RfaH into a translation factor. Cell 150:291–303
11. Tuinstra RL, Peterson FC, Kutlesa S et al (2008) Interconversion between two unrelated protein folds in the lymphotactin native state. Proc Natl Acad Sci U S A 105:5057–5062
12. Littler DR, Harrop SJ, Fairlie WD et al (2004) The intracellular chloride ion channel protein clic1 undergoes a redox-controlled structural transition*. J Biol Chem 279:9298–9305
13. Bryan PN, Orban J (2010) Proteins that switch folds. Curr Opin Struct Biol 20:482–488
14. Chen N, Das M, LiWang A et al (2020) Sequence-based prediction of metamorphic behavior in proteins. Biophys J 119:1380–1390
15. Li W, Godzik A (2006) Cd-hit: a fast program for clustering and comparing large sets of protein or nucleotide sequences. Bioinformatics 22:1658–1659
16. Katoh K, Standley DM (2013) MAFFT multiple sequence alignment software version 7: improvements in performance and usability. Mol Biol Evol 30:772–780
17. Kabsch W, Sander C (1983) Dictionary of protein secondary structure: Pattern recognition of hydrogen-bonded and geometrical features. Biopolymers 22:2577–2637
18. Shindyalov IN, Bourne PE (1998) Protein structure alignment by incremental combinatorial extension (CE) of the optimal path. Protein Eng 11:739–747
19. Waterhouse AM, Procter JB, Martin DMA et al (2009) Jalview Version 2—a multiple sequence

alignment editor and analysis workbench. Bioinformatics 25:1189–1191
20. Chang Y-G, Cohen SE, Phong C et al (2015) Circadian rhythms. A protein fold switch joins the circadian oscillator to clock output in cyanobacteria. Science 349:324–328
21. Gammons MV, Renko M, Johnson CM et al (2016) Wnt signalosome assembly by DEP domain swapping of dishevelled. Mol Cell 64: 92–104
22. Altschul SF, Gish W, Miller W et al (1990) Basic local alignment search tool. J Mol Biol 215:403–410
23. Steinegger M, Söding J (2017) MMseqs2 enables sensitive protein sequence searching for the analysis of massive data sets. Nat Biotechnol 35:1026–1028
24. Notredame C, Higgins DG, Heringa J (2000) T-coffee: a novel method for fast and accurate multiple sequence alignment. J Mol Biol 302: 205–217
25. DePristo MA, de Bakker PIW, Blundell TL (2004) Heterogeneity and inaccuracy in protein structures solved by X-ray crystallography. Structure 12:831–838
26. Li Z, Jaroszewski L, Iyer M et al (2020) FATCAT 2.0: towards a better understanding of the structural diversity of proteins. Nucleic Acids Res 48:W60–W64
27. Zhang Y, Skolnick J (2004) Scoring function for automated assessment of protein structure template quality. Proteins 57:702–710

Chapter 11

Transition State-Based Computational Enzyme Design

Thomas Gaillard and Thomas Simonson

Abstract

Our approach to computational protein design is physics-based. We develop a software called Proteus, allowing both physics-based energy evaluation and sequence-conformation exploration. Unlike knowledge-based models, a physics-based energy function facilitates the consideration of unusual chemical entities, such as novel ligands or transition states. Additionally, the adaptive landscape flattening method allows direct sampling on free energy difference between two states. We show here how these ingredients combined can benefit enzyme design. As an illustration, we revisit the stereospecificity inversion of tyrosyl-tRNA synthetase. Following a tutorial presentation, we explore various design criteria that can be related to enzyme experimental parameters. Our model is able to retrieve the native sequence when targeting L-tyrosine. Mutations predicted to favor D-tyrosine are proposed.

Keywords Protein, Enzyme, Sequence, Design, CPD, Transition state, Aminoacyl-tRNA synthetase, D-amino acid

1 Introduction

Amino acids are mostly found in nature in their L form. The D isomers can have useful applications, for example, to make therapeutic peptides resistant to protease degradation. In protein synthesis, amino acids are loaded on their cognate tRNA with the assistance of aminoacyl-tRNA synthetase enzymes. The first step of amino acid loading is the adenylation reaction, which converts ATP and amino acid reactants into aminoacyl-adenylate and pyrophosphate. One way to make D-amino acid peptides in vivo is to modify aminoacyl-tRNA synthetases so that they recognize D- instead of L-amino acids. Computational protein design (CPD), which aims at predicting new proteins or modifying existing ones to achieve a desired goal, can be used for this task. TyrRS is a good candidate for stereospecificity inversion, as it naturally has a non-negligible activity for D-Tyr [1, 2].

Shahid M. Khan and Florencio Pazos (eds.), *Protein Evolution: Methods and Protocols*, Methods in Molecular Biology, vol. 2979, https://doi.org/10.1007/978-1-0716-4828-5_11,

Our group develops a CPD software called Proteus [3–5], available at the address https://proteus.polytechnique.fr. We follow a structure- and physics-based approach. The protein is modeled as a fixed backbone and an ensemble of sidechain rotamers. The energy is calculated with an all-atom molecular mechanics model, completed by an implicit model for solvent polar and nonpolar effects. The sequence and conformation landscape is explored with different algorithms, including Monte Carlo sampling. Proteus has been applied to diverse problems including the prediction of sidechain conformations [6, 7], stability mutants [6, 8], p*K*as [9–13], as well as fold recognition [14, 15], full sequence design [3, 16–18], and active site engineering [19–26].

E. coli TyrRS stereospecificity inversion has already been investigated by our group [4, 21–23]. In the 2013 article presenting Proteus [4] and in more details later in a 2016 combined modeling and experimental study [22], four positions were targeted (D81, Y175, Q179, and Q201) by automatic design. Sequences and conformations were selected by their folding energy, and then by their ability to better bind D-Tyr than the native sequence. D, E, H, and K were found at position 81; Y and W at position 175; and E and D at positions 179 and 201. A selection of mutants was then made manually (RYQQ, KYEQ, KYED, HYED, NYQQ, and NDYQQ in which the D41N mutation was added). The mutants were tested by alchemical free energy simulations and experimentally. The D81R mutant (RYQQ) was found experimentally to have inverted stereospecificity; however, the activity was weak. In another 2016 study [21], the binding free energy difference between D- and L-Tyr ligands was estimated by a constant-activity Monte Carlo approach, for the native TyrRS and six mutants. A molecular dynamics + GBSA approach was considered for comparison. Automated design of individual positions 81, 179, and 201 was performed with an updated physical model, and with a mutation space limited to polar types. Sequences were ranked according to their ligand binding energy. With the D-Tyr ligand, R, E, and H were respectively predicted at positions 81, 179, and 201. At last, in a 2017 article [23], native and mutant TyrRS complexes were simulated by molecular dynamics. Alchemical free energy simulations were used to compute relative L/D binding free energies. The D81Q mutant was tested but found to reinforce L-Tyr specificity.

In current work, we revisit the TyrRS stereospecificity inversion problem by exploiting two new ingredients not available in previous works: a transition state model of the amino acid adenylation reaction and the adaptive landscape flattening method. CPD explicitly targeting transition states has not been often performed [24, 26]. Empirical energy functions often used in other CPD programs cannot easily model a transition state. In addition, the adaptive landscape flattening method [5, 27] allows direct sampling on the free energy difference between two states. The two ingredients combined allow us to design on criteria directly related to experimentally measured enzyme parameters. We present here the theory of the adaptive landscape flattening method application to enzyme design, the details of application of our CPD approach to the design goal, followed by the results presented as a tutorial (available on Proteus website).

2 Theory

The adaptive landscape flattening method allows direct Monte Carlo sampling on the free energy difference between two states. It has been presented and applied to protein systems by our group [5, 12, 24–28], with protein-peptide binding, acid-base competition, enzyme-substrate or transition state binding, and enzyme substrate vs. antibiotic binding, as free energy differences.

The adaptive landscape flattening method can be used in two ways to optimize free energy differences in the context of protein design. In the first, direct approach, one of the two states, called the reference state, is chosen to develop a bias potential, taking the form of position-specific amino acid energies, that flattens the amino acid landscape. That is, the sequences of the reference state are sampled with comparable probabilities when the bias potential is applied. The second state is then sampled with the bias potential of the reference state applied, directly producing sequences populated according to the free energy difference between the two states (Fig. 1a). A practical difficulty with this approach is that the sequence space coverage for the second state can be limited.

In the second, indirect approach, the two states are sampled with their own bias potential applied (Fig. 1b). As the bias leads to approximate equiprobability of amino acids, sequence space coverage is high for both states. This gives more flexibility in analyzing the results, such as the possibility of multi-criteria filtering. For two sequences i and j sampled in two states X and Υ, the relative (inter-sequence) free energy difference between the two states can be calculated a posteriori from the biased Monte Carlo simulation results as

$$\begin{aligned} \Delta\Delta G_{X\to\Upsilon} &= \Delta G_{X\to\Upsilon}(j) - \Delta G_{X\to\Upsilon}(i) = G(\Upsilon,j) - G(X,j) - G(\Upsilon,i) + G(X,i) \\ &= -RT\ln\frac{p(\Upsilon,j)}{p(\Upsilon,i)} + RT\ln\frac{p(X,j)}{p(X,i)} - B(\Upsilon,j) + B(X,j) + B(\Upsilon,i) - B(X,i) \end{aligned} \quad (1)$$

where $G(X,i)$ is the free energy of sequence i in state X, $p(X,i)$ is the population of sequence i in the biased simulation of state X, and $B(X,i)$ is the bias potential for sequence i in state X.

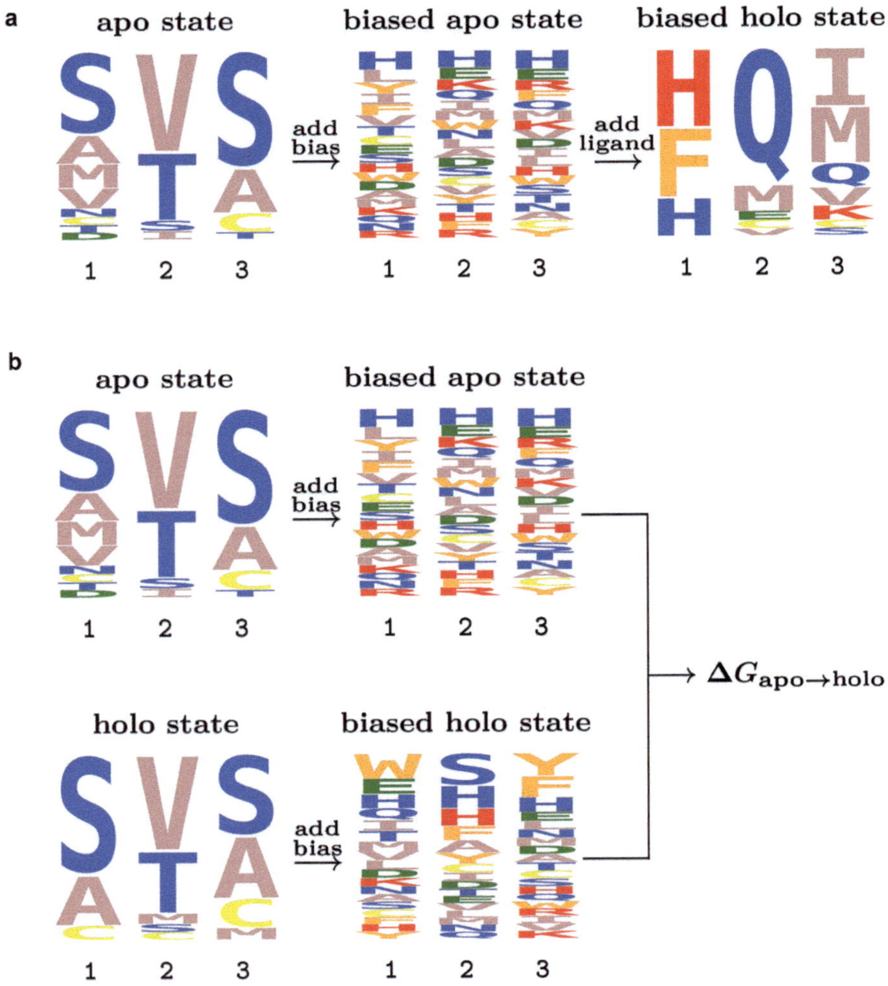

Fig. 1 Adaptive landscape flattening method. (**a**) In the direct approach, one of the two states, here the apo state, is chosen to develop a bias potential that flattens the sequence landscape. When the bias potential is added to the apo state, the sequences are sampled with comparable probabilities. The second state, here the holo state, is then sampled with the bias potential of the apo state applied, directly producing sequences populated according to the free energy difference between the two states. (**b**) In the indirect approach, the two states are sampled with their own bias potential applied. The free energy difference between the two states is calculated a posteriori from the biased populations, according to Eq. 1

In the following, we recall how the adaptive landscape flattening method can be applied to enzyme design, with stability, affinity, catalytic rate, catalytic efficiency, and substrate competition as design criteria. Some of these criteria require the consideration of the transition state. In Fig. 2, we provide a schematic view of enzyme energetics, along with the notations that we use. In the following, we refer to enzyme, substrate, and transition state, as E, S, and $S^{\ddagger}$. When $\Delta\Delta G$ quantities are considered, the first Δ indicates a difference between two sequences of the enzyme.

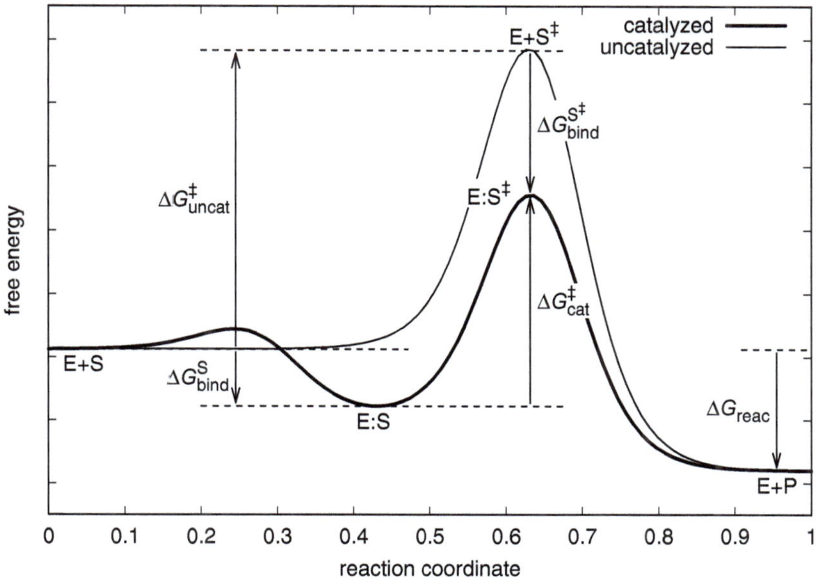

Fig. 2 Free energy along the reaction coordinate. The uncatalyzed (thin line) and catalyzed (thick line) cases are compared. Enzyme, substrate, transition state, and product are noted E, S, S‡, and P. The free energies of binding the substrate and the transition state are indicated as ΔG^{S}_{bind} and $\Delta G^{S^‡}_{bind}$. The activation barriers in the uncatalyzed and catalyzed cases are indicated as $\Delta G^{‡}_{uncat}$ and $\Delta G^{‡}_{cat}$

Stability Design In stability design, the optimization criterion is the folding free energy, that is, the difference between the folded and unfolded state free energies:

$$\Delta G_{fold} = G_{folded} - G_{unfolded} \tag{2}$$

The unfolded state free energy is calculated as a sum of amino acid reference energies, independent of the conformation. As a consequence, only the folded state needs to be simulated, and Eq. 1 applied to the relative folding free energy takes the following simplified form:

$$\begin{aligned}\Delta\Delta G_{fold} &= \Delta G_{fold}(j) - \Delta G_{fold}(i) = G_{folded}(j) - G_{unfolded}(j) - G_{folded}(i) + G_{unfolded}(i) \\ &= -RT \ln \frac{p(j)}{p(i)} - B(j) + B(i) - G_{unfolded}(j) + G_{unfolded}(i)\end{aligned} \tag{3}$$

Affinity Design In affinity design, the optimization criterion is the difference of free energy between the substrate complex and the apoenzyme:

$$\Delta G_{E \to E:S} = G(E:S) - G(E) \tag{4}$$

In Michaelis-Menten kinetics, the free energy of binding the substrate is related to Michaelis constant K_M by the following equation:

$$\Delta G^{\mathrm{S}}_{\mathrm{bind}} = G(\mathrm{E:S}) - G(\mathrm{E}) - G(\mathrm{S}) = RT \ln K_{\mathrm{M}} \tag{5}$$

When two sequences of the enzyme are compared, that is, when the relative free energy of binding the substrate is considered, the $G(\mathrm{S})$ term disappears as it is independent of the enzyme sequence:

$$\Delta\Delta G_{\mathrm{E}\rightarrow\mathrm{E:S}} = \Delta\Delta G^{\mathrm{S}}_{\mathrm{bind}} = RT\Delta \ln K_{\mathrm{M}} \tag{6}$$

Catalytic Rate Design In catalytic rate design, the optimization criterion is the free energy difference between the transition state and substrate complexes. This difference of free energy, the activation barrier, is related to the catalytic rate constant, k_{cat}. Following Eyring equation, the following relationship is obtained between the activation barrier and the catalytic rate constant:

$$\Delta G_{\mathrm{E:S}\rightarrow\mathrm{E:S}^{\ddagger}} = \Delta G^{\ddagger}_{\mathrm{cat}} = G(\mathrm{E:S}^{\ddagger}) - G(\mathrm{E:S}) \propto - RT \ln k_{\mathrm{cat}} \tag{7}$$

Catalytic Efficiency Design In catalytic efficiency design, the optimization criterion is the free energy difference between the transition state complex and the apoenzyme:

$$\Delta G_{\mathrm{E}\rightarrow\mathrm{E:S}^{\ddagger}} = G(\mathrm{E:S}^{\ddagger}) - G(\mathrm{E}) \tag{8}$$

When comparing two sequences of the enzyme, this difference of free energy is related to the $k_{\mathrm{cat}}/K_{\mathrm{M}}$ ratio, alternatively called catalytic efficiency or specificity constant. Indeed, the free energy of binding the transition state is

$$\Delta G^{\mathrm{S}^{\ddagger}}_{\mathrm{bind}} = G(\mathrm{E:S}^{\ddagger}) - G(\mathrm{E}) - G(\mathrm{S}^{\ddagger}) \tag{9}$$

When the relative free energy of binding the transition state is considered, the $G(\mathrm{S}^{\ddagger})$ term disappears as it is independent of the enzyme sequence. As a consequence, we obtain the following relationship:

$$\Delta\Delta G_{\mathrm{E}\rightarrow\mathrm{E:S}^{\ddagger}} = \Delta\Delta G^{\mathrm{S}^{\ddagger}}_{\mathrm{bind}} = \Delta\Delta G^{\ddagger}_{\mathrm{cat}} + \Delta\Delta G^{\mathrm{S}}_{\mathrm{bind}} \propto -RT\Delta \ln \frac{k_{\mathrm{cat}}}{K_{\mathrm{M}}} \tag{10}$$

Substrate Competition Design In substrate competition design, the optimization criterion is the free energy difference between the transition state complexes of substrates S1 and S2:

$$\Delta G_{\mathrm{E:S1}^{\ddagger}\rightarrow\mathrm{E:S2}^{\ddagger}} = G(\mathrm{E:S2}^{\ddagger}) - G(\mathrm{E:S1}^{\ddagger}) \tag{11}$$

When comparing two enzyme sequences, this difference of free energy is related to the ratio of S1 and S2 catalytic efficiency constants, that is, $k^{\mathrm{S2}}_{\mathrm{cat}}/K^{\mathrm{S2}}_{\mathrm{M}} \cdot K^{\mathrm{S1}}_{\mathrm{M}}/k^{\mathrm{S1}}_{\mathrm{cat}}$. This allows direct design on substrate competition. Indeed, when considering the relative $\Delta\Delta G_{E:S1^\ddagger \to E:S2^\ddagger}$, the following relationship holds, thanks to the canceling of $G(S1^\ddagger)$ and $G(S2^\ddagger)$ terms, which are independent of the enzyme sequence:

$$\Delta\Delta G_{E:S1^\ddagger \to E:S2^\ddagger} = \Delta\Delta G_{bind}^{S2^\ddagger} - \Delta\Delta G_{bind}^{S1^\ddagger} \propto -RT\Delta \ln \frac{k_{cat}^{S2}}{K_M^{S2}} \cdot \frac{K_M^{S1}}{k_{cat}^{S1}} \tag{12}$$

3 Materials and Methods

3.1 Transition State Model

A model of the adenylation reaction transition state was prepared. The α phosphate is assumed to adopt a pentavalent, trigonal bipyramidal geometry (*see* Fig. 3), based on preliminary quantum calculations. One of the axial oxygen connects the $\beta\gamma$ diphosphate group, and the other axial oxygen connects the Tyr amino acid. One of the equatorial oxygen connects the adenosine group. Two atom types were introduced, for the pentavalent phosphate (P5) and its axial oxygens (OA). The equilibrium P5-OA distance was set to 2.4 Å. Other parameters were attributed by analogy with existing AMBER parameters. A conformation of the transition state model was built with internal coordinates at their equilibrium values. RESP charges were calculated, following the AMBER procedure [29].

3.2 Structure Preparation

Structural models of the TyrRS protein, alone and in complex with L- or D-Tyr reactants or transition states, were prepared. A list of the states modeled is given in Table 1. Representation of the TyrRS in complex with L-Tyr transition state is shown in Fig. 4.

TyrRS The TyrRS model was based on the 1VBM PDB entry [30], an *E. coli* TyrRS structure in complex with an L-Tyr-AMP

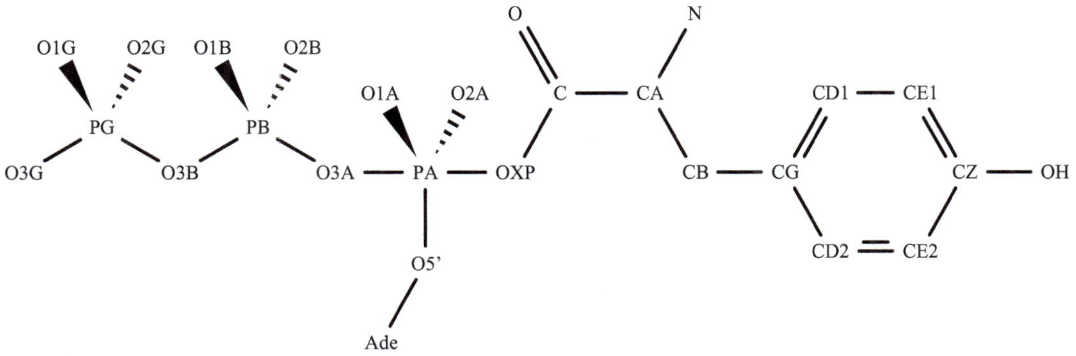

Fig. 3 Transition state model of the tyrosine adenylation reaction. 2D representation without hydrogens. Atom names are indicated. The PA phosphate adopts a trigonal bipyramidal geometry. The adenosine group is abbreviated (Ade)

Table 1
Enzyme states considered. The name of the state is given along with a description, the ligand if present, and the corresponding tutorial folder

State	Description	Ligand	Folder
E	Apoenzyme	–	`apo`
E:L	Complex of enzyme and L reactants	ATP+ L-Tyr + Mg^{2+}	`reacL`
E:D	Complex of enzyme and D reactants	ATP + D-Tyr + Mg^{2+}	`reacD`
E:L‡	Complex of enzyme and L transition state	[PP··· AMP··· L-Tyr]‡ + Mg^{2+}	`tsL`
E:D‡	Complex of enzyme and D transition state	[PP··· AMP··· D-Tyr]‡ + Mg^{2+}	`tsD`

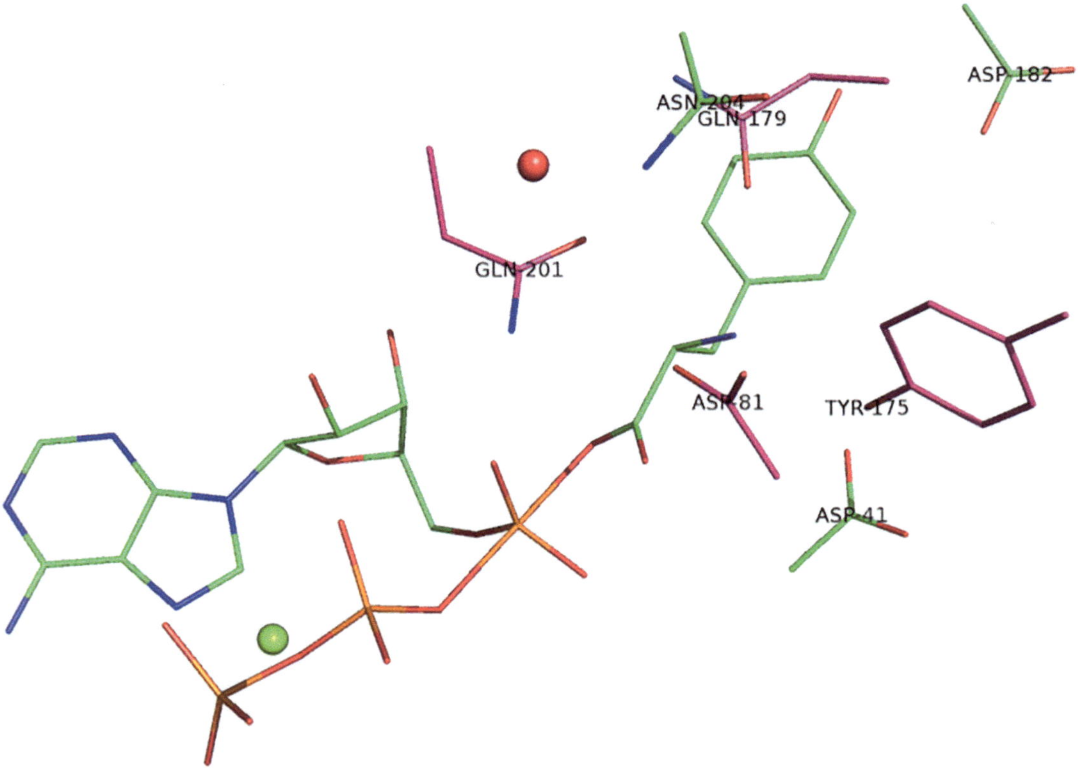

Fig. 4 Transition state complex structure of the L-tyrosine adenylation reaction. 3D representation without hydrogens. TyrRS amino acids close to the tyrosine part of the ligand are shown and labeled. Carbon atoms of amino acids allowed to mutate in design are colored in magenta, and other carbons are in green. The magnesium ion and water oxygen are shown as spheres

analog. Only the A chain was used, stripped of water molecules, ions, and ligands. A buried water molecule (residue number 3041 in 1VBM) is conserved in several crystal structures and was included in the model, as in previous works [21–23].

Ligands The four ligands considered were the L and D reactants of the adenylation reaction, that is, ATP plus either L- or D-Tyr, as well as the L and D transition states. Ligand conformations were prepared, starting from the L-Tyr-AMP analog (residue code YSA) present in the 1VBM structure. The analog was first converted into a real L-Tyr-AMP by modifying the appropriate atoms. L-Tyr and AMP molecules were separated. ATP was then built by combining the AMP extracted from 1VBM with the second and third phosphates extracted from the ATP molecule present in the 1H3E PDB entry [31], after superposition on the AMP. 1H3E is a *T. thermophilus* TyrRS structure in complex with tRNA, ATP, and tyrosinol. At last, the L transition state structure was built by combining ATP and L-Tyr, and modifying the first phosphate (PA, O1A, O2A, O5' atoms) geometry to make it trigonal bipyramidal (PA, O1A, O2A, O5' in the same plane; OXP-PA-O3A axis perpendicular to the plane). The distance of PA to OXP and O3A was 2.2 Å, while the distance of PA to O1A, O2A, and O5' was 1.5 Å, close to the equilibrium values. The D-Tyr version of the ligands was prepared by rotating the N atom by $-120°$ around the CA-C axis. A magnesium ion was included in all ligand models, positioned between the second and third phosphates.

Hydrogens Hydrogen atoms were added to all models with the Reduce program [32]. Protonation state of residues was their standard one at pH 7, except for Asp 182, which was considered protonated in previous studies [21–23]. A PROPKA [33, 34] analysis confirmed this with a predicted p*K*a of 7.8.

Minimization The models were built and minimized with the XPLOR program [35]. The AMBER ff99SB force field [36, 37] was employed, along with Carlson parameters [38] for ATP, Horn parameters for zwitterionic amino acids [39], and our transition state parameters. The TIP3P model was used for the buried water molecule. A GB HCT solvent model [40–42] was used. Energy minimizations were performed with Powell's conjugate gradient algorithm until an energy gradient norm of 0.5 kcal/mol/Å was reached.

3.3 Sequence Design

We redesigned the type of selected amino acid positions close to the Tyr ligand backbone. Residues 81, 175, 179, and 201, whose native sequence is DYQQ, were chosen, as in previous studies [21–23]. The type of the other positions was kept native. The conformation of the amino acids whose sidechain is within 8 Å of the ligand Tyr part was allowed to vary.

Energy Matrix Calculation An interaction energy matrix was calculated for each model, using our Proteus program. We followed the same protocol as in previous works [7, 17], where details are given. Briefly, the rotamer library was prepared from the 1995 library of Tuffery et al. [43], the energy function was of the MMGBSA form, and the molecular mechanics force field was AMBER ff99SB [36, 37]. An internal dielectric constant of 8 was used, along with compatible surface energy coefficients that were previously optimized [17].

Sequence-Conformation Exploration Designed sequences and conformations were generated from the energy matrices by Monte Carlo simulations implemented in the Proteus program. The Monte Carlo move probabilities were 0.9, 0.1, 0.9, and 0.1, respectively, for single rotamer, single mutation, a pair of rotamers, and mutation-rotamer moves. The thermal energy RT was 0.6 kcal/mol, and the trajectory length was 10^9. For stability design, the reference energies were those optimized previously [17].

4 Tutorial and Results

We present the results of TyrRS sequence redesign, in the presence of L- or D-Tyr reactants or transition states, in the form of a tutorial (available on Proteus website). The states considered are summarized in Table 1. Different design criteria are tested: stability, affinity, catalytic rate, catalytic efficiency, and stereospecificity, as summarized in Table 2. For design criteria other than stability, we only keep the sequences having a stability at least as good as the native sequence in the apoenzyme. For stereospecificity design, we apply

Table 2
Enzyme design criteria considered. The criterion is given along with states A and B taken into account in the application of the adaptive landscape flattening method, and the enzyme parameter that is optimized. [LD] indicates L or D

Criterion	State A	State B	$\Delta G_{A \to B}$ = or is related to
Stability	Unfolded	Folded	ΔG_{fold}
Affinity	E	E:[LD]	$K_M^{[LD]}$
Catalytic rate	E:[LD]	E:[LD]‡	$k_{cat}^{[LD]}$
Catalytic efficiency	E	E:[LD]‡	$k_{cat}^{[LD]}/K_M^{[LD]}$
Stereospecificity	E:[DL]‡	E:[LD]‡	$\frac{k_{cat}^{[LD]}}{K_M^{[LD]}} \cdot K_M^{[DL]}/k_{cat}^{[DL]}$

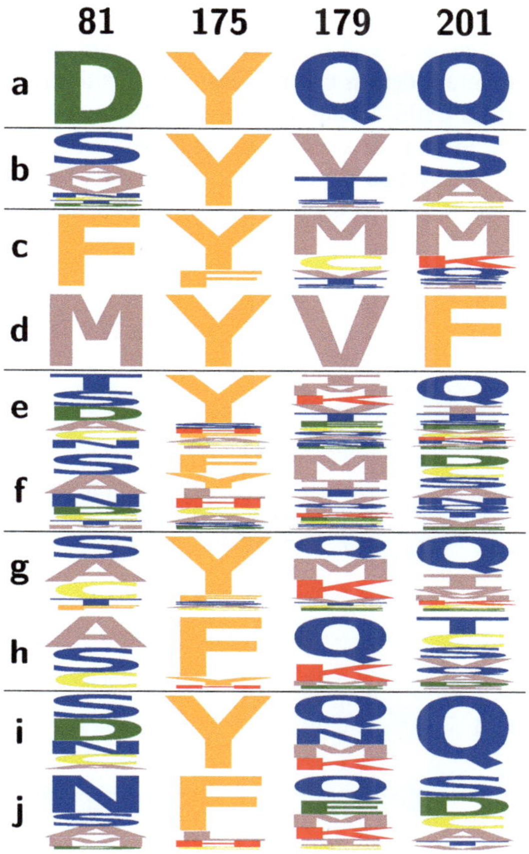

Fig. 5 Sequence logos obtained in design. Sequences are filtered (*see* Table 3 legend). (**a**) Native sequence. (**b**) Stability design of apo TyrRS. (**c**) L- and (**d**) D-Tyr affinity design. (**e**) L- and (**f**) D-Tyr catalytic rate design. (**g**) L- and (**h**) D-Tyr catalytic efficiency design. (**i**) L-Tyr vs. D-Tyr and (**j**) D-Tyr vs. L-Tyr stereospecificity design

a multi-criteria filter explained below. Sequence logo presentations of the results are given in Fig. 5. The native sequence is recalled in Fig. 5a. The best designed sequences and their relative free energies with respect to the native sequence for the different design criteria are given in Table 3. More detailed results are given in Tables 4 and 5.

4.1 Downloading Proteus and the Tutorial Files

Proteus 3.0 can be obtained from https://proteus.polytechnique.fr. It is free to academic and government scientists, while scientists from companies should contact the corresponding author. The distribution includes source code, extensive test cases, and detailed documentation. No compilation is needed for standard use, as static binaries are included that should run on any Linux system. It is sufficient to include the location of scripts and binaries in the PATH environment variable. The "tuto_TyrRS"

Table 3
Best designed sequences and their relative free energies (kcal/mol) with respect to the native sequence for different design criteria. Designed positions are 81, 175, 179, and 201, for which the native sequence is DYQQ. Design criteria are affinity, catalytic rate, catalytic efficiency, and stereospecificity. Sequences are filtered. For affinity, rate, and efficiency, the stability of the reference state is at least as good as the native. For stereospecificity, stability of the apo state, affinity, rate, and efficiency are at least as good as the native. Results are given first for the L-Tyr ligand and then for the D-Tyr

Ligand	Affinity		Rate		Efficiency		Stereospecificity	
L	FYMM	−7.27	TYEI	−1.18	CYQQ	−1.97	DYQQ	0.00
L	FYMK	−7.10	DYIQ	−1.14	AYQQ	−1.95	SYNQ	0.20
L	FYMQ	−6.92	TYKQ	−0.86	SYKQ	−1.94	NYMQ	0.29
L	FYCM	−6.83	FAMQ	−0.85	CYKQ	−1.92	SYQQ	0.70
L	FFMM	−6.76	NYVQ	−0.78	SYKI	−1.87	CYQQ	0.82
L	FYMS	−6.60	TYCI	−0.73	AYMQ	−1.85	CYKQ	0.86
L	FFCM	−6.43	CYKQ	−0.71	FYMQ	−1.82	AYKQ	0.90
L	FYTM	−6.34	SYKQ	−0.66	SYQQ	−1.82	SYKQ	1.36
L	FFVM	−6.20	TYTE	−0.64	SYMQ	−1.81		
L	FYCI	−6.16	TYMQ	−0.63	CYMQ	−1.76		
D	MYVF	−10.30	SFMN	−2.84	CFQT	−4.68	MLQS	−6.13
D	CFQT	−5.78	AFMN	−2.65	AFQV	−4.64	SFQD	−5.89
D	AFQM	−5.77	NYMD	−2.56	AFQT	−4.61	NFES	−5.71
D	CFQS	−5.62	SYMN	−2.49	SFQT	−4.56	NFKA	−5.67
D	CFQA	−5.50	ALVA	−2.37	CFQC	−4.55	NFKS	−5.64
D	CFQC	−5.39	NYKD	−2.31	AFQC	−4.54	AFQD	−5.62
D	SFQT	−5.31	MAMS	−2.28	AFKQ	−4.50	NFEA	−5.58
D	AFQT	−5.29	NLIS	−2.26	SFQV	−4.45	NFIC	−5.43
D	SFQM	−5.28	AYMD	−2.23	SFQC	−4.41	NFCD	−5.41
D	CYKM	−5.23	SYMD	−2.18	AFKT	−4.39	NFMT	−5.25

tutorial files can also be downloaded from the same address. The following commands need to be executed.

```
tar xzf Proteus_3.0.tar.gz
export PATH="$(pwd)/Proteus_3.0/protMC:$PATH"
tar xzf tuto_TyrRS.tar.gz
cd tuto_TyrRS
export PATH="$(pwd)/bin:$PATH"
```

Table 4
Best designed sequences according to a given design criterion (bold font) and their relative free energies (kcal/mol) with respect to the native sequence for different criteria. L ligand criteria sequences. *See* Table 3 legend

seq.	aff.L	rateL	eff.L	stereo.L	aff.D	rateD	eff.D	stereo.D
FYMM	**−7.27**	6.62	−0.65	−1.21	−1.33	1.89	0.56	1.21
FYMK	**−7.10**	6.46	−0.64	−3.10	2.22	0.24	2.46	3.10
FYMQ	**−6.92**	5.10	−1.82	−0.49	0.33	−1.67	−1.33	0.49
FYCM	**−6.83**	7.59	0.77	−0.55	–	–	1.32	0.55
FFMM	**−6.76**	8.79	2.03	2.15	−1.41	1.29	−0.11	−2.15
FYMS	**−6.60**	6.53	0.03	0.30	0.73	−1.00	−0.28	−0.30
FFCM	**−6.43**	8.60	2.17	1.65	−1.70	2.22	0.52	−1.65
FYTM	**−6.34**	7.48	1.14	−1.95	0.02	3.07	3.09	1.95
FFVM	**−6.20**	8.35	2.15	0.45	−0.66	2.36	1.70	−0.45
FYCI	**−6.16**	6.50	0.34	−0.60	–	–	0.94	0.60
TYEI	0.10	**−1.18**	−1.08	−3.48	−0.59	2.99	2.40	3.48
DYIQ	3.04	**−1.14**	1.91	−0.41	2.14	0.18	2.32	0.41
TYKQ	−1.18	**−0.86**	−2.03	−0.10	−2.54	0.61	−1.94	0.10
FAMQ	−1.51	**−0.85**	−2.35	−0.08	−0.78	−1.50	−2.28	0.08
NYVQ	1.81	**−0.78**	1.03	−0.56	1.34	0.24	1.58	0.56
TYCI	1.52	**−0.73**	0.79	–	0.40	–	–	–
CYKQ	−1.21	**−0.71**	−1.92	0.86	−3.45	0.66	−2.78	−0.86
SYKQ	−1.28	**−0.66**	−1.94	1.36	−3.49	0.20	−3.30	−1.36
TYTE	3.67	**−0.64**	3.04	−2.40	3.84	1.60	5.44	2.40
TYMQ	0.09	**−0.63**	−0.54	−2.39	−0.08	1.93	1.84	2.39
CYQQ	−1.88	−0.10	**−1.97**	0.82	−3.42	0.62	−2.79	−0.82
AYQQ	−2.21	0.27	**−1.95**	0.34	−3.08	0.79	−2.29	−0.34
SYKQ	−1.28	−0.66	**−1.94**	1.36	−3.49	0.20	−3.30	−1.36
CYKQ	−1.21	−0.71	**−1.92**	0.86	−3.45	0.66	−2.78	−0.86
SYKI	−2.73	0.86	**−1.87**	1.41	−4.25	0.97	−3.28	−1.41
AYMQ	−2.62	0.78	**−1.85**	−0.05	−1.87	0.08	−1.80	0.05
FYMQ	−6.92	5.10	**−1.82**	−0.49	0.33	−1.67	−1.33	0.49
SYQQ	−1.63	−0.18	**−1.82**	0.70	−2.97	0.45	−2.52	−0.70
SYMQ	−2.13	0.33	**−1.81**	0.18	−1.64	−0.34	−1.99	−0.18
CYMQ	−2.05	0.28	**−1.76**	−0.34	−1.99	0.57	−1.42	0.34

(continued)

Table 4
(continued)

seq.	aff.L	rateL	eff.L	stereo.L	aff.D	rateD	eff.D	stereo.D
DYQQ	0.00	0.00	0.00	**0.00**	0.00	0.00	0.00	0.00
SYNQ	−0.27	−0.01	−0.28	**0.20**	−1.15	0.66	−0.48	−0.20
NYMQ	−1.28	−0.19	−1.48	**0.29**	−0.87	−0.90	−1.77	−0.29
SYQQ	−1.63	−0.18	−1.82	**0.70**	−2.97	0.45	−2.52	−0.70
CYQQ	−1.88	−0.10	−1.97	**0.82**	−3.42	0.62	−2.79	−0.82
CYKQ	−1.21	−0.71	−1.92	**0.86**	−3.45	0.66	−2.78	−0.86
AYKQ	−1.14	−0.47	−1.60	**0.90**	−2.81	0.31	−2.50	−0.90
SYKQ	−1.28	−0.66	−1.94	**1.36**	−3.49	0.20	−3.30	−1.36

4.2 Monte Carlo Simulations

We follow the indirect approach to the adaptive landscape flattening method, described in the Theory section (Fig. 1b). For each enzyme state considered, the first step is to develop a bias potential, which when applied, will lead to approximately equal probabilities of amino acids at each designed position. This is achieved by the following command, to be executed in each state folder:

```
protMC.exe < adapt.conf > adapt.log
```

The last iteration of the bias potential present in the `bias.dat` file produced needs to be extracted and saved in a `bias.in` file. Then, a Monte Carlo simulation is run with the bias potential applied. The goal is to sample as many sequences as possible.

```
protMC.exe < MC.conf > MC.log
protMC.exe < POST.conf > POST.log
```

The following `analyze_seq.py` script will then read the Monte Carlo simulation result files `proteus.seq` and `proteus.rich` and produce a `proteus.dat` file, containing the sampled sequences and associated populations.

```
analyze_seq.py proteus.seq proteus.rich 81,175,179,201 > proteus.dat
```

4.3 Stability Design

Stability design is performed for the apo TyrRS and the four complexes. This is achieved by the following command, to be executed in each state folder:

```
stability.py proteus.dat bias.in -rf DYQQ -p 81 175 179 201 \
  > stability.dat
```

Table 5
Best designed sequences according to a given design criterion (bold font) and their relative free energies (kcal/mol) with respect to the native sequence for different criteria. D ligand criteria sequences. *See* Table 3 legend

seq.	aff.L	rateL	eff.L	stereo.L	aff.D	rateD	eff.D	stereo.D
MYVF	16.24	−4.26	11.98	15.82	**−10.30**	6.46	−3.85	−15.82
CFQT	−2.19	2.55	0.36	5.04	**−5.78**	1.10	−4.68	−5.04
AFQM	−3.83	3.92	0.10	3.34	**−5.77**	2.53	−3.24	−3.34
CFQS	−2.38	3.15	0.77	5.08	**−5.62**	1.31	−4.31	−5.08
CFQA	−2.67	3.36	0.70	5.06	**−5.50**	1.13	−4.37	−5.06
CFQC	−2.72	3.10	0.38	4.93	**−5.39**	0.84	−4.55	−4.93
SFQT	−1.71	2.57	0.85	5.41	**−5.31**	0.75	−4.56	−5.41
AFQT	−2.23	2.56	0.33	4.94	**−5.29**	0.68	−4.61	−4.94
SFQM	−2.85	4.12	1.26	3.91	**−5.28**	2.64	−2.64	−3.91
CYKM	−3.17	2.13	−1.04	0.69	**−5.23**	3.50	−1.73	−0.69
SFMN	3.01	0.38	3.39	3.54	2.69	**−2.84**	−0.15	−3.54
AFMN	1.68	0.80	2.48	3.05	2.08	**−2.65**	−0.57	−3.05
NYMD	2.46	0.95	3.41	3.29	2.67	**−2.56**	0.12	−3.29
SYMN	2.15	−0.34	1.81	0.93	3.37	**−2.49**	0.88	−0.93
ALVA	−0.69	3.02	2.33	1.61	3.08	**−2.37**	0.71	−1.61
NYKD	1.89	−0.08	1.81	3.43	0.69	**−2.31**	−1.62	−3.43
MAMS	–	–	–	5.72	–	**−2.28**	–	−5.72
NLIS	3.48	2.67	6.15	5.01	3.40	**−2.26**	1.14	−5.01
AYMD	−0.10	1.42	1.32	2.43	1.12	**−2.23**	−1.11	−2.43
SYMD	0.41	1.11	1.52	2.44	1.25	**−2.18**	−0.93	−2.44
CFQT	−2.19	2.55	0.36	5.04	−5.78	1.10	**−4.68**	−5.04
AFQV	−3.08	2.79	−0.28	4.35	−4.85	0.22	**−4.64**	−4.35
AFQT	−2.23	2.56	0.33	4.94	−5.29	0.68	**−4.61**	−4.94
SFQT	−1.71	2.57	0.85	5.41	−5.31	0.75	**−4.56**	−5.41
CFQC	−2.72	3.10	0.38	4.93	−5.39	0.84	**−4.55**	−4.93
AFQC	−2.81	3.21	0.40	4.94	−4.83	0.29	**−4.54**	−4.94
AFKQ	−1.27	0.34	−0.93	3.57	−4.32	−0.18	**−4.50**	−3.57
SFQV	−2.46	2.75	0.29	4.74	−4.69	0.24	**−4.45**	−4.74
SFQC	−2.20	2.85	0.65	5.06	−4.87	0.47	**−4.41**	−5.06
AFKT	−1.21	1.71	0.50	4.89	−4.82	0.44	**−4.39**	−4.89

(continued)

Table 5
(continued)

seq.	aff.L	rateL	eff.L	stereo.L	aff.D	rateD	eff.D	stereo.D
MLQS	2.95	2.92	5.87	6.13	−0.17	−0.09	−0.26	**−6.13**
SFQD	−0.34	2.20	1.86	5.89	−3.32	−0.71	−4.03	**−5.89**
NFES	2.11	2.16	4.27	5.71	−1.36	−0.08	−1.44	**−5.71**
NFKA	0.85	1.92	2.76	5.67	−2.63	−0.28	−2.91	**−5.67**
NFKS	0.65	1.79	2.44	5.64	−3.09	−0.10	−3.19	**−5.64**
AFQD	−0.83	2.31	1.48	5.62	−3.47	−0.67	−4.14	**−5.62**
NFEA	1.22	2.64	3.86	5.58	−1.54	−0.17	−1.72	**−5.58**
NFIC	2.51	2.18	4.69	5.43	−0.63	−0.12	−0.74	**−5.43**
NFCD	2.18	2.23	4.42	5.41	−0.27	−0.72	−0.99	**−5.41**
NFMT	1.63	1.84	3.47	5.25	−1.18	−0.60	−1.77	**−5.25**

The `stability.py` script takes as inputs the sequence populations (`proteus.dat`) and bias (`bias.in`) of the state considered. A reference sequence can be provided so that relative stabilities are calculated, and here the native sequence is given. The design positions also need to be provided. The output file (`stability.dat`) contains sequences and associated folding free energies.

Results are presented in the form of sequence logos (Fig. 5b for apo TyrRS). The profiles are very similar for the five systems (not shown), indicating that the ligands tested have little influence on stability. Ser, Tyr, and Ser are the most frequent amino acids at positions 81, 175, and 201. Thus we retrieve the native type at position 175. As Tyr is found with high frequency at this position in all states, this suggests that it has an important role for TyrRS stability, and not for ligand recognition. Val and Thr are found at position 179, with a slight preference for Val for apo and L-Tyr complexes (reactant and transition state), while Thr is preferred for D-Tyr complexes. Compared to previous works by our group [22], where the folding free energy was used as design criterion, Tyr at position 175 is found in common, while other positions differ.

4.4 Affinity Design

Applying the adaptive landscape flattening method with the apo TyrRS as reference state and the L or D reactant complex as second state, we obtain designs based on the affinity of the reactants for the TyrRS. That is, when $\Delta G_{\mathrm{E}\rightarrow\mathrm{E:S}}$ is minimized, K_{M} is minimized.

This is achieved by the following command, to be executed in `reacL` or `reacD` folders:

```
affinity.py ../apo/proteus.dat proteus.dat \
 --bias_unbound ../apo/bias.in --bias_bound bias.in \
 -rf DYQQ -p 81 175 179 201 > affinity.dat
```

The `affinity.py` script takes as inputs the sequence populations (`proteus.dat`) and biases (`bias.in`) of the unbound and bound states. A reference sequence can be provided so that relative affinities are calculated, and here the native sequence is given. The design positions also need to be provided. The output file (`affinity.dat`) contains sequences and associated binding free energies.

Results are shown in the form of sequence logos in Fig. 5c, d, for the L and D reactant affinities. Tyr is found again at position 175 for L and D affinities, suggesting that this position is not able to discriminate the stereoisomers in terms of affinity. At the other positions, the designs differ. Phe, Met, and Met are obtained for L affinity at positions 81, 179, and 201. For D affinity, we find Met, Val, and Phe at positions 81, 179, and 201. Interestingly, Phe and Met at positions 81 and 201 appear to be inverted when switching between the L and D reactants.

4.5 Catalytic Rate Design

The adaptive landscape flattening method is now applied with the L or D reactant complex as reference state and the corresponding L or D transition state complex as second state, yielding designs based on catalytic rate. That is, when $\Delta G_{E:S \rightarrow E:S^{\ddagger}}$ is minimized, k_{cat} is maximized. This is achieved by the following command, to be executed in `tsL` or `tsD` (in the latter case, replace `reacL` by `reacD` in the command line) folders. Despite its name, the same Python script as in affinity design can be used, as it is applicable to any free energy difference between two states.

```
affinity.py ../reacL/proteus.dat proteus.dat \
 --bias_unbound../reacL/bias.in --bias_bound bias.in \
 -rf DYQQ -p 81 175 179 201 > rate.dat
```

Results are shown as sequence logos in Fig. 5e, f, for the L- and D-Tyr adenylation catalytic rate designs. The logos show more diversity for catalytic rate than for stability and affinity. For L catalytic rate, we find Thr/Ser/Asp, Tyr, Ile/Met, and Gln at positions 81, 175, 179, and 201. The native types are thus retrieved for three (81, 175, and 201) out of four positions. This could indicate that these positions have been selected by evolution for catalytic rate. For D catalytic rate, we obtain quite different results, with Ser/Ala/Asn, Phe/Tyr, Met, and Asp/Cys/Ser/Ala at positions 81, 175, 179, and 201.

4.6 Catalytic Efficiency Design

Applying the adaptive landscape flattening method by taking the apo TyrRS as reference state, as in affinity design, but this time with the L or D transition state complexes as second state, we obtain designs based on catalytic efficiency. That is, when $\Delta G_{E \rightarrow E:S^{\ddagger}}$ is minimized, k_{cat}/K_M is maximized. The following command is used, to be executed in `tsL` or `tsD` folders:

```
affinity.py ../apo/proteus.dat proteus.dat \
 --bias_unbound ../apo/bias.in --bias_bound bias.in \
 -rf DYQQ -p 81 175 179 201 > efficiency.dat
```

Results as sequence logos are shown in Fig. 5g, h, for L- and D-Tyr adenylation catalytic efficiency designs. For L catalytic efficiency, we obtain Ser/Ala/Cys, Tyr, Gln/Met/Lys, and Gln/Ile at positions 81, 175, 179, and 201. The native types are thus retrieved for three positions (175, 179, and 201) out of four. For D catalytic efficiency, we obtain Ala/Ser/Cys, Phe, Gln/Lys, and Thr/Cys/Ser at positions 81, 175, 179, and 201.

4.7 Stereospecificity Design

For stereospecificity design in favor of L-Tyr, we apply the adaptive landscape flattening method with the D transition state complex as reference state and the L transition state complex as second state. When $\Delta G_{E:D^{\ddagger} \rightarrow E:L^{\ddagger}}$ is minimized, $k_{cat}^{L}/K_{M}^{L} \cdot K_{M}^{D}/k_{cat}^{D}$ is maximized. This is achieved by the following command, to be executed in `tsL` folder. Inverting L and D provides stereospecificity design in favor of D-Tyr.

```
affinity.py ../tsD/proteus.dat proteus.dat \
 --bias_unbound ../tsD/bias.in --bias_bound bias.in \
 -rf DYQQ -p 81 175 179 201 > stereospecificity.dat
```

Results are shown in Fig. 5i for L-Tyr stereospecificity design. We obtain Ser/Asp, Tyr, Gln, Gln at positions 81, 175, 179, and 201. Remarkably, the native types are thus retrieved at the four designed positions, suggesting that stereospecificity was a selection criterion in the evolution of TyrRS. Figure 5j shows D-Tyr stereospecificity design results. We obtain Asn, Phe, Gln, and Ser/Asp at positions 81, 175, 179, and 201. As in affinity design, we obtain somewhat inverted results between positions 81 and 201 when switching between L- and D-Tyr.

4.8 Multi-criteria Filtering

Protein sequences optimized according to a single criterion can be irrelevant for other criteria. For example, it can be pertinent to filter out from affinity designs the sequences having a poor stability. This can be achieved by the following command, to be executed in `reacL` or `reacD` folders:

```
filter_sequences.py affinity.dat ../apo/stability.dat 0.0 \
  > affinity.filtered.dat
```

The `filter_sequences.py` script takes as inputs the sequences and associated quantities to be filtered (here `affinity.dat`), the sequences and associated quantities to be used as filter (here `apo/stability.dat`), as well as a threshold value for filtering. As quantities have been calculated relatively to the native sequence, a threshold value of 0 will select only sequences as good as the native for the filtering criterion. The output file (`affinity.filtered.dat`) contains the filtered sequences and associated affinities. Stability filtering was applied to all design results presented above.

Multi-criteria filtering is also possible, by chaining the filters. This can be obtained by the following command, to be executed in `tsL` folder. The resulting `stereospecificity.filtered.dat` file will contain sequences and associated stereospecificities that have been selected to be as good as the native sequence for apo stability, affinity, catalytic rate, and catalytic efficiency.

```
filter_sequences.py stereospecificity.dat \
  ../apo/stability.dat 0.0| \
  filter_sequences.py - ../reacL/affinity.dat 0.0| \
  filter_sequences.py - rate.dat 0.0| \
  filter_sequences.py - efficiency.dat 0.0 \
  > stereospecificity.filtered.dat
```

5 Concluding Remarks

Our goal was to modify the TyrRS enzyme to adenylate the D form of Tyr instead of the natural L form. For this, we performed different sequence designs on a TyrRS model, considering four ligands: the reactants and the transition states of the adenylation reaction, in L and D forms. Stability, affinity, catalytic rate, catalytic efficiency, and stereospecificity designs were performed with our Proteus CPD software. While stability and affinity were considered in previous studies addressing TyrRS stereospecificity inversion, catalytic rate, catalytic efficiency, and a direct stereospecificity criterion were not. This illustrates the versatility of the adaptive landscape flattening method applied to enzyme design. We were able to retrieve the native amino acid at the four positions designed. Asp was retrieved at position 81 among the best candidates in designs based on catalytic rate and stereospecificity in favor of L-Tyr. Tyr was retrieved at position 175 in stability design of all forms and retained in most other designs when a stability filter was applied. Gln was retrieved at positions 201 in design based on the catalytic rate of the L-Tyr adenylation, and at positions 179 and 201 in designs based on catalytic efficiency as well as stereospecificity in favor of L-Tyr. Stereospecificity design in favor of L-Tyr thus retrieves the four amino acids of the native sequence, while catalytic rate and catalytic efficiency designs in favor of L-Tyr retrieve three native sequence amino acids. The capability of our approach to retrieve the native sequence is in favor of the validity of the model and shows the interest of considering the transition state. The results suggest that discriminating L-Tyr from D-Tyr was among TyrRS evolution criteria. To achieve our goal of TyrRS stereospecificity inversion, we recommend the mutations found by stereospecificity design in favor of D-Tyr: Asn, Phe, and Ser at positions 81, 175, and 201. Gln179 seems to be important for catalytic efficiency of both L- and D-Tyr and thus should not be mutated.

6 Notes

1. In the Theory section and in Fig. 1, we presented two ways of using the adaptive landscape flattening method. While the direct approach is more intuitive and readily produces sequences sampled according to the design criterion, we recommend the indirect approach. Multi-criteria filtering (e.g., affinity+stability) is indeed often impossible in practice with the direct approach, as the sequence spaces sampled according to different design criteria do not necessarily overlap.
2. In the application presented, we have redesigned four amino acid positions of the TyrRS enzyme. It has to be noted that increasing the number of designed positions requires longer Monte Carlo simulations to reach convergence. We recommend targeting 1000 visits per sequence variant. With four mutated positions and a mutation space of 20 amino acids, a trajectory length of 10^9 is sufficient.
3. Convergence of the bias potentials should be assessed by looking at the amino acid frequencies obtained from the biased Monte Carlo simulations. It is expected that the proportions of amino acids at each position are roughly equal (see middle sequence logos in Fig. 1). Perfect equipartition is not needed though.
4. We have sometimes observed difficulties in the positioning of certain amino acids, often of aromatic type, leading to mutants with abnormal energies. This is caused by the limits of the rotamer library. A simple solution is to remove the amino acid type involved from the mutation space of the position. If this is not acceptable, we recommend considering more rotamers and/or increasing the number of minimization steps.

5. Note that the bias potentials (bias.in) and results of biased Monte Carlo simulations (proteus.dat) are provided in the tutorial files. It is thus possible to bypass Subheading 4.2 above.
6. If the Monte Carlo simulations are performed, note that they can be run in parallel over the states.

References

1. Calendar R, Berg P (1966) The catalytic properties of tyrosyl ribonucleic acid synthetases from *Escherichia coli* and *Bacillus subtilis*. Biochemistry 5:1690–1695
2. Soutourina J, Plateau P, Blanquet S (2000) Metabolism of d-aminoacyl-tRNAs in *Escherichia coli* and *Saccharomyces cerevisiae* cells. J Biol Chem 275:32535–32542
3. Schmidt am Busch M, Lopes A, Mignon D et al (2008) Computational protein design: software implementation, parameter optimization, and performance of a simple model. J Comput Chem 29:1092–1102
4. Simonson T, Gaillard T, Mignon D et al (2013) Computational protein design: the Proteus software and selected applications. J Comput Chem 34:2472–2484
5. Mignon D, Druart K, Michael E et al (2020) Physics-based computational protein design: an update. J Phys Chem A 124:10637–10648
6. Lopes A, Alexandrov A, Bathelt C et al (2007) Computational sidechain placement and protein mutagenesis with implicit solvent models. Proteins 67:853–867
7. Gaillard T, Panel N, Simonson T (2016) Protein side chain conformation predictions with an MMGBSA energy function. Proteins 84: 803–819
8. Schmidt am Busch M, Lopes A, Amara N et al (2008) Testing the Coulomb/accessible surface area solvent model for protein stability, ligand binding, and protein design. BMC Bioinform 9:148
9. Aleksandrov A, Polydorides S, Archontis G et al (2010) Predicting the acid/base behavior of proteins: a constant-pH Monte Carlo approach with Generalized Born solvent. J Phys Chem B 114:10634–10648
10. Polydorides S, Simonson T (2013) Monte Carlo simulations of proteins at constant pH with Generalized Born solvent. J Comput Chem 34:2742–2756
11. Villa F, Mignon D, Polydorides S et al (2017) Comparing pairwise-additive and many-body generalized born models for acid/base calculations and protein design. J Comput Chem 38: 2396–2410
12. Villa F, Simonson T (2018) Protein pKa's from adaptive landscape flattening instead of constant-pH simulations. J Chem Theory Comput 14:6714–6721
13. Michael E, Polydorides S, Simonson T et al (2020) Hybrid MC/MD for protein design. J Chem Phys 153:054113
14. Schmidt am Busch M, Mignon D, Simonson T (2009) Computational protein design as a tool for fold recognition. Proteins 77:139–158
15. Schmidt am Busch M, Sedano A, Simonson T (2010) Computational protein design: validation and possible relevance as a tool for homology searching and fold recognition. PLOS ONE 5:e10410
16. Mignon D, Panel N, Chen X et al (2017) Computational design of the Tiam1 PDZ domain and its ligand binding. J Chem Theory Comput 13:2271–2289
17. Gaillard T, Simonson T (2017) Full protein sequence redesign with an MMGBSA energy function. J Chem Theory Comput 13:4932–4943
18. Opuu V, Sun YJ, Hou T et al (2020) A physics-based energy function allows the computational redesign of a PDZ domain. Sci Rep 10: 11150
19. Lopes A, Schmidt am Busch M, Simonson T (2010) Computational design of protein-ligand binding: modifying the specificity of asparaginyl-tRNA synthetase. J Comput Chem 31:1273–1286
20. Polydorides S, Amara N, Aubard C et al (2011) Computational protein design with a Generalized Born solvent model: application to asparaginyl-tRNA synthetase. Proteins 79: 3448–3468
21. Druart K, Palmai Z, Omarjee E et al (2016) Protein:ligand binding free energies: a stringent test for computational protein design. J Comput Chem 37:404–415
22. Simonson T, Ye-Lehmann S, Palmai Z et al (2016) Redesigning the stereospecificity of tyrosyl-tRNA synthetase. Proteins 84:240–253
23. Druart K, Le Guennec M, Palmai Z et al (2017) Probing the stereospecificity of tyrosyl- and glutaminyl-tRNA synthetase with molecular dynamics. J Mol Graph Model 71:192–199

24. Opuu V, Nigro G, Gaillard T et al (2020) Adaptive landscape flattening allows the design of both enzyme: substrate binding and catalytic power. PLOS Comput Biol 16:e1007600
25. Michael E, Saint-Jalme R, Mignon D et al (2023) Computational protein design repurposed to explore enzyme vitality and help predict antibiotic resistance. Front Mol Biosci 9: 905588
26. Opuu V, Nigro G, Lazennec-Schurdevin C et al (2023) Redesigning methionyl-tRNA synthetase for β-methionine activity with adaptive landscape flattening and experiments. Protein Sci 32:e4738
27. Villa F, Panel N, Chen X et al (2018) Adaptive landscape flattening in amino acid sequence space for the computational design of protein: peptide binding. J Chem Phys 149:072302
28. Panel N, Villa F, Opuu V et al (2021) Computational design of PDZ-peptide binding. in: Borg JP (ed) Methods in molecular biology: PDZ mediated interactions. Humana, New York
29. Cieplak P, Cornell WD, Bayly C et al (1995) Application of the multimolecule and multiconformational RESP methodology to biopolymers: charge derivation for DNA, RNA, and proteins. J Comput Chem 16:1357–1377
30. Kobayashi T, Takimura T, Sekine R et al (2005) Structural snapshots of the KMSKS loop rearrangement for amino acid activation by bacterial tyrosyl-tRNA synthetase. J Mol Biol 346: 105–117
31. Yaremchuk A, Kriklivyi I, Tukalo M et al (2002) Class I tyrosyl-tRNA synthetase has a class II mode of cognate tRNA recognition. EMBO J 21:3829–3840
32. Word JM, Lovell SC, Richardson JS et al (1999) Asparagine and glutamine: using hydrogen atom contacts in the choice of side-chain amide orientation. J Mol Biol 285:1735–1747
33. Li H, Robertson AD, Jensen JH (2005) Very fast empirical prediction and interpretation of protein pKa values. Proteins 61:704–721
34. Olsson MHM, Søndergard CR, Rostkowski M et al (2011) PROPKA3: consistent treatment of internal and surface residues in empirical pKa predictions. J Chem Theory Comput 7:525–537
35. Brünger AT (1992) X-Plor Version 3.1. A system for X-ray crystallography and NMR. Yale University Press, New Haven
36. Wang J, Cieplak P, Kollman PA (2000) How well does a restrained electrostatic potential (RESP) model perform in calculating conformational energies of organic and biological molecules? J Comput Chem 21:1049–1074
37. Hornak V, Abel R, Okur A et al (2006) Comparison of multiple Amber force fields and development of improved protein backbone parameters. Proteins 65:712–725
38. Meagher KL, Redman LT, Carlson HA (2003) Development of polyphosphate parameters for use with the AMBER force field. J Comput Chem 24:1016–1025
39. Horn AHC (2014) A consistent force field parameter set for zwitterionic amino acid residues. J Mol Model 20:2478
40. Hawkins GD, Cramer CJ, Truhlar DG (1995) Pairwise solute descreening of solute charges from a dielectric medium. Chem Phys Lett 246:122–129
41. Hawkins GD, Cramer CJ, Truhlar DG (1996) Parametrized models of aqueous free energies of solvation based on pairwise descreening of solute atomic charges from a dielectric medium. J Phys Chem 100:19824–19839
42. Moulinier L, Case DA, Simonson T (2003) Reintroducing electrostatics into protein X-ray structure refinement: bulk solvent treated as a dielectric continuum. Acta Crystallogr Sect D 59:2094–2103
43. Tufféry P, Etchebest C, Hazout S (1997) Prediction of protein side chain conformations: a study on the influence of backbone accuracy on conformation stability in the rotamer space. Protein Eng 10:361–372

Part III

3D Structure Analysis: Molecular Recognition

Chapter 12

Quantification of Conformational Changes of Kinase Regulators, Kinases, and Regulator–Kinase Complexes Using the TSR Algorithm

Tarikul I. Milon, Poorya Khajouie, Feng Chen, Tyler A. Borel, Kathleen D. Knierim, Vijay Raghavan, and Wu Xu

Abstract

This chapter provides step-by-step instructions on applying a triangular spatial relationship (TSR) algorithm for precise molecular recognition studies, with a focus on kinase–ligand interactions. Given the central role of kinases in cellular processes and that their dysregulation is a hallmark of cancer, the development of kinase inhibitors is a major focus in drug discovery efforts. In particular, understanding the conformational changes involved in kinase signaling and drug binding is essential for elucidating regulatory mechanisms and designing effective therapeutics. Using three-dimensional structures from the Protein Data Bank, we demonstrate the utility of the TSR algorithm in characterizing and quantifying conformational changes in protein–ligand complexes. Our findings include: (i) the identification of a unique substructure specific to adenosine triphosphate (ATP) molecules, distinguishing them from related compounds such as adenosine diphosphate (ADP) and adenylyl imidodiphosphate (ANP); (ii) the discovery of a backbone substructure that differentiates bound from unbound forms of cyclic adenosine monophosphate-dependent kinases (cAMPDKs); and (iii) the observation that glutamate and asparagine residues within molecular recognition interfaces exhibit distinctive spatial geometries. These results support the existence of recognition codes—structural patterns that govern the specificity of interactions between cAMPDKs and their regulators (ATP, ADP, and ANP). The TSR algorithm and its variants are broadly applicable to other protein families, their ligands, and the complexes they form, beyond kinase–ligand interactions, offering a powerful framework for molecular recognition analysis.

Key words TSR-based method, 3D structure, Kinase, Kinase inhibitor, Kinase activator, Conformational change, Molecular recognition, Binding site

---

Supplementary Information: The online version contains supplementary material available at https://doi.org/10.1007/978-1-0716-4828-5_12.

Shahid M. Khan and Florencio Pazos (eds.), *Protein Evolution: Methods and Protocols*, Methods in Molecular Biology, vol. 2979, https://doi.org/10.1007/978-1-0716-4828-5_12,

1 Introduction

Biomolecular recognition, for example, protein and protein interactions, inhibitor/activator and protein interactions, deoxyribonucleic acid (DNA) and protein interactions, is a process in which biomolecules recognize and physically bind to their targets. Such recognition events typically occur with high affinity, high specificity, and reversibility, often through noncovalent interactions [1]. Noncovalently controlled phenomena are poorly understood, probably because the repertoire of noncovalent interactions underlying molecular recognition has not yet been fully defined [2]. Therefore, it is challenging to design a small molecule from first principles that will bind tightly and specifically to a pocket in a protein of known structure, despite widespread interest in rational drug design [2]. Biomolecular recognition often leads to the formation of binding complexes that are accompanied by small-scale to large-scale conformational changes. This process is fundamental to biomolecular functions at the molecular and cellular levels. Uncovering the molecular mechanisms of biomolecular recognition through the tracking of key conformational changes is vital for understanding these functions [3]. For example, defining and quantifying conformational changes of catalytically important structural elements of kinases is important for dictating inhibitor binding modes and affinities [4]. The lack of widely applicable methods for tracking molecular conformational changes has hindered our understanding of protein regulation and our ability to design conformationally selective inhibitors [4].

In response to this lack, the triangular spatial relationship (TSR)-based algorithm was developed. The first version was designed for comparing individual protein three-dimensional (3D) structures and motif discovery using alpha carbon (C_α) atoms [5]. The first version has limitations in certain situations, such as when quantifying homologous protein structures, comparing a small protein with a large protein, distinguishing stereometric differences, and quantifying large-scale protein structural comparisons. By introducing the amino acid grouping algorithm [6], the size-filtering algorithm [7], the mirror image algorithm [8], and the parallelization [9] strategy, the later versions are able to overcome or partially overcome those problems. Although the C_α-based TSR algorithm can quantify the conformational changes [10] and define the specific geometry of metal ion binding sites [11], the C_α-based algorithms cannot precisely define and quantify conformational changes at the atomic level due to the fact that side chains of amino acids play a critical role in molecular recognition [12–18]. To extend the capacity of the original C_α-based TSR algorithm for quantifying drug and target interactions, and protein and protein interactions, we have developed the all-atom TSR algorithms for inhibitors and targets [19] as well as for individual amino acids [20]. The TSR algorithm can be integrated with experimental approaches [21–23] and theoretical approaches (e.g., molecular dynamics [MD] simulations) [5, 21, 22]. The TSR-based method can be effectively used in a laboratory setting to educate undergraduate students on big data analysis, as it offers a simple approach for comparing and analyzing molecular structures without requiring prior structural alignment [24].

The purpose of this book chapter is to clearly describe and illustrate the steps involved in applying the TSR-based algorithm to allow for accurate implementation of molecular recognition studies. To achieve this objective, we have chosen the interactions between kinases and their inhibitors or activators as an example. It is worth noting that the method can be applied to any other biomolecular recognition. A large body of publicly available 3D structures of two-molecule complexes, and ideally of each component alone in the Protein Data Bank (PDB), provide a structural foundation for studying conformational changes upon molecular recognition events. Kinases are among the most diverse and largest class of proteins with 500+ members [25, 26]. They are enzymes that biochemically catalyze the transfer of a phosphate group to their targeted substrates of either small molecules [27] or proteins [27, 28] using adenosine triphosphate (ATP) [29, 30]. Kinase signaling pathways have been shown to be essential for many cellular processes [31, 32], and dysregulation of kinase signaling drives many of the hallmark phenotypes of tumor biology [32–35]. In addition, dysregulation of kinases has been demonstrated in many other human disorders including immune and neurological disorders and in infectious diseases [36–40]. Kinase inhibitors now account for a quarter of all current drug discovery research and development efforts [36]. Kinase inhibitors have revolutionized the treatment of a select group of diseases [35, 41]. The Food and Drug Administration (FDA) has approved 35 drugs (31 for the treatment of cancers) targeting kinases [36]. Some examples of such drugs are the following: imatinib for treatment of chronic myelogenous and acute lymphoblastic leukemia through inhibiting breakpoint cluster region–Abelson murine leukemia viral oncogene homolog 1 (BCR–ABL1) [42], crizotinib for treatment of cancers driven by anaplastic lymphoma kinase (ALK) fusions [43], vemurafenib for treatment of B-rapidly accelerated fibrosarcoma (BRAF) mutant tumors [44], lapatinib for treatment of erythroblastic oncogene B2/ human epidermal growth factor receptor 2 (ERBB2/ HER2)-amplified tumors [45], gefitinib and erlotinib for treatment of epidermal growth factor receptor (EGFR)-mutated tumors [45], and fasudil for treatment of cerebral vasospasms [46]. In each of these cases, superior clinical benefit was observed for patients who received the treatment compared with those receiving the prior standard-of-care regimens [47].

This chapter demonstrates the capacity of the TSR algorithm for defining and quantifying conformational changes of proteins and small inhibitors by addressing three questions: (i) Can we quantify the conformational changes of the same inhibitors in the same type of targets under different conditions or in different targets (*see* Fig. 1a)? (ii) Can we quantify the conformational changes of protein targets induced by the binding of inhibitors (*see* Fig. 1b)? (iii) Can we quantify the conformational changes of an amino acid induced by the binding of an inhibitor (*see* Fig. 1c)? To help with understanding the approach used for addressing these three questions, each question was further divided into three specific subquestions (*see* Fig. 1a–c). All three questions and their corresponding subquestions aim to address whether there is a molecular recognition code (*see* Fig. 1). Despite the successful development of kinase inhibitors [36, 48], the problems with drug resistance, toxicity, and compromised efficacy present critical challenges in understanding how to fully exploit kinase inhibitors for a therapeutic cure [36, 47, 49]. More than 2000 kinase structures are available in the PDB. Therefore, kinases and kinase regulators (inhibitors and activators) are well suited to be used as the examples for addressing the questions and subquestions. Key conformational changes play a critical role in understanding both the regulation mechanisms of kinase signaling and the mechanisms of drug and target interactions [4]. However, a quantitative understanding of these phenomena has been hampered by a lack of accessible techniques for defining and quantifying these conformational changes [4]. The TSR-based method, along with other methods developed in the field, is shedding new light on kinase regulation and drug development.

To provide a step-by-step guide for addressing the three questions, we have divided the Procedures and Results section into three subsections: (i) dataset preparation and annotation for studying conformational changes of kinase inhibitors or activators using the all-atom TSR algorithm, (ii) dataset preparation and annotation for studying local and global conformational changes of kinases using the C_α TSR algorithm, and (iii) dataset preparation and annotation for studying conformational changes of amino acids (glutamate [Glu] and asparagine [Asn]) of kinases using the amino acid TSR algorithm. In the remainder of the chapter, using the different versions of the TSR algorithm, we have demonstrated the following:

1. A signature and characteristic substructure exclusively belonging to ATPs, but not to ATP-related small molecules, adenosine diphosphates (ADPs) and adenylyl imidodiphosphates (ANPs), was identified.

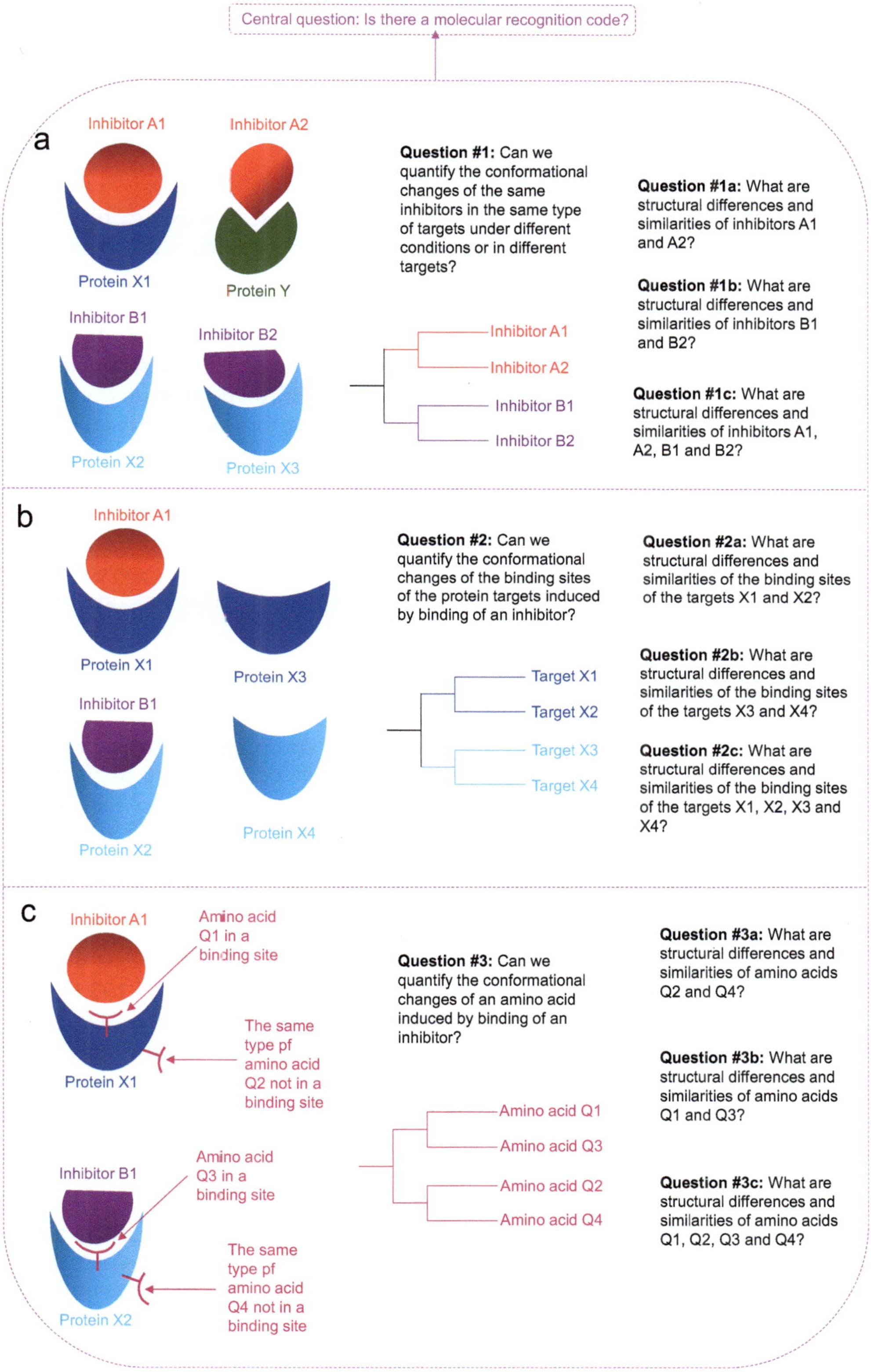

Fig. 1 The diagram shows the central question, three main questions, and three subquestions for each main question that will be addressed. (**a**) Four protein inhibitors (A1, A2, B1, and B2) bind to their targets (X1, Y, X2, and X3) and have different shapes. Four protein targets have different shapes. (**b**) Two protein inhibitors

2. A backbone substructure was identified for distinguishing the bound structures from the unbound structures of cyclic adenosine monophosphate-dependent kinases (cAMPDKs).
3. Glutamate and asparagine residues in the interfaces of molecular recognition sites were found to have their specific and characteristic geometries.

Taken together, our results suggest the existence of recognition codes that govern the interactions between cAMPDKs and their regulators (ADP, ANP, and ATP). This means there are likely specific features or patterns in these molecules that dictate how they interact and bind with each other. The TSR algorithm and its variants that were employed to investigate these interactions are applicable to other protein families, their inhibitors or activators, and the resulting complexes. We wish to alert the reader, however, that the formulation of precise and comprehensive molecular recognition codes faces challenges due to the inherent complexity and variability of molecular recognition events. The complexity of molecular recognition arises from the dynamic nature of interactions, the influence of the environment (e.g., solvent), and the potential for conformational changes in the interacting molecules.

2 Procedures and Results

2.1 Dataset Preparation and Annotation for Studying Conformational Changes of Kinase Inhibitors or Activators

The procedure for studying the conformational changes of kinase inhibitors or activators can be divided into three steps: (i) dataset preparation and annotation of kinase inhibitors and activators, (ii) hierarchical clustering of kinase inhibitors and activators, and (iii) identification of common and specific substructures. Both kinase inhibitors and activators are referred to as kinase regulators in this book chapter.

2.1.1 Dataset Preparation and Annotation of Kinase Regulators

We have prepared a large kinase dataset containing 2527 kinase 3D structures (Supplementary File 1) [10] from the PDB [50]. We have developed an in-house algorithm for outputting all small molecules that closely interact with every protein kinase in the kinase dataset (Python Code 1). We carefully examined the output

Fig. 1 (continued) (A1 and B1) bind to their targets (X1 and X2) and have different shapes. Two proteins are in the bound states or forms (X1 and X2), and two proteins are in the unbound states or forms (X3 and X4). (**c**) Two protein inhibitors (A1 and B1) bind to their targets (X1 and X2) and have different shapes. Four same amino acids (Q1, Q2, Q3, and Q4) have different shapes. Two amino acids are in the binding sites (Q1 and Q3) and two are not in the binding sites (Q2 and Q4). (**a–c**) Different shapes present different 3D structures of the molecules. The central question, three main questions, and their associated three subquestions are shown. A hypothetical clustering map illustrates the structure relationships of inhibitors (**a**), proteins (**b**), and amino acids (**c**) file for all the kinase regulators. Most of the regulators have very few instances of regulator–kinase complexes. To provide accurate and reliable quantifications of conformational changes, more types of kinase regulators and more instances of each type of kinase regulator are needed. To meet these criteria, we have chosen ADP, ANP, and ATP as examples representing kinase regulators for studying their conformational changes (Supplementary File 2). We have identified 40 ADP 3D structures with identical total atom numbers (a total of 27 atoms) (hydrogen atoms are excluded from the study) and embedded in protein kinases. Similarly, we have identified 45 ATP 3D structures (a total of 31 atoms). Initially, 61 ANP 3D structures were identified. However, they have different total atom numbers. These 61 ANPs can be divided into two groups: One group has a total of 27 atoms and the other group has a total of 31 atoms, although all are annotated as ANP from their PDB files. We refer to the ANPs with 27 atoms as ANPA and the ANPs with 31 atoms as ANPB. There are 6 and 55 instances for ANPA and ANPB, respectively.

2.1.2 Hierarchical Clustering Showing the Diversity of Conformations of Kinase Regulators

The key generation method for kinase regulators uses all atoms except hydrogen atoms [19]. MaxDist and Theta were reported previously [5]. Three vertices of triangle i are labeled as l_{i1}, l_{i2}, and l_{i3} that are determined using a rule-based formula. MaxDist is defined as the distance of the longest edge of a triangle. Theta is defined as the angle that is $<90°$ between the line from the midpoint of the edge of l_{i1} and l_{i2} to the opposite vertex l_{i3} and half of the l_{i1}—l_{i2} edge. The bin boundaries used for MaxDist and Theta were the same as those we reported for the ligand TSR algorithm [19]. To generate TSR keys for kinase regulators, information on PDB identifier (ID), chain, regulator name, and ID is needed. Each type of atom was assigned an integer. The TSR algorithm formulates MaxDist, Theta, and three atoms of a triangle into an integer (a TSR key). If two triangles have an identical TSR key, it means that they have similar geometries, and three corresponding atoms are identical. If two triangles have different TSR keys, it means that they have different geometries or at least one corresponding atom pair is different. The key generation Python code for kinase regulators is available (Python Code 2). The generalized Jaccard coefficient measure [51] was used for calculating pairwise similarities between any two regulator structures in the dataset (Supplementary File 2) [5]. The distance matrix is derived from the similarity matrix (Python Code 3) [5]. The hierarchical clustering of kinase regulators is visualized based on the average linkage clustering (Python Code 4) [52]. Structural images were prepared using the visual molecular dynamics (VMD) package [53].

The hierarchical clustering result is shown in Fig. 2a. Out of 146 kinase regulators, no two have identical 3D structures,

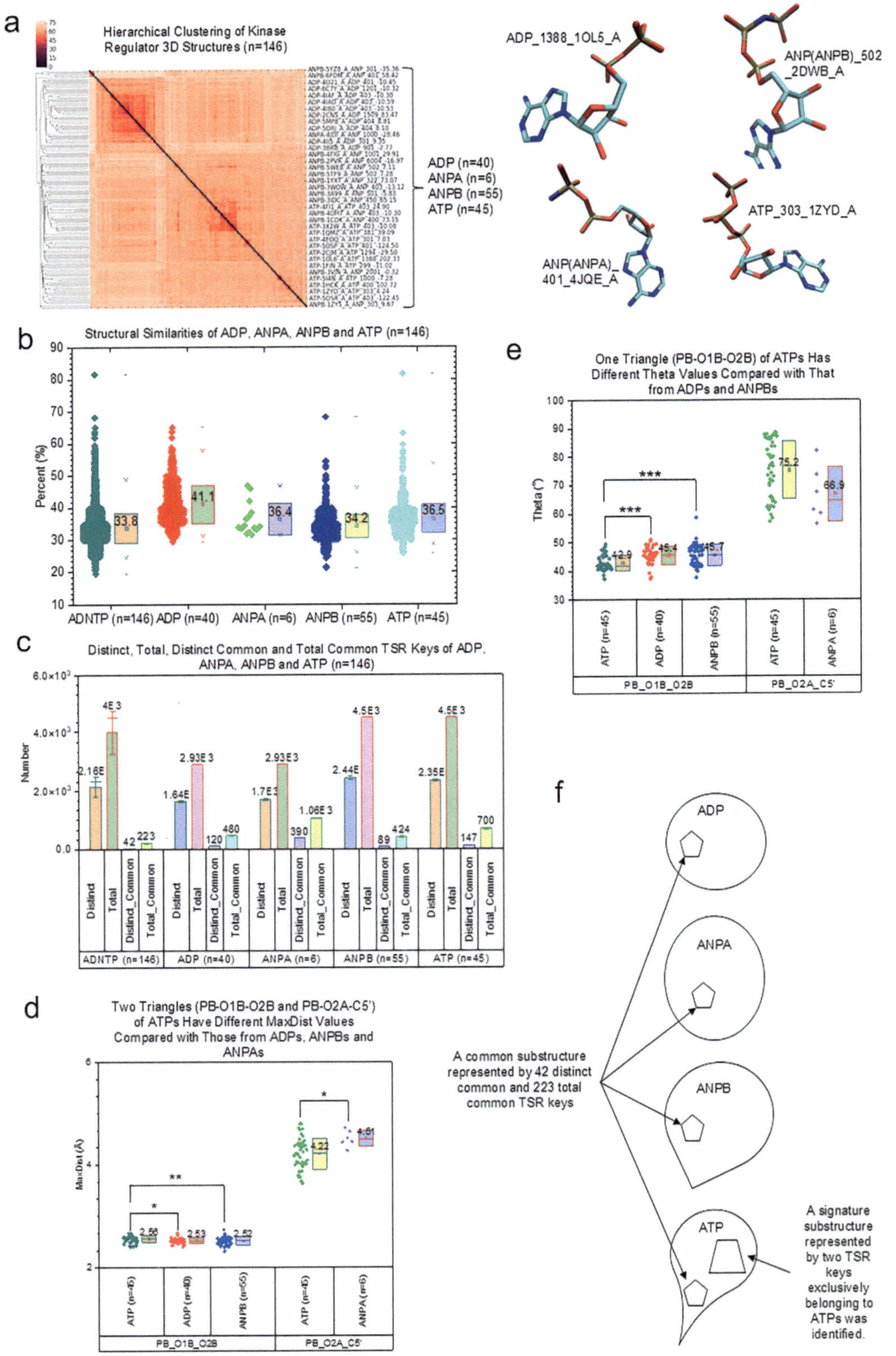

Fig. 2 Kinase regulator adenosine triphosphate (ATP) has its specific and characteristic geometries that are different from the related kinase regulators, adenosine diphosphate (ADP) and ANP. (**a**) The hierarchical clustering result using all-atom TSR keys of kinase regulators is shown. The numbers of kinase regulators,

although each type of regulator, ADP, ANPA, ANPB, and ATP, has its identical formula, respectively. We did not observe a distinct and clear cluster for ADP structures well separated from ANPAs, ANPBs, and ATPs, although most ADPs are clustered together. Similar scenarios were observed for ANPAs, ANPBs, and ATPs. The results demonstrate a high diversity of conformations of kinase regulators. On average, ADPs have the highest pairwise structural similarity (41.1%), while ANPBs have the lowest pairwise structural similarity (34.2%) (*see* Fig. 2b). ATPs and ANPAs are in the middle between ADPs and ANPBs, and ATPs (36.5%) have a slightly higher structural similarity than ANPAs (36.4%) (*see* Fig. 2b).

2.1.3 Identification of Common and Specific Substructures for Further Understanding of Structural Relationships of Kinase Regulators

The uniqueness of the TSR algorithm lies in its ability to interpret clustering results using *common* and *specific* TSR keys and to offer valuable insights into the underlying hierarchical relationships of kinase regulator structures within the dataset. *Common* substructures shared by each regulator in the dataset can be used to infer a similar portion of the binding sites. *Specific* substructures exclusively belonging to a particular type of kinase regulators can be considered as their conformational characteristics. Both *common* and *specific* substructures could serve as structural foundations for understanding kinase regulation modes. The TSR keys and their occurrence frequencies were calculated for each kinase regulator in the dataset. An intersection of distinct TSR keys of all kinase regulators in the dataset is named distinct *common* keys. If the key occurrence frequency for the distinct *common* key of each regulator is counted, they are referred to as total *common* keys (Python Code 5). Distinct and total *common* TSR keys can be calculated for the whole dataset of all types of kinase regulators as well as only for ADPs ($n = 40$) that are considered as a subdataset of the whole kinase regulator dataset ($n = 146$). Similarly, distinct and total

Fig. 2 (continued) ADP, ANPA, ANPB, and ATP, are labeled. One representative 3D structure for each type of kinase regulators, ADP, ANPA, ANPB, and ATP, is shown. Name, residue number, PDB, and chain for each kinase regulator are indicated. (**b**) Pairwise structural similarities between all kinase regulators and between individual types of kinase regulators were calculated and are presented. (**c**) Distinct, total, distinct common, and total common TSR keys were calculated and are presented for all kinase regulators and individual types of kinase regulators. (**d–e**) MaxDist (**d**) and Theta (**e**) values of two triangles (PB-01B-02B and PB-02A-C5') of kinase regulators, ATPs, ADPs, ANPBs and ANPAs, were calculated and are presented. A *t*-test was used to identify statistical differences between the test group and the control group. A threshold of $p < 0.05$ was used to determine significance. This applies to all other figures. The symbol * means that a p value <0.05 (**d**). ** means that a p value <0.01 (**d**). *** means that a p value <0.001 (**e**). (**f**) Each type of kinase regulator, ADP, ANPA, ANPB, and ATP, is represented by a different shape. The pentagon shape represents the common substructure of four types of kinase regulators. The isosceles trapezoid shape represents the specific and signature substructure for ATPs. The numbers of common TSR keys and specific triangles are indicated. (**b–c**) The numbers of total kinase regulators and individual types of kinase regulators are labeled. (**b–e**) The average values are shown, and standard deviations (SDs) represented by error bars are labeled *common* TSR keys can be calculated for the subdatasets of ANPAs ($n = 6$), ANPBs ($n = 55$), and ATPs ($n = 45$). *Specific* TSR keys for ADPs mean that those keys exclusively belong to ADPs but are not found in ANPAs, ANPBs, and ATPs. A similar calculation is used to find *specific* TSR keys for ANPAs, ANPBs, and ATPs (Python Code 6).

ANPBs and ATPs have higher numbers of distinct and total TSR keys than ADPs and ANPAs, probably because ANPBs and ATPs have more atoms than ADPs and ANPAs (*see* Fig. 2c). The order of the numbers of instances from high to low is ANPB ($n = 55$) > ATP ($n = 45$) > ADP ($n = 40$) > ANPA ($n = 6$) (*see* Fig. 2c). We observed that the kinase regulators with fewer instances have larger numbers of distinct and total *common* TSR keys on average, probably because a given regulator type with more instances has a higher conformational diversity than one with fewer instances. Therefore, the order of average numbers of distinct and total *common* keys from large to small is ANPA > ADP > ATP > ANPB (*see* Fig. 2c). All the kinase regulators ($n = 146$) share a total of 42 distinct *common* keys and an average of 233 total *common* keys (*see* Fig. 2c). Two triangles (PB-O1-B-O2B and PB-O2A-C5′) of ATPs were identified as a signature substructure that can be used to distinguish ATP structures from the structures of ADPs, ANPAs, and ANPBs because one or two triangles of ATPs have different MaxDist (*see* Fig. 2d) or Theta (*see* Fig. 2e) values compared with those from ADPs, ANPAs, and ANPBs. In summary, the common substructure shared by ADPs, ANPAs, ANPBs, and ATPs, as well as the signature substructure exclusively belonging to ATPs, were identified to understand their structural relationships (*see* Fig. 2f).

2.2 Dataset Preparation and Annotation for Studying Local and Global Conformational Changes of Kinases

2.2.1 Dataset Preparation and Annotation of Kinases

Quantitatively tracking the conformational changes upon the formation of a two-molecule complex is still one of the greatest challenges for researchers studying molecular interactions [54]. A direct way to tackle this problem is to study the differences between the unbound and the bound structures of the same protein [55–57] or the differences between alternative conformations either in bound or unbound proteins under different conditions [58, 59]. As discussed in the previous section, 146 kinase regulators (ADP: $n = 40$; ANPA: $n = 6$; ANPB: $n = 55$; and ATP: $n = 45$) were identified. We decided to examine the kinase targets of these 146 kinase regulators and found that most kinase regulators bind cAMP-dependent protein kinases (cAMPDKs) [60] and casein kinases (CKs) [61]. We have followed criteria similar to those for kinase regulators; more instances of kinase targets will have a higher chance for accurate and reliable studies of conformational changes. Thus, cAMPDKs were chosen as a representative example to show the application of the TSR-based method in defining and quantifying conformations of global protein structures and local binding sites of protein regulators. A total of 31 instances of cAMPDKs were found to have a binding regulator (bound form): an ADP, an ANP, or an ATP. For convenience of explanation, the term "ADNTP" in this book chapter refers to an ADP, an ANP, or an ATP kinase regulator. Only 4 instances of cAMPDKs were identified without a kinase regulator or a peptide inhibitor (unbound form). A dataset containing 35 cAMPDK structures was prepared and annotated for showing conformational diversity and demonstrating their structure–function relationships (Supplementary File 3).

2.2.2 Hierarchical Clustering Showing the Diversity of Conformations of cAMPDKs

The process of C_α TSR key generation begins with extracting C_α atoms from the PDB files of each protein under analysis. The three side lengths and angles of all triangles constructible from these C_α atoms were calculated. Each of the 20 amino acids was labeled or assigned with one of 20 consecutive unique integer identifiers [5]. The C_α TSR algorithm converts l_{i1}, l_{i2}, and l_{i3} (vertex labels), Theta, and MaxDist to an integer (a TSR key). This design ensures that TSR keys can effectively capture scale changes in protein structures and remain invariant to rotations and translations, making them suitable for alignment-free, pairwise comparison of 3D structures (Python Code 7). Similar to kinase regulators, the generalized Jaccard coefficient measure [51] was used for calculating pairwise similarities between any two cAMPDK structures in the dataset (Python Code 8). The hierarchical clustering of kinase regulators is visualized based on the average linkage clustering (Python Code 9) [52]. The complexity of the multiple dimensional relations among 3D structures is reduced and represented by the multidimensional scaling (MDS) method (Python Code 10) [62].

Fourteen amino acids were identified as candidates for the binding of ADNTP to cAMPDKs. Two distinct clusters were observed from the hierarchical clustering of these 14 amino acids (*see* Fig. 3a). One larger cluster includes the cAMPDK structures with an ADNTP (bound structures), and the smaller cluster contains the cAMPDK structures without an ADNTP (unbound structures), suggesting that the binding of ADNTP induced local conformational changes in cAMPDKs. To further confirm that the local structural differences between bound and unbound forms of cAMPDKs are due to the binding of an ADNTP, we performed hierarchical cluster analysis of entire protein structures using all C_α atoms of cAMPDKs (global structural comparison). We also observed two clusters similar to the local structural comparisons (14 amino acids) (*see* Fig. 3b). To examine the global structural differences between the bound and unbound forms, we compared amino acid sequences using the sequence alignment algorithm. The sequence alignment result shows that there is a deletion in the unbound forms of cAMPDKs (*see* Fig. 3c), suggesting that the global structural differences between the bound and unbound forms (*see* Fig. 3b) are mainly due to the deletion. We also

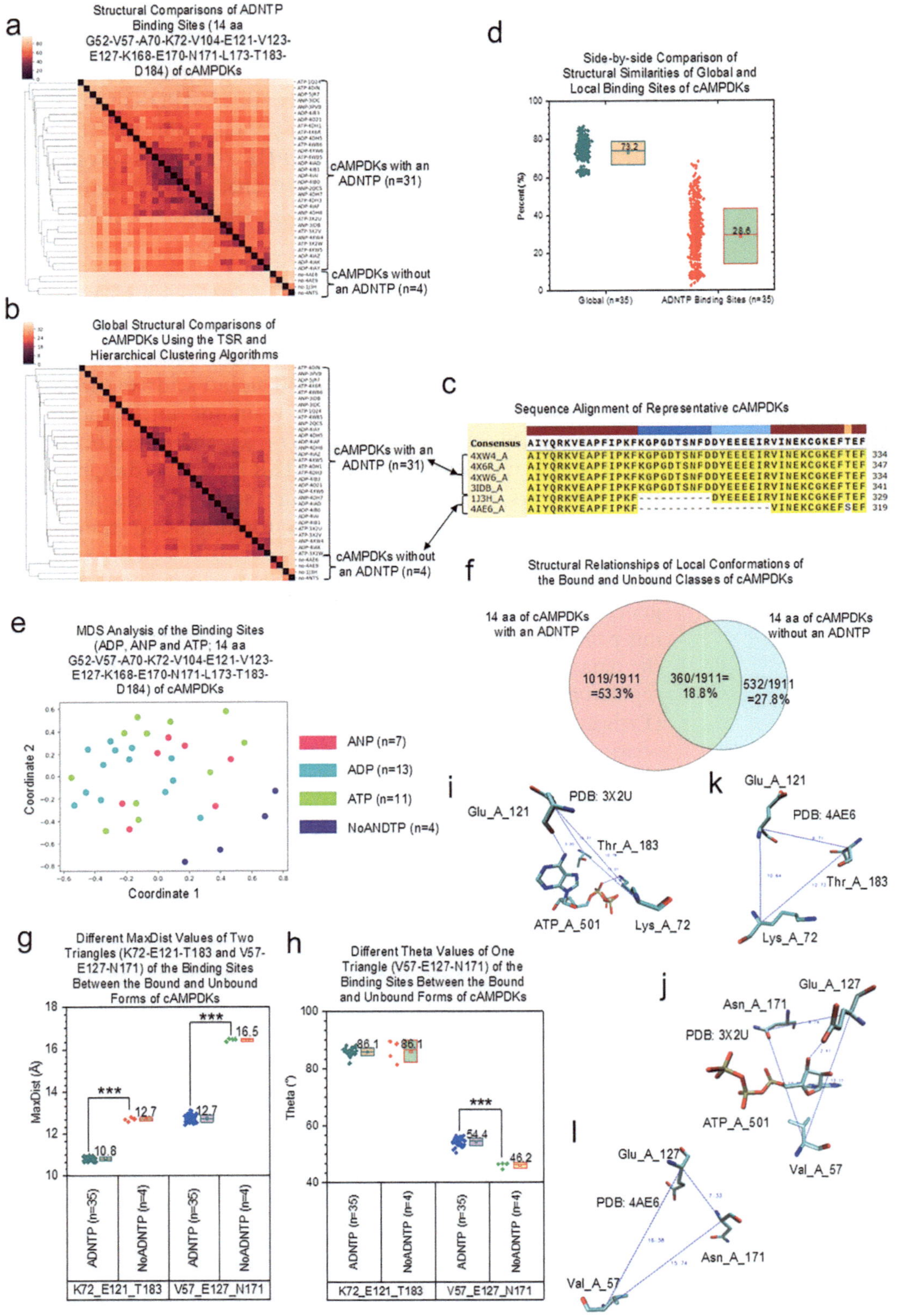

Fig. 3 The regulator binding sites of the bound states of cAMPDKs have their specific and characteristic conformations that are different from those of the unbound states. (**a**–**b**) The hierarchical clustering results of local (**a**) and global (**b**) cAMPDK structures using the alpha carbon (C_{α}) TSR keys are illustrated. The numbers

suspect that the local structural differences (14 amino acids) between the bound and unbound forms (*see* Fig. 3a) may also be largely due to the remote effect of the deletion and may not be primarily due to the binding of ADNTP. Finding the deletion using the sequence alignment algorithm emphasizes the critical role of accurately annotating the datasets for using the TSR algorithm. The global structural similarity is much higher than the local structural similarity (*see* Fig. 3d). The MDS analysis indicates that the binding sites of ADPs, ANPs, and ATPs are indistinguishable from each other (*see* Fig. 3e).

2.2.3 Identification of Specific Substructures for Further Understanding of Local Structural Relationships of cAMPDKs

After C_α TSR keys were calculated for the 14 amino acids of each protein, a union of TSR keys for each bound structure is considered the total distinct key for the bound group, and a union of TSR keys for each unbound is considered the total distinct key for the unbound group. The bound group has a total of 1379 distinct TSR keys, and the unbound group has a total of 892 distinct TSR keys. It is worth pointing out that it does not mean that every local structure in the bound group will have such 1379 distinct keys. This argument will be applied to the unbound group. A total of 360 distinct TSR keys are shared by the bound and unbound groups (*see* Fig. 3f). Therefore, 1019 distinct TSR keys were exclusively found in the bound group, whereas 532 distinct TSR keys were found only in the unbound group (Python Code 11) (*see* Fig. 3f). One uniqueness of the TSR algorithm is that the TSR keys can be mapped back to the geometries of three or more amino

Fig. 3 (continued) of the bound states (cAMPDKs with an ADNTP) and unbound states (cAMPDKs without an ADNTP) are labeled. The local cAMPDK structures are represented by 14 amino acids (G52, V57, A70, K72, V104, E121, V123, E127, K168, E170, N171, L173, T183, and D184). (**c**) Protein sequence alignment of representative cAMPDKs using the SnapGene MUSCLE algorithm shows the deletions in some of cAMPDKs. cAMPDKs without the deletion are correlated with the larger cluster from the hierarchical clustering analysis of global cAMPDK structures, and cAMPDKs with the deletion are correlated with the smaller cluster from the hierarchical clustering analysis of global cAMPDK structures. (**d**) Pairwise structural similarities for local (the ADNTP binding sites) and global cAMPDK structures were calculated and are presented. The numbers of local (the ADNTP binding sites) and global cAMPDK structures and average values are labeled. The error bars represent SDs. (**e**) MDS analysis result of the binding sites (ADP, ANPA, ANPB, and ATP; 14 amino acids) of cAMPDKs is presented. The numbers of kinase regulators, ADP, ANPA, ANPB, and ATP, are labeled. (**f**) Structural relationships of local conformations (14 amino acids) of the bound and unbound classes of cAMPDKs are illustrated by a Venn diagram. Percentages for each section were calculated and are shown. (**g–h**) MaxDist (**g**) and Theta (**h**) values of two triangles (K72-E121-T183 and V57-E127-N171) of both bound and unbound local cAMPDK structures were calculated and are presented. *** means that a p value <0.001. The average values are shown, and SDs represented by error bars are labeled. The numbers of the bound and unbound states are labeled. (**i**) A representative bound triangle K72-E121-T183 is illustrated. (**j**) A representative bound triangle V57-E127-N171 is illustrated. (**k**) A representative unbound triangle K72-E121-T183 is illustrated. (**l**) A representative unbound triangle V57-E127-N171 is illustrated. (**i–j**) Kinase regulators and their numbers are labeled. (**i–l**) PDB identifiers, amino acids, chains, and amino acid positions are labeled acids. Two triangles (K72-E121-T183 and V57-E127-N171) of the cAMPDKs were identified as a signature substructure for distinguishing the bound forms from the unbound forms because the two triangles have different MaxDist (*see* Fig. 3g) and Theta (*see* Fig. 3h) values between the bound and unbound forms. One representative triangle (K72-E121-T183) of the bound form is illustrated in Fig. 3i, and one representative triangle (V57-E127-N171) of the bound form is shown in Fig. 3j. Two representative triangles of the unbound form are shown in Fig. 3k for K72-E121-T183 and Fig. 3l for V57-E127-N171. In conclusion, the TSR keys or their corresponding triangles can be used for quantifying the conformations of both bound and unbound forms as well as for distinguishing the bound structures from the unbound structures or vice versa.

2.3 Dataset Preparation and Annotation for Studying Conformational Changes of Amino Acids of Kinases

The use of the C_α TSR algorithm as a measure of structural differences makes it difficult to judge the extent of local movements taking place because side chains, especially bulky side chains, contribute a significant percentage of the intermolecular contacts. Experimentally determined 3D molecular structures deposited in the PDB have been the source for studies of the side-chain conformations in proteins in general because of an encouraging growth factor of the PDB [63]. Studies suggested that there is a recognition code because a significant bias in amino acid composition was found among interface residues [55]. The hot spots are particularly enriched in tyrosine, tryptophan, and arginine [64]. For example, tyrosine residues are exceptionally versatile for mediating contacts at interfaces [65]. This bias found in natural biomolecular interfaces can enable accurate prediction of natural binding partners and provide guidelines for designing and engineering new proteins for specific functions that do not exist in nature [65].

2.3.1 Dataset Preparation and Annotation of Glutamate and Asparagine Residues

To study the conformations of amino acids, we identified the amino acids of cAMPDKs that closely interact with ADP, ANP, and ATP. The top abundant three amino acids are glutamate, lysine, and aspartate (*see* Fig. 4a). All three amino acids are charged under physiological conditions. Glutamate residues were chosen as representative amino acids for studying their roles in molecular recognition because they are the most abundant residues in the binding sites for all three kinase regulators (ADP, ANP, and ATP). A total of 878 glutamate residues were found from the cAMPDK structures that were examples of conformational studies using the C_α TSR algorithm discussed in Subheading 2.2. These 878 glutamate residues were annotated into three groups (Supplementary File 4): the ADNTP group where glutamate residues closely interact with either an ADP, an ANP, or an ATP, the phosphoserine (SEP)

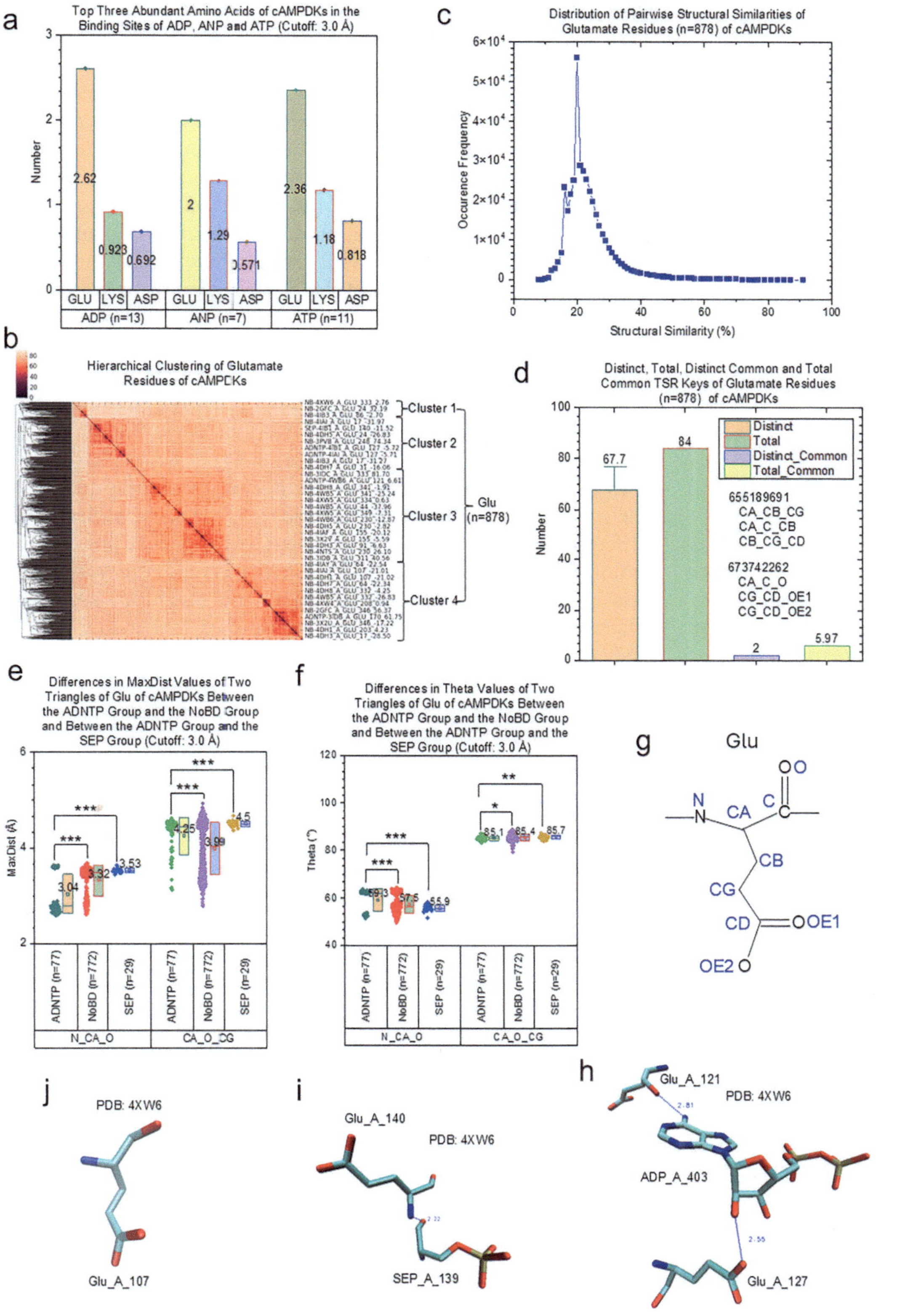

Fig. 4 Glutamate (Glu) residues in the binding sites of cAMPDK regulators have their specific and characteristic conformations that are different from those not in the binding sites. (**a**) The occurrences of 20 amino acids in the binding sites of cAMPDK regulators, ADP, ANP, and ATP, were calculated. The top three abundant amino

group where glutamate residues closely interact with a phosphoserine (SEP), and the no binding (NoBD) group where glutamate residues do not closely interact with either a kinase regulator (ADP, ANP, or ATP) or a SEP. Since a significant percentage of cAMPDKs were phosphorylated at the serine sites, we also examined the SEP binding sites of cAMPDKs. The most abundant residue in the SEP binding sites of cAMPDKs is asparagine. A total of 568 asparagine residues were identified from the cAMPDK structures. These 568 asparagine residues can be divided into two main groups (Supplementary File 5): the SEP group where asparagine residues closely interact with a SEP and the NoSEP group where asparagine residues do not closely interact with a SEP. The SEP group can be further categorized into two subgroups: the N1134 group where Asn113 residues or Asn114 residues interact with a SEP and the N3401 group where Asn340 residues or Asn341 residues interact with a SEP.

2.3.2 Hierarchical Clustering Showing the Diversity of Conformations of Glutamate Residues and Structure–Function Relationships of Glutamate Residues of cAMPDKs

The TSR concept was used to develop an algorithm for quantifying the structures of different amino acids and the same amino acids at different positions. All atoms except hydrogen atoms of every amino acid were used for TSR key generation. Each type of atom was assigned an integer. The bin boundaries used for MaxDist and Theta were the same as those we discussed in Subheading 2.1. The TSR algorithm formulates MaxDist, Theta, and three atoms of a

Fig. 4 (continued) acids for each type of regulators are presented. The average values are shown, and SDs represented by error bars are labeled. The numbers of each type of kinase regulator are labeled. The cutoff value for the close binding is labeled. (**b**) The hierarchical clustering result of glutamate residues from the cAMPDK structures using the amino acid TSR keys is illustrated. The major clusters and the total number of glutamate residues are labeled. (**c**) The distribution of pairwise structural similarities of glutamate residues from cAMPDK structures is illustrated. The total number of glutamate residues is indicated. (**d**) Distinct, total, distinct common, and total common TSR keys were calculated and are presented for glutamate residues. The total number of glutamate residues and the average values are labeled. SDs represented by error bars are indicated. The two distinct common TSR keys and their corresponding triangles are shown. (**e–f**) MaxDist (**e**) and Theta (**f**) values of two triangles (N-CA-O and CA-O-CG) of glutamate residues were calculated and are presented. * means that a p value <0.05 (**f**). ** means that a p value <0.01 (**f**). *** means that a p value <0.001 (**e–f**). The average values are shown. SDs represented by error bars and cutoff value for a binding site are labeled. The numbers of glutamate residues in the kinase regulator binding sites, in the phosphoserine (SEP) binding sites, not in the kinase regulator binding sites and not in the SEP binding sites are labeled. (**g**) A two-dimensional (2D) structure of glutamate residue is shown. The atom names are labeled. (**h**) A representative Glu121, a representative Glu127, and a representative ADP are illustrated. The hydrogen bond interactions between ADP and Glu121 and between ADP and Glu127 are shown with dash lines. ADP, chain, and its number are labeled. (**i**) A representative Glu140 and a representative SEP are illustrated. The hydrogen bond interaction between Glu140 and SEP is shown using a dashed line. SEP, chain, and its number are labeled. (**j**) A representative glutamate residue that does not interact with a kinase regulator or a SEP is illustrated. PDB identifiers, amino acids, chains, and amino acid positions are labeled. (**h–i**) PDB identifiers and chains for amino acids are labeled triangle of an amino acid into an integer (a TSR key) (Python Code 12). Similar to the all-atom and C_α TSR algorithms, the generalized Jaccard coefficient measure [51] was used for calculating pairwise similarities between any two amino acids in the datasets [5]. The hierarchical clustering of amino acids is visualized based on the average linkage clustering [52]. Structural images were prepared using the VMD package [53].

The hierarchical clustering shows relationships of 878 glutamate conformations (*see* Fig. 4b). Four major clusters were observed, although their boundaries are not clear (*see* Fig. 4b). No two glutamate residues have identical conformations (*see* Fig. 4b), and the highest pairwise structural similarity is 90.9%. Nearly all glutamate residues interacting with a SEP were clustered together (Supplementary File 6), suggesting that those glutamate residues have specific conformation characteristics in common. Some, but not all, glutamate residues interacting with an ADNTP are clustered together (Supplementary File 6). The highest peak of pairwise structural similarities is ~20%, indicating a high diversity of glutamate conformations (*see* Fig. 4c). To further reveal conformational relationships of glutamate residues, we calculated distinct, total, distinct *common*, and total *common* TSR keys for 878 glutamate residues from the selected cAMPDK structures. The definitions of distinct, total, distinct *common*, and total *common* TSR keys for kinase regulators discussed in Subheading 2.1 are the same for amino acids (e.g., glutamate and asparagine). All 878 glutamate residues share two distinct *common* keys (655189691 and 673742262) (*see* Fig. 4d), indicating that the triangles of all the glutamates associated with those two keys have similar geometries. Three possible triangles correspond to each of the two keys (655189691: CA-CB-CG, CA-C-CB, and CB-CG-CD; 673742262: CA-C-O, CG-CD-OE1, and CG-CD-OE2) (*see* Fig. 4d). To further characterize the relationships of the conformations, we investigated specific triangles for the glutamate residues that interact with an ADNTP. Two specific triangles (N-CA-O and CA-O-CG) were identified as a signature substructure for the glutamate residues that interact with an ADNTP because they have different MaxDist (*see* Fig. 4e) and Theta (*see* Fig. 4f) values between the ADNTP group and the NoBD group and between the ADNTP group and the SEP group. A 2D glutamate structure and its atom names are shown in Fig. 4g. Representative examples are shown in Fig. 4h for two glutamate residues interacting with an ADP, Fig. 4i for one glutamate interacting with a SEP, and Fig. 4j for one glutamate not interacting with a kinase regulator nor with a SEP. Glu121 and Glu127 hydrogen-bond to an ADP (*see* Fig. 4h), and Glu140 hydrogen-bonds to a SEP (*see* Fig. 4i). In conclusion, the relationships of glutamate conformations are established through the identification of common and specific triangles that can also be used for interpreting the clustering results.

2.3.3 Hierarchical Clustering Showing the Diversity of Conformations of Asparagine Residues and Structure–Function Relationships of Asparagine Residues of cAMPDKs

Studies showed that certain types of amino acids are frequently found in the interfaces in molecular recognition sites. The TSR algorithm was able to demonstrate the specific signature geometries for the glutamate residues that recognize and interact with a kinase regulator. The objective of this section is to test whether the asparagine residues interacting with a SEP have specific and characteristic conformations. To achieve this objective, we have performed similar analyses for asparagine as for glutamate. The hierarchical cluster analysis shows two major clusters (*see* Fig. 5a). Some, but not all, asparagine residues interacting with a SEP are clustered together (Supplementary File 7). The highest peak of pairwise structural similarities is ~17%, indicating a high diversity of asparagine conformations, and only two asparagine residues have identical conformations (100% structural similarity) (*see* Fig. 5b). All 568 asparagine residues share two distinct *common* keys (655189691 and 673742262) (*see* Fig. 5c), indicating that the triangles of all the asparagines associated with those two keys have similar geometries. Two possible triangles correspond to each of the two keys (655189691: CA-CB-CG and CA-C-CB; 673742262: CA-C-O and CB-CG-OD1) (*see* Fig. 5c). One specific triangle (CA-O-CG) was identified as a signature substructure for the asparagine residues that interact with a SEP because they have different MaxDist values between the SEP group and the NoSEP group (*see* Fig. 5d). One specific triangle (C-O-CB) was identified as a signature substructure for Asn113 or Asn114 that interacts with a SEP because they have different MaxDist (*see* Fig. 5d) and Theta (*see* Fig. 5e) values between the N1134 subgroup and the NoSEP group and between the N1134 subgroup and the N3401 subgroup. A 2D asparagine structure and its atom names are illustrated in Fig. 5f. A representative example of the asparagine structures interacting with a SEP is shown in Fig. 5g. Asn113 hydrogen-bonds to SEP338, and Asn340 also hydrogen-bonds to SEP338 (*see* Fig. 5g). A representative asparagine that does not interact with a SEP is shown in Fig. 5h. Similar to the analysis of glutamate conformations, we conclude that the relationships of asparagine conformations are established through the identification of common and specific triangles that can also be used for interpreting the clustering results.

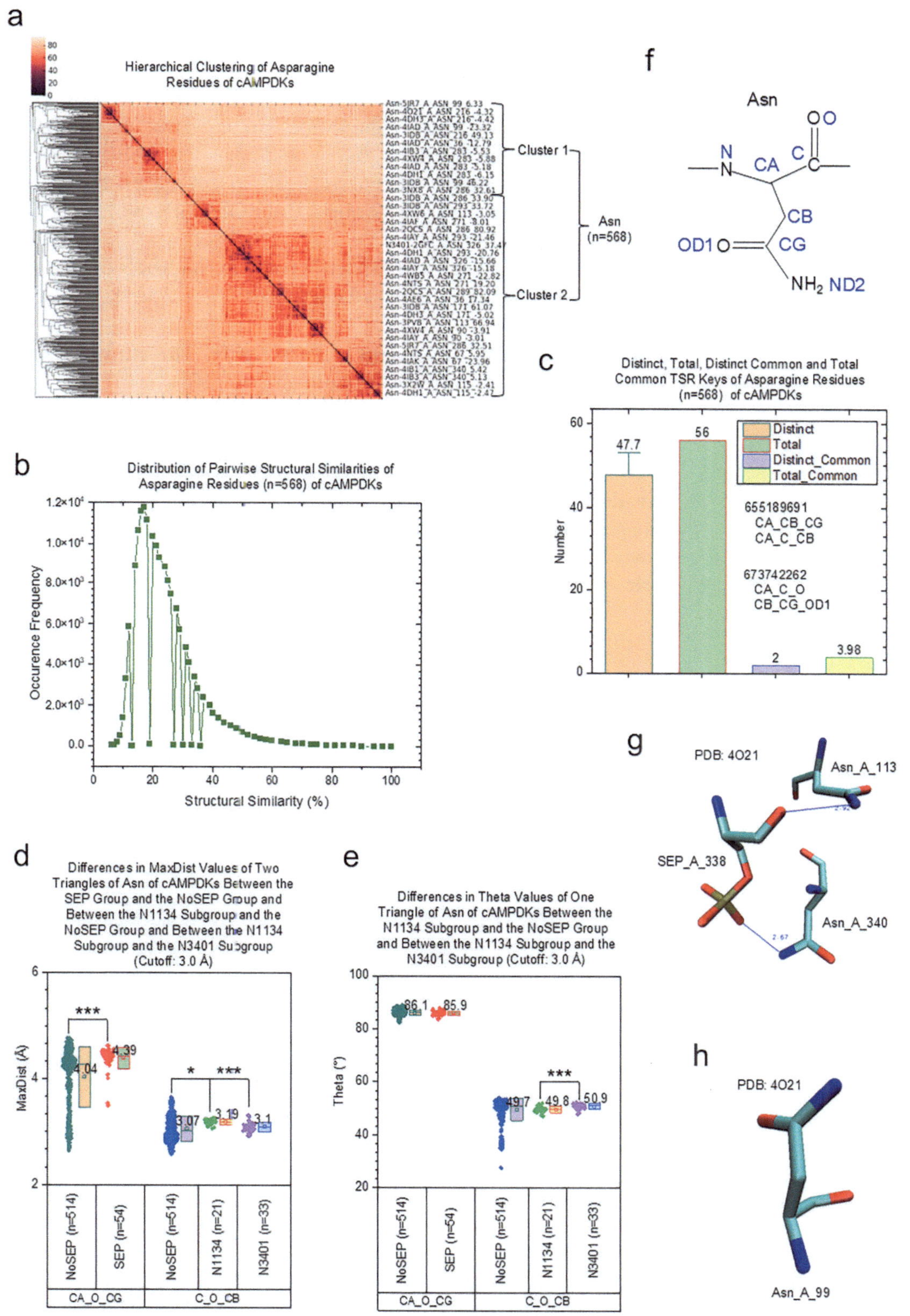

Fig. 5 Asparagine (Asn) residues in the SEP binding sites of cAMPDKs have their specific and characteristic conformations that are different from those not in the binding sites. (**a**) The hierarchical clustering result of asparagine residues from the cAMPDK structures using the amino acid TSR keys is illustrated. The major

Acknowledgments

Most of this research was conducted with high-performance computational resources provided by the Louisiana Optical Network Infrastructure (http://www.loni.org). Shahd A. Muthana, Josee E. Robinson, Lauren E. Savoy, and Sophia R. LeBlanc helped to prepare the datasets. Here, we want to appreciate the LONI support team, especially Jianxiong Li and Oleg Starovoytov.

Declarations The authors explicitly confirm that the work is original and has not been published elsewhere.

Ethics Approval and Consent to Participate This study does not involve human subjects and/or animals.

Consent for Publication Not applicable.

Availability of Data and Materials These data were derived from the following resource available in the public domain: https://www.rcsb.org/. The supplementary files and source codes are available for academic users on GitHub.

Competing Interests The authors declare no conflict of interest.

Authors' Contributions WX proposed and designed this study. TIM, PK, TAB, and WX carried out the study, collected data, and prepared the figures. TIM, PK, and FC wrote the Python code. VR and WX supervised students. KDK contributed to the discussion. All authors wrote the manuscript, read it, and revised it.

Fig. 5 (continued) clusters and the total number of asparagine residues are labeled. (**b**) The distribution of pairwise structural similarities of asparagine residues from cAMPDK structures is illustrated. The total number of asparagine residues is indicated. (**c**) Distinct, total, distinct common, and total common TSR keys were calculated and are presented for asparagine residues. The total number of asparagine residues and the average values are labeled. SDs represented by error bars are indicated. The two distinct common TSR keys and their corresponding triangles are shown. (**d–e**) MaxDist (**d**) and Theta (**e**) values of two triangles (CA-O-CG and C-O-CB) of asparagine residues were calculated and are presented. * means that a p value <0.05 (**d**). *** means that a p value <0.001 (**d–e**). The average values are shown. SDs represented by error bars and cutoff value for a binding site are labeled. The numbers of asparagine residues in the SEP binding sites and not in the SEP binding sites are labeled. The numbers of Asn113 or Asn114 and Asn340 or Asn341 are also labeled. (**f**) A 2D structure of asparagine residue is shown. The atom names are labeled. (**g**) A representative Asn113, a representative Asn340, and a representative SEP338 are illustrated. The hydrogen bond interactions between Asn113 and SEP338 and between Asn340 and SEP338 are shown with dash lines. SEP and chain are labeled. (**h**) A representative asparagine residue that does not interact with a SEP is illustrated. (**g–h**) PDBs and chains for amino acids are labeled

Funding

This study is supported by NIH NIGMS (1R15GM144944-01).

References

1. Kahn K, Plaxco KW (2010) Principles of biomolecular recognition. In: Zourob M (ed) Recognition receptors in biosensors. Springer, New York, pp 3–45
2. Gellman SH (1997) Introduction: molecular recognition. Chem Rev 97:1231–1232. https://doi.org/10.1021/cr970328j
3. Chu WT, Yan Z, Chu X, Zheng X, Liu Z, Xu L, Zhang K, Wang J (2021) Physics of biomolecular recognition and conformational dynamics. Rep Prog Phys 84. https://doi.org/10.1088/1361-6633/ac3800
4. Baker ZD, Rasmussen DM, Levinson NM (2024) Exploring the conformational landscapes of protein kinases: perspectives from FRET and DEER. Biochem Soc Trans 52: 1071–1083. https://doi.org/10.1042/bst20230558
5. Kondra S, Sarkar T, Raghavan V, Xu W (2021) Development of a TSR-based method for protein 3-D structural comparison with its applications to protein classification and motif discovery. Front Chem 8. https://doi.org/10.3389/fchem.2020.602291
6. Sarkar T, Raghavan VV, Chen F, Riley A, Zhou S, Xu W (2021) Exploring the effectiveness of the TSR-based protein 3-D structural comparison method for protein clustering, and structural motif identification and discovery of protein kinases, hydrolase, and SARS-CoV-2's protein via the application of amino acid grouping. Comput Biol Chem:107479. https://doi.org/10.1016/j.compbiolchem.2021.107479
7. Kondra S, Chen F, Chen Y, Chen Y, Collette CJ, Xu W (2022) A study of a hierarchical structure of proteins and ligand binding sites of receptors using the triangular spatial relationship-based structure comparison method and development of a size-filtering feature designed for comparing different sizes of protein structures. Proteins 90:239–257. https://doi.org/10.1002/prot.26215
8. Sarkar T, Chen Y, Wang Y, Chen Y, Chen F, Reaux CR, Moore LE, Raghavan V, Xu W (2023) Introducing mirror-image discrimination capability to the TSR-based method for capturing stereo geometry and understanding hierarchical structure relationships of protein receptor family. Comput Biol Chem 103: 107824. https://doi.org/10.1016/j.compbiolchem.2023.107824
9. Chen F, Milon IT, Khajouie P, Myers A, Xu W (2024) A parallel implementation for large-scale TSR-based 3D structural comparisons of protein and amino acid. Curr Bioinforma 19: 1–16. https://doi.org/10.2174/0115748936306625240724102438
10. Milon TI, Rauniyar K, Furman S, Orthi KH, Wang Y, Raghavan V, Xu W (2024) Representing and quantifying conformational changes of kinases and phosphatases using the TSR-based algorithm. Kinases Phosphatases 2:315–339
11. Sarkar T, Reaux CR, Li J, Raghavan VV, Xu W (2022) The specific applications of the TSR-based method in identifying Zn2+ binding sites of proteases and ACE/ACE2. Data Brief 45:108629. https://doi.org/10.1016/j.dib.2022.108629
12. Luscombe NM, Laskowski RA, Thornton JM (2001) Amino acid–base interactions: a three-dimensional analysis of protein–DNA interactions at an atomic level. Nucleic Acids Res 29: 2860–2874. https://doi.org/10.1093/nar/29.13.2860
13. Armstrong CT, Mason PE, Anderson JLR, Dempsey CE (2016) Arginine side chain interactions and the role of arginine as a gating charge carrier in voltage sensitive ion channels. Sci Rep 6:21759. https://doi.org/10.1038/srep21759
14. Vaitheeswaran S, Thirumalai D (2008) Interactions between amino acid side chains in cylindrical hydrophobic nanopores with applications to peptide stability. Proc Natl Acad Sci 105:17636–17641. https://doi.org/10.1073/pnas.0803990105
15. Spassov VZ, Yan L, Flook PK (2007) The dominant role of side-chain backbone interactions in structural realization of amino acid code. ChiRotor: a side-chain prediction algorithm based on side-chain backbone interactions. Protein Sci 16:494–506. https://doi.org/10.1110/ps.062447107
16. Baker CM, Grant GH (2007) Role of aromatic amino acids in protein-nucleic acid recognition. Biopolymers 85:456–470. https://doi.org/10.1002/bip.20682
17. Madhusudan Makwana K, Mahalakshmi R (2015) Implications of aromatic-aromatic

interactions: from protein structures to peptide models. Protein Sci 24:1920–1933. https://doi.org/10.1002/pro.2814

18. Diarra dit Konté N, Krepl M, Damberger FF, Ripin N, Duss O, Šponer J, FHT A (2017) Aromatic side-chain conformational switch on the surface of the RNA recognition motif enables RNA discrimination. Nat Commun 8:654. https://doi.org/10.1038/s41467-017-00631-3
19. Milon TI, Wang Y, Fontenot RL, Khajouie P, Villinger F, Raghavan V, Xu W (2024) Development of a novel representation of drug 3D structures and enhancement of the TSR-based method for probing drug and target interactions. Comput Biol Chem 112:108117. https://doi.org/10.1016/j.compbiolchem.2024.108117
20. Milon TI, Sarkar T, Chen Y, Grider JM, Chen F, Ji J-Y, Jois SD, Kousoulas KG, Raghavan V, Xu W (2025) Development of the TSR-based computational method to investigate spike and monoclonal antibody interactions. Front Chem 13. https://doi.org/10.3389/fchem.2025.1395374
21. Luo L, Daigle J, Chen Y, Duan X, Wang Y, Chistoserdov A, Wang Y, Xu W (2023) Structural and functional studies of a eukaryotic type Ser/Thr kinase, Slr0599, of Synechocystis sp. PCC 6803 using a combination of experimental and computational approaches. In: Hou HJM, Allakhverdiev SI (eds) Photosynthesis from plants to nanomaterials. Academic Press
22. Xu W, Xie X-J, Faust A, Liu M, Li X, Chen F, Naquin A, Walton A, Kishbaugh P, Ji J-Y (2020) All-atomic molecular dynamic studies of human and drosophila CDK8: insights into their kinase domains, the LXXLL motifs, and drug binding site. Int J Mol Sci 21
23. Luo L, Milon TI, Tandoh EK, Galdamez WJ, Chistoserdov AY, Yu J, Kern J, Wang Y, Xu W (2025) Development of a TSR-based method for understanding structural relationships of cofactors and local environments in photosystem I. BMC Bioinformatics 26:15. https://doi.org/10.1186/s12859-025-06038-y
24. Reaux CR, Meche SA, Grider JM, Dhanabal S, Milon TI, Chen F, Xu W (2025) Design of a TSR-based project learning strategy for biochemistry undergraduate teaching and research labs: a case study. Front Educ 9. https://doi.org/10.3389/feduc.2024.1455173
25. Kanev GK, de Graaf C, de Esch IJP, Leurs R, Würdinger T, Westerman BA, Kooistra AJ (2019) The landscape of atypical and eukaryotic protein kinases. Trends Pharmacol Sci 40:818–832. https://doi.org/10.1016/j.tips.2019.09.002
26. Hunter T (1995) Protein kinases and phosphatases: the Yin and Yang of protein phosphorylation and signaling. Cell 80:225–236. https://doi.org/10.1016/0092-8674(95)90405-0
27. Matte A, Tari LW, Delbaere LTJ (1998) How do kinases transfer phosphoryl groups? Structure 6:413–419. https://doi.org/10.1016/S0969-2126(98)00043-4
28. Burnett G, Kennedy EP (1954) The enzymatic phosphorylation of proteins. J Biol Chem 211:969–980. https://doi.org/10.1016/S0021-9258(18)71184-8
29. Pawson T, Scott JD (2005) Protein phosphorylation in signaling – 50 years and counting. Trends Biochem Sci 30:286–290. https://doi.org/10.1016/j.tibs.2005.04.013
30. Ware TB, Hsu K-L (2021) Advances in chemical proteomic evaluation of lipid kinases – DAG kinases as a case study. Curr Opin Chem Biol 65:101–108. https://doi.org/10.1016/j.cbpa.2021.06.007
31. Tonks NK (2006) Protein tyrosine phosphatases: from genes, to function, to disease. Nat Rev Mol Cell Biol 7:833–846. https://doi.org/10.1038/nrm2039
32. Schwartz PA, Murray BW (2011) Protein kinase biochemistry and drug discovery. Bioorg Chem 39:192–210. https://doi.org/10.1016/j.bioorg.2011.07.004
33. Hanahan D, Weinberg RA (2011) Hallmarks of cancer: the next generation. Cell 144:646–674. https://doi.org/10.1016/j.cell.2011.02.013
34. Yesilkanal AE, Johnson GL, Ramos AF, Rosner MR (2021) New strategies for targeting kinase networks in cancer. J Biol Chem 297:101128. https://doi.org/10.1016/j.jbc.2021.101128
35. Knight ZA, Lin H, Shokat KM (2010) Targeting the cancer kinome through polypharmacology. Nat Rev Cancer 10:130–137. https://doi.org/10.1038/nrc2787
36. Bhullar KS, Lagarón NO, McGowan EM, Parmar I, Jha A, Hubbard BP, Rupasinghe HPV (2018) Kinase-targeted cancer therapies: progress, challenges and future directions. Mol Cancer 17:48. https://doi.org/10.1186/s12943-018-0804-2
37. Sato S, Sanjo H, Takeda K, Ninomiya-Tsuji J, Yamamoto M, Kawai T, Matsumoto K, Takeuchi O, Akira S (2005) Essential function for the kinase TAK1 in innate and adaptive immune responses. Nat Immunol 6:1087–1095. https://doi.org/10.1038/ni1255
38. Mueller BK, Mack H, Teusch N (2005) Rho kinase, a promising drug target for

neurological disorders. Nat Rev Drug Discov 4:387–398. https://doi.org/10.1038/nrd1719

39. Chong ZZ, Shang YC, Wang S, Maiese K (2012) A critical kinase cascade in neurological disorders: PI 3-K, Akt, and mTOR. Future Neurol 7:733–748. https://doi.org/10.2217/fnl.12.72

40. Tabit CE, Shenouda SM, Holbrook M, Fetterman JL, Kiani S, Frame AA, Kluge MA, Held A, Dohadwala MM, Gokce N, Farb MG, Rosenzweig J, Ruderman N, Vita JA, Hamburg NM (2013) Protein kinase C-β contributes to impaired endothelial insulin signaling in humans with diabetes mellitus. Circulation 127:86–95. https://doi.org/10.1161/circulationaha.112.127514

41. Zhang J, Yang PL, Gray NS (2009) Targeting cancer with small molecule kinase inhibitors. Nat Rev Cancer 9:28–39. https://doi.org/10.1038/nrc2559

42. Kantarjian H, Sawyers C, Hochhaus A, Guilhot F, Schiffer C, Gambacorti-Passerini C, Niederwieser D, Resta D, Capdeville R, Zoellner U, Talpaz M, Druker B, Goldman J, O'Brien SG, Russell N, Fischer T, Ottmann O, Cony-Makhoul P, Facon T, Stone R, Miller C, Tallman M, Brown R, Schuster M, Loughran T, Gratwohl A, Mandelli F, Saglio G, Lazzarino M, Russo D, Baccarani M, Morra E (2002) Hematologic and cytogenetic responses to imatinib mesylate in chronic myelogenous leukemia. N Engl J Med 346:645–652. https://doi.org/10.1056/NEJMoa011573

43. Shaw AT, Kim DW, Mehra R, Tan DS, Felip E, Chow LQ, Camidge DR, Vansteenkiste J, Sharma S, De Pas T, Riely GJ, Solomon BJ, Wolf J, Thomas M, Schuler M, Liu G, Santoro A, Lau YY, Goldwasser M, Boral AL, Engelman JA (2014) Ceritinib in ALK-rearranged non-small-cell lung cancer. N Engl J Med 370:1189–1197. https://doi.org/10.1056/NEJMoa1311107

44. Chapman PB, Hauschild A, Robert C, Haanen JB, Ascierto P, Larkin J, Dummer R, Garbe C, Testori A, Maio M, Hogg D, Lorigan P, Lebbe C, Jouary T, Schadendorf D, Ribas A, O'Day SJ, Sosman JA, Kirkwood JM, Eggermont AM, Dreno B, Nolop K, Li J, Nelson B, Hou J, Lee RJ, Flaherty KT, McArthur GA (2011) Improved survival with vemurafenib in melanoma with BRAF V600E mutation. N Engl J Med 364:2507–2516. https://doi.org/10.1056/NEJMoa1103782

45. Arteaga CL, Engelman JA (2014) ERBB receptors: from oncogene discovery to basic science to mechanism-based cancer therapeutics. Cancer Cell 25:282–303. https://doi.org/10.1016/j.ccr.2014.02.025

46. Shibuya M, Suzuki Y, Sugita K, Saito I, Sasaki T, Takakura K, Nagata I, Kikuchi H, Takemae T, Hidaka H et al (1992) Effect of AT877 on cerebral vasospasm after aneurysmal subarachnoid hemorrhage. Results of a prospective placebo-controlled double-blind trial. J Neurosurg 76:571–577. https://doi.org/10.3171/jns.1992.76.4.0571

47. Gross S, Rahal R, Stransky N, Lengauer C, Hoeflich KP (2015) Targeting cancer with kinase inhibitors. J Clin Invest 125:1780–1789. https://doi.org/10.1172/jci76094

48. Garraway LA, Lander ES (2013) Lessons from the cancer genome. Cell 153:17–37. https://doi.org/10.1016/j.cell.2013.03.002

49. Fabbro D, Cowan-Jacob SW, Moebitz H (2015) Ten things you should know about protein kinases: IUPHAR review 14. Br J Pharmacol 172:2675–2700. https://doi.org/10.1111/bph.13096

50. Bernstein FC, Koetzle TF, Williams GJ, Meyer EF Jr, Brice MD, Rodgers JR, Kennard O, Shimanouchi T, Tasumi M (1977) The protein data bank: a computer-based archival file for macromolecular structures. J Mol Biol 112:535–542. https://doi.org/10.1016/s0022-2836(77)80200-3

51. Jaccard P (1901) Etude comparative de la distribution florale dans une portion des Alpes et des Jura. Bull Soc Vaudoise Sci Nat 37:547–579

52. Ackerman M, Ben-David S (2016) A characterization of linkage-based hierarchical clustering. J Mach Learn Res 17:8182–8198

53. Humphrey W, Dalke A, Schulten K (1996) VMD: visual molecular dynamics. J Mol Graph 14:33–38. https://doi.org/10.1016/0263-7855(96)00018-5

54. Ruvinsky AM, Kirys T, Tuzikov AV, Vakser IA (2011) Side-chain conformational changes upon protein-protein association. J Mol Biol 408:356–365. https://doi.org/10.1016/j.jmb.2011.02.030

55. Lo Conte L, Chothia C, Janin J (1999) The atomic structure of protein-protein recognition sites. J Mol Biol 285:2177–2198. https://doi.org/10.1006/jmbi.1998.2439

56. Betts MJ, Sternberg MJ (1999) An analysis of conformational changes on protein-protein association: implications for predictive docking. Protein Eng 12:271–283. https://doi.org/10.1093/protein/12.4.271

57. Guharoy M, Janin J, Robert CH (2010) Side-chain rotamer transitions at protein-protein

interfaces. Proteins 78:3219–3225. https://doi.org/10.1002/prot.22821

58. Davis IW, Arendall WB 3rd, Richardson DC, Richardson JS (2006) The backrub motion: how protein backbone shrugs when a sidechain dances. Structure 14:265–274. https://doi.org/10.1016/j.str.2005.10.007
59. Bhardwaj N, Gerstein M (2009) Relating protein conformational changes to packing efficiency and disorder. Protein Sci 18:1230–1240. https://doi.org/10.1002/pro.132
60. Taylor SS, Buechler JA, Yonemoto W (1990) cAMP-dependent protein kinase: framework for a diverse family of regulatory enzymes. Annu Rev Biochem 59:971–1005. https://doi.org/10.1146/annurev.bi.59.070190.004543
61. Knippschild U, Gocht A, Wolff S, Huber N, Löhler J, Stöter M (2005) The casein kinase 1 family: participation in multiple cellular processes in eukaryotes. Cell Signal 17:675–689. https://doi.org/10.1016/j.cellsig.2004.12.011
62. Kruskal JB, Wish M (1978) Multidimensional scaling. Thousand Oaks
63. Berman H, Henrick K, Nakamura H (2003) Announcing the worldwide protein data Bank. Nat Struct Biol 10:980. https://doi.org/10.1038/nsb1203-980
64. Bogan AA, Thorn KS (1998) Anatomy of hot spots in protein interfaces. J Mol Biol 280:1–9. https://doi.org/10.1006/jmbi.1998.1843
65. Koide S, Sidhu SS (2009) The importance of being tyrosine: lessons in molecular recognition from minimalist synthetic binding proteins. ACS Chem Biol 4:325–334. https://doi.org/10.1021/cb800314v

Chapter 13

Identifying Functional Water Encapsulation Sites Within the Transmembrane Domain of Sensor Kinases

Rahmi Yusuf, Robert J. Lawrence, Fadhael Alrahman Hasan, Peltine A. Ndifon, Joy Jun Yan 雋恩 Yau 邱, Monabel May Birao, Maetrese Beley, George Baker-Valla, and Roger R. Draheim

Abstract

Although many decades of research have been conducted on bacterial sensor kinases, their mechanism of transmembrane signal transduction remains unclear. Here, we have compiled a set of methodologies that we believe would be beneficial to any research group starting studies on their sensor kinase of interest. Alternatively, these methods would be beneficial to those monitoring evolution of sensor kinases or potentially to those engineering sensor kinases for a particular function. To accomplish this, we began by comparing the previously determined orientation of the transmembrane domain (TMD) in native bacterial membranes to a structure generated in silico after atomistic molecular dynamics (MD) simulation. This comparison showed that the in silico structure was an accurate representation of the EnvZ TMD in native bacterial membranes. Due to this similarity, we mapped previously existing data that assessed the change in EnvZ signal output from a set of ~60 Cys substitutions onto the structure generated in silico and found that water is present near where the sensor kinase becomes activated by Cys mutations. For brevity, we solely focus on functional water encapsulation within the TMD of EnvZ, but other parameters such as individual helical position, tilt, or rotation and their role in signal transduction could also be assessed. We hope that other research groups find this workflow useful and decide to employ it to assess their sensor kinases of interest.

Key words Sensor kinase, Atomistic molecular dynamics simulation, Transmembrane water encapsulation, Transmembrane communication, EnvZ

1 Introduction

Sensor kinases are the predominant signaling system found in bacteria, and they have been extensively studied for decades. However, compared to other membrane-spanning protein complexes, namely G protein-coupled receptors (GPCRs) [1, 2] and multimeric ion channels [3], little is known about how they coordinate water within their transmembrane domain (TMD) and whether

Shahid M. Khan and Florencio Pazos (eds.), *Protein Evolution: Methods and Protocols*, Methods in Molecular Biology, vol. 2979, https://doi.org/10.1007/978-1-0716-4828-5_13,

that water plays a functional role in transmembrane communication.

Although limited, some experimental results support the role of water in transmembrane signaling by sensor kinases. *Escherichia coli* PhoQ was shown to require a hydrophilic residue within its TMD [4], leading to the proposal that it possesses a cytoplasmic-facing water-filled hemichannel. In addition, the high-resolution crystal structure of a TMD-containing *E. coli* NarQ fragment was shown to possess water within its TMD. Finally, a preprint examining atomistic molecular dynamics (MD) simulations of a membrane-embedded *E. coli* PhoQ has shown a high level of hydration within the TMD, and it has been suggested that this represents an intermediate signaling state [5]. Although direct analysis of water within sensor kinase TMDs might be lacking, several decades of other experimental results from sensor kinases, and the related bacterial chemoreceptors, exist in great volume.

Here, we provide an overview of using computational methods to generate structures of sensor kinases and to use preexisting data to confirm their appropriateness. More precisely, we employ preexisting sulfhydryl-reactivity data from *E. coli* EnvZ to confirm the structure of the transmembrane domain from EnvZ generated in silico. We then employ this structure in atomistic molecular dynamics simulations to monitor water dynamics within the TMD, leading to the identification of a TM1-TM1′ water encapsulation pocket. We then superimpose preexisting signal output data from point mutations across the TMD onto the simulated structures, which suggests that the proposed cytoplasmic-facing water-filled hemichannel might exist, along with two other potential TM1-TM2′ water encapsulation pockets. These results provide a detailed example of generating new computational data and comparing it to preexisting datasets. Unfortunately, it also provides an example of understanding the limitations of computational experimentation and how to decide whether to invest significant resources to identify low-probability but functionally important events. Although these studies monitoring water dynamics within sensor kinases are in their infancy, we hope that other groups find them useful and start looking at water dynamics within their sensor kinase of interest.

2 Materials

Here, our focus is on the analysis of computational datasets; therefore, only the required hardware is listed below. These datasets will be compared to preexisting results generated within intact bacteria, and those materials are not described here but can be found within the citations provided.

2.1 Required Computational Resources

1. A graphics processing unit (GPU)-enabled workstation.
2. Access to a GPU-containing high-performance computing (HPC) cluster or supercomputer.

3 Methods

3.1 Introduction

Below, the focus is on identifying and monitoring the position and dynamics of water within the TMD of a canonical sensor kinase, EnvZ from *Escherichia coli*. However, before that was possible, previous results from a few well-established protocols were required, which will be referenced and noted here but not described in detail. Initially, a library of ~60 single-Cys-containing EnvZ variants that composed the EnvZ TMD was made [6]. These were then used for sulfhydryl-reactivity experimentation that determined the orientation of the helices within the TMD while present in intact *E. coli* cells [7, 8]. The primary sequence of EnvZ was then subjected to AlphaFold v2.3.0 multimer [9, 10] (*see* **Note 1**), and the resultant protein coordinates were used to build a protein–membrane system ($EnvZ_2$/G-IM) composed of a homodimeric EnvZ and a model gram-negative inner membrane [11] with Membrane Builder [12–15] and CHARMM-GUI [16]. Three replicates of the $EnvZ_2$/G-IM system (*see* Fig. 1) were equilibrated and simulated with nanoscale molecular dynamics (NAMD) [17], which, after examination with MDAnalysis [18], resulted in quantification of inter-residue distance between analogous positions within the homodimeric TMD, i.e., Ser^{11} and $Ser^{11'}$. Finally, signal output from the library of single-Cys-containing EnvZ variants within intact bacterial cells was assessed [7, 8]. Possessing results from these different methods beforehand was necessary to confirm

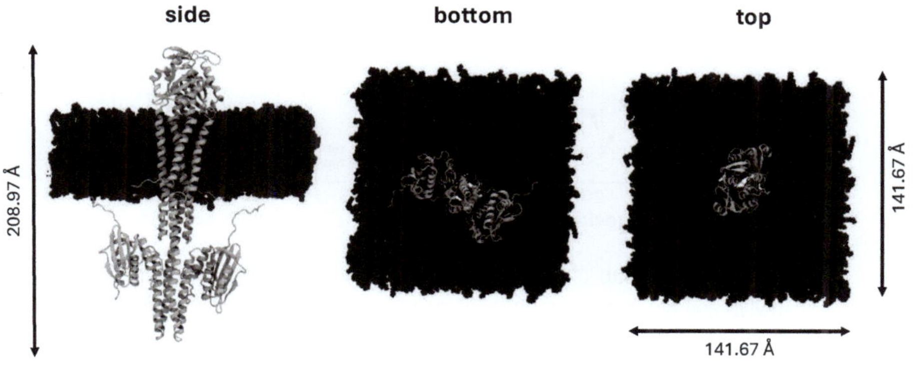

leaflet	peri	cyto
PMPE	141	138
POPE	39	39
QMPE	36	36
OYPE	24	24
PMPG	30	30
PYPG	27	27
$PVCL_2$	6	6

other	
H_2O	102149
K^+	415
Cl^-	281

Fig. 1 Composition and dimensions of the $EnvZ_2$/G-IM system. The dimensions of the system are provided in three panels (side view indicating the perspective as viewed from the side, bottom view indicating the perspective as viewed from the cytoplasm, and top view indicating the perspective as viewed from the periplasm). The composition of the periplasmic and cytoplasmic leaflets and the other components of the system are also provided

the validity of the computational output and to identify potentially functional water encapsulation within the EnvZ TMD.

3.2 Comparing TMD Structure in Bacterial Membranes to Simulated TMD Structures

1. Densitometrically calculate the ratio of dimeric to monomeric EnvZ species after in vivo sulfhydryl-reactivity analysis as described previously in [7, 8] (*see* upper panels of Fig. 2).
2. Using the *mdanalysis.lib.mdamath.norm* mathematical helper function of MDAnalysis [18], measure the interatomic distance between the β-carbons from analogous homodimeric positions, i.e., Ser^{11} and $Ser^{11'}$, over the residue range present in the single-Cys-containing library. If a Gly residue is present, such as Gly^{173} and Gly^{174} in *E. coli* EnvZ, measure the distance between the α-carbons. Complete this distance analysis for every frame of the simulations performed and plot the average distance and standard deviation for all replicates (*see* lower panels of Fig. 2).
3. To confirm that the simulated TMDs are an accurate representation of the TMD structure in living cells, localized, i.e., within one helical turn, maxima in sulfhydryl reactivity should correlate with localized minima in inter-residue distance observed in the simulations. These correlations are highlighted by the red dashed lines in Fig. 2 and indicate that the experimental results from intact bacterial cells can be overlaid onto the EnvZ TMD structures generated in silico.

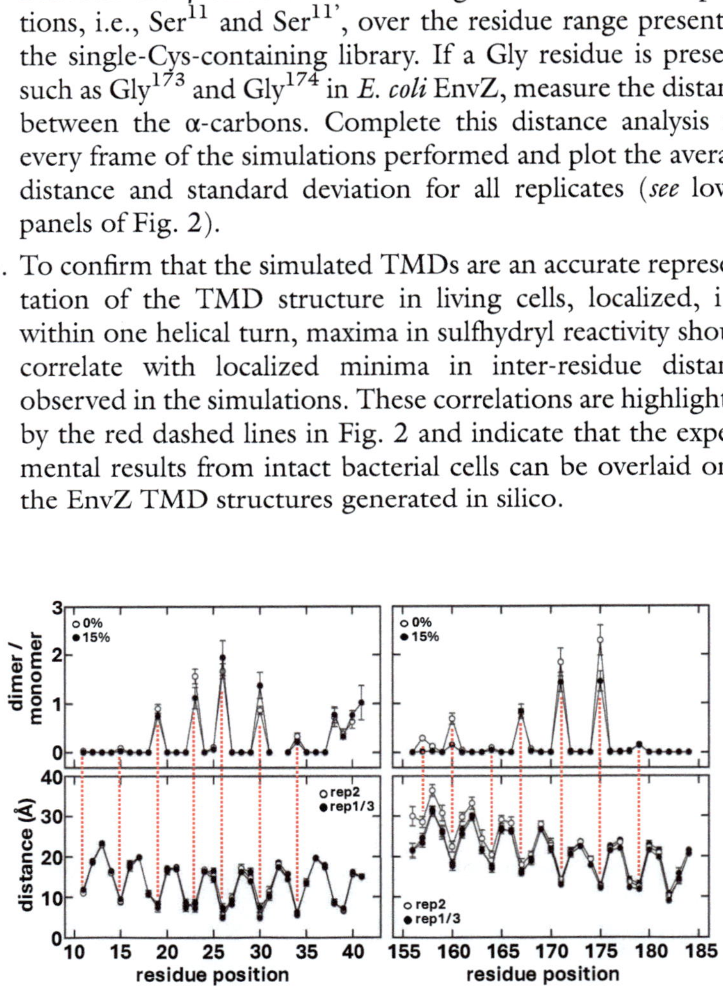

Fig. 2 Comparison of inter-residue distances by sulfhydryl-reactivity analysis in vivo and within the three replicates of the $EnvZ_2$/G-IM simulation. (top) Dimer–monomer ratio estimates inter-residue distances between the side chains within an EnvZ homodimer, i.e., Cys^{11} and $Cys^{11'}$. The TM1-TM1′ and TM2-TM2′ data were previously available [7, 8]. (bottom) The average inter-residue distance between the same side chains within the EnvZ homodimer is shown from three replicate simulations of the same $EnvZ_2$/G-IM system. These orthogonal methods show (red dashed lines) that localized crosslinking maxima correlate with the localized distance minima in the simulations

3.3 Monitoring Water Encapsulation Within the Atomistic MD Simulations

1. Open visual molecular dynamics (VMD) [19] and load the relevant system and trajectory data.
2. To quickly analyze the trajectories in a visual manner, create two new graphical representations. The first is a selection of "all not water" with a "NewCartoon" style that will display only the homodimeric sensor kinase in a cartoon style, and the second, a section of "water within 5 of segid PROA PROB and (resid 11–41 or resid 156–184)" with a style of "VDW" will highlight any water within 5 Å of the residues selected to encompass the TMD, i.e., 11–41 and 156–184.
3. With these graphical representations enabled, watch the trajectories in a frame-by-frame manner to determine whether any water is observed within the TMD. From the three replicates performed here, water was only observed within the TMD of the second replicate and was absent in the TMD of the first and third replicates.
4. Should any water be observed, a quantitative visualization can be achieved with the VolMap graphical user interface (GUI) plugin present in VMD under Extensions → Analysis → VolMap tool. Before performing this analysis, the trajectory needs to be aligned under Extensions → Analysis → RSMD Trajectory Tool → Align.
5. The VolMap GUI plugin can be employed to create a volumetric map of water present within the TMD of the homodimeric sensor kinase. To accomplish this, employ the same selection used during the initial visualization, namely "water within 5 of segid PROA PROB and (resid 11–41 or resid 156–184)." The density VolMap type should be selected, along with the default 1.0 Å resolution, 1.0x radius atom size, and mass-based weighting. Select the "compute for all frames option, and combine using: avg" and append to molecule: "same."
6. The result can be visualized, and in the case of the second replicate a single water density can be volumetrically shown as existing between residues Ser^{26}, Thr^{30}, Ser$^{26'}$, and Thr$^{30'}$ (*see* green in Fig. 3). Using these default settings, a similar density representing water within the TMD was not observed in the first and third replicates.

3.4 Superimposition of Signal Output from the Single-Cys-Containing Variants onto the Simulated TMD Structure

1. Baseline signal output from EnvZ, measured in intact *E. coli* cells as the ratio of cyan fluorescent protein (CFP) to yellow fluorescent protein (YFP) fluorescence, is 0.3 as previously shown in [7, 8]. Single-Cys-containing EnvZ moieties that have greater than 100% signal output over baseline, i.e., a CFP/YFP of ≥0.6, were previously assigned to the elevated category, while those resulting in less than 75% signal output of baseline, i.e., CFP/YFP ≤ 0.23, were previously assigned as

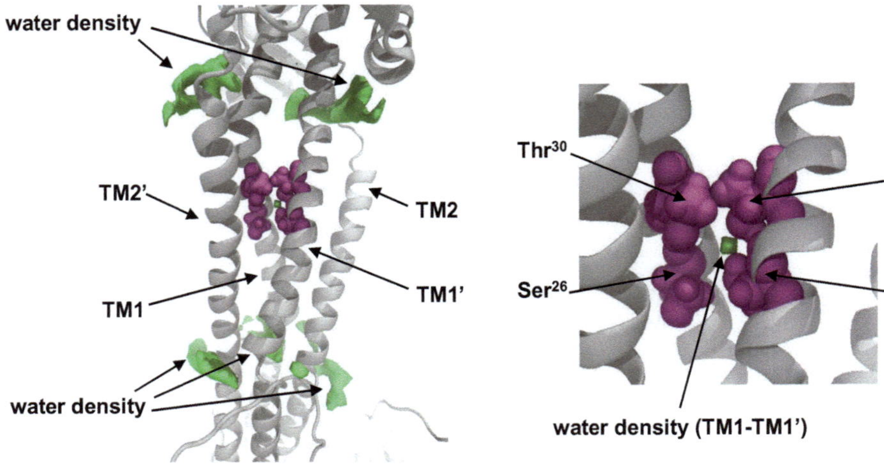

Fig. 3 Overlay of VolMap output over the $EnvZ_2$ structure from the second replicate. (left) The VolMap results are shown in green and represent water that is found near the cytoplasmic and periplasmic transmembrane boundaries. In addition, one density is found flanked by Ser^{26} and Thr^{30} (purple) from each EnvZ monomer. (right) A magnified view of the four residues (purple) flanking the single water density (green) found within the transmembrane domain

possessing reduced signal output. The final frame of the second replicate was visualized with VMD, and residue positions where a Cys substitution resulted in elevated or reduced signal output were subjected to a graphical representation with a "Surf" style and a green or red color, respectively (*see* Fig. 4a).

2. These elevated and reduced residue positions were then compared to areas that have been observed to encapsulate water via VolMap analysis in Subheading 3.3, **step 6**. However, when the visual inspection was performed in Subheading 3.3, **step 2**, a few frames were detected where a water molecule was in close proximity to $Thr^{164'}$, thereby forming a TM1-TM2′ encapsulation pocket with residue Thr^{30} (*see* Fig. 5). Although observed visually, the events were infrequent and did not appear in the default density-averaged VolMap result. In addition, Cys substitutions at positions 23, 27, 164, 167, and 168 are shown to elevate signal output from EnvZ. Taken together, the discrepancy in these results from different methodologies suggests the possibility of two additional TM1-TM2′ water encapsulation pockets that warrant further examination with more intense computational analysis (*see* Fig. 5). Also note that based on signal output analysis, a cytoplasmic-facing water-filled hemichannel is also possibly present. Again, water can be observed within this cavity when examined visually as performed in Subheading 3.3, **step 2** and again might be a relatively low-probability event that will require additional computational analysis (*see* Subheading 3.5).

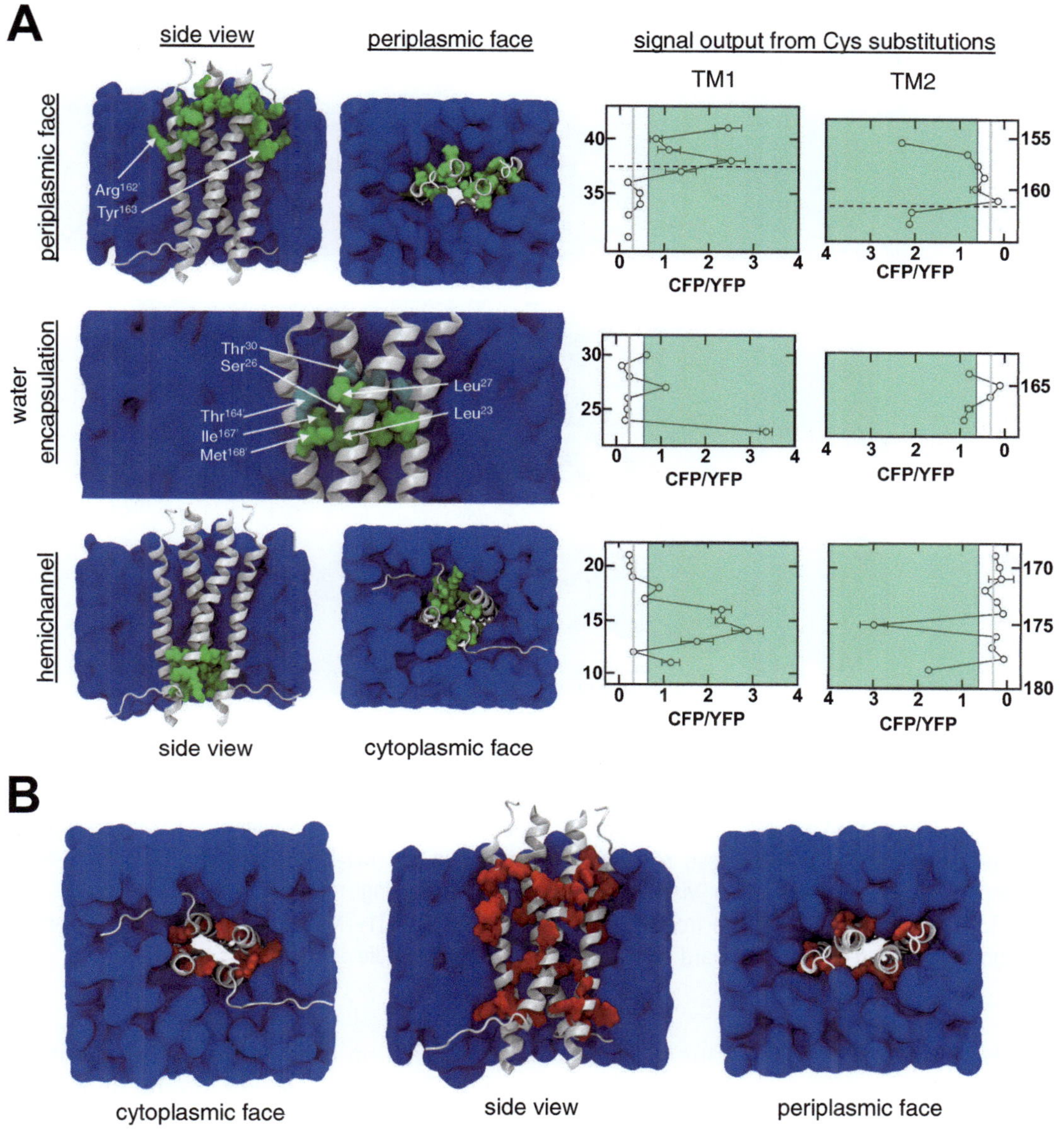

Fig. 4 Facilitation of water encapsulation governs EnvZ signal output. (**a**) Alterations of the three water coordination motifs modulate signal output. Residue positions where Cys substitutions activate signal output twofold above baseline (CFP/YFP $\geq$ 0.6) are highlighted in green. Three motifs have been identified where this occurs: when the Cys residues face into the cytoplasmic-facing water-filled hemichannel (lower third), when the water encapsulation pockets are expanded (middle third), or when one of two outward-facing members of the periplasmic anchoring belt are substituted (upper third). (**b**) Residue positions where a Cys substitution reduces signal output to below 75% of baseline output are highlighted in red. Please note that none of these side chains face into the transmembrane bundle core

3. This methodology also showed that all Cys substitutions that reduce EnvZ signal output below 75% of baseline output face outward from the helical core of the EnvZ TMD (*see* Fig. 4b).

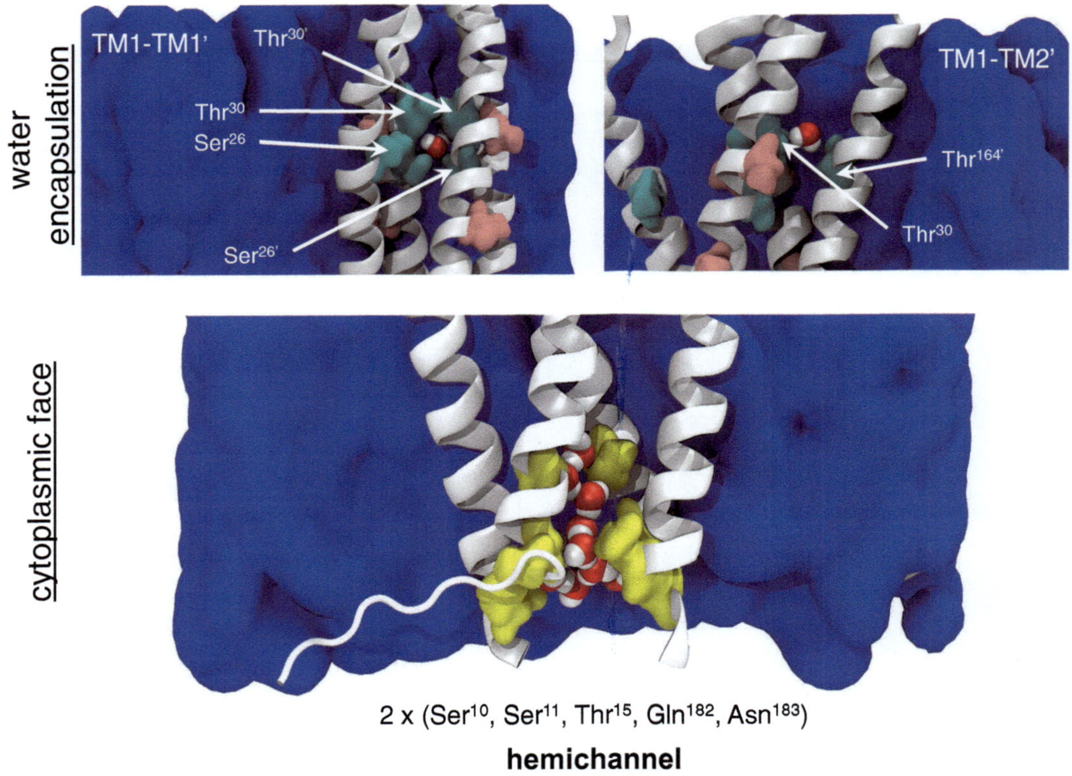

Fig. 5 Water coordination motifs identified visually within the EnvZ transmembrane domain (TMD). At the cytoplasmic face of the EnvZ TMD, a water-filled hemichannel composed of ten polar residues was visualized (yellow). Water within 5 Å of these residues is also shown. A TM1-TM1′ and two TM1-TM2′ water encapsulation pockets were also visualized, and the participating residues are shown (cyan). The TM1-TM1′ pocket was much more frequently occupied than the TM1-TM2′ pockets. Other polar residues are highlighted (pink) but face outward from the core TM1-TM1′ bundle

3.5 Potential Further Computational Optimizations

1. Due to the visual determination of water in the proximity of Thr-164′ in Subheading 3.3, **step 2** (*see* Fig. 5) and its absence in the default VolMap analysis (*see* Subheading 3.3, **step 6** and Fig. 3) coupled with changes in signal output that might support TM1-TM2′ water encapsulation pockets, a more intensive set of simulations should be performed. Also lending credence to further simulation is the faint appearance of a water density representing the potential cytoplasmic-facing water-filled hemichannel in the VolMap analysis (*see* Fig. 3). In both cases, the signal output analysis coupled with the visual detection of water molecules supports further simulation.
2. For further simulations, the "dcdfreq" parameter, or the number of steps occurring before writing a subsequent set of coordinates to the trajectory file, will be reduced from the CHARMM-GUI default, which is 50,000 frames, to

10,000 frames. This change facilitates capturing a greater number of events with smaller molecules, i.e., water, at the expense of larger file size (*see* **Note 2**).

3. In addition, the overall length of the simulations will be extended; however, this will dramatically increase the time frame required to complete the computational analysis and the overall file size, which also need to be considered (*see* **Note 3**).
4. Based on the signal output analysis of single-Cys-containing receptors and the potential water encapsulation motifs present, the TM1-TM1′ water encapsulation pocket, the potential TM1-TM2′ water encapsulation pockets, and the water-filled hemichannel, other $EnvZ_2$/G-IM systems should be created in CHARMM-GUI that possess Cys mutations at one of the following positions: 15, 23, 27, 168, and 175. This will allow for potential differences in water densities provided by the VolMap GUI plugin to be assessed.
5. Overall, it is critical to assess the number, length, and size of the simulations required to capture potentially infrequent events, or events with small molecules, such as water encapsulation. Furthermore, comparing these requirements to the computational resources available will guide the direction of the project toward streamlining computational requirements or toward gaining access to national-level HPC facilities (*see* **Note 4**).

4 Notes

1. AlphaFold 3 [20] is now available at https://alphafoldserver.com.
2. The $EnvZ_2$/G-IM system produced is approximately 400,000 atoms. Using the default "dcdfreq" parameter of 50,000 resulted in a 10 GB file size for 200 ns of simulation. Reducing the "dcdfreq" parameter to 10,000 will result in files that are five times the size, i.e., 25 GB per 100 ns of simulation. Therefore, local storage and backup should be considered before proceeding.
3. Extending the duration of the simulation to 1 μs while reducing the "dcdfreq" results in a single trajectory that is approximately 250 GB. Thus, considering the infrastructure requirements, in terms of TBs of data to be transferred and stored, is critical during the decision-making process about workflow.
4. In addition to file size, it remains important to consider required time to complete a simulation. With our $EnvZ_2$/G-IM system (400,000 atoms), a high-spec workstation with two

Nvidia RTX A4000 video cards can complete 12–14 ns of a single simulation per day. The local HPC (SCIAMA HPC at the University of Portsmouth) possesses 10 Nvidia A100 GPUs. Using a full node with GPU offloading and NAMD 3.0.1 allows completion of approximately 30 ns of simulation per day per node available. Thus, completing 1 μs of simulation for a single replicate requires a full node being occupied continuously for approximately 30 days.

Acknowledgments

R.Y. was generously supported by the Indonesia Endowment Fund for Education, Ministry of Finance (S-4833/LPDP.3/2015). R.R.D. was supported with start-up funding from the Faculty of Science and from the Institute of Biomedical and Biomolecular Science (IBBS) at the University of Portsmouth. Numerical computations were carried out on the SCIAMA high-performance computing (HPC) cluster which is supported by the ICG and the University of Portsmouth.

References

1. Lesca E, Panneels V, Schertler GFX (2018) The role of water molecules in phototransduction of retinal proteins and G protein-coupled receptors. Faraday Discuss 207:27–37. https://doi.org/10.1039/C7FD00207F
2. Venkatakrishnan AJ, Ma AK, Fonseca R et al (2019) Diverse GPCRs exhibit conserved water networks for stabilization and activation. Proc Natl Acad Sci 116:3288–3293. https://doi.org/10.1073/pnas.1809251116
3. Paulino J, Yi M, Hung I et al (2020) Functional stability of water wire–carbonyl interactions in an ion channel. Proc Natl Acad Sci 117: 11908–11915. https://doi.org/10.1073/pnas.2001083117
4. Goldberg SD, Clinthorne GD, Goulian M, DeGrado WF (2010) Transmembrane polar interactions are required for signaling in the Escherichia coli sensor kinase PhoQ. Proc Natl Acad Sci 107:8141–8146. https://doi.org/10.1073/pnas.1003166107
5. Lazaridi S, Yuan J, Lemmin T (2024) Atomic insights into the signaling landscape of E. coli PhoQ histidine kinase from molecular dynamics simulations. bioRxiv 2024.04.19.590235. https://doi.org/10.1101/2024.04.19.590235
6. Yusuf R, Lawrence RJ, Eke LV, Draheim RR (2018) Tuning chemoreceptor signaling by positioning aromatic residues at the lipid–aqueous interface. In: Manson MD (ed) Bacterial chemosensing: methods and protocols. Springer, New York, pp 147–158
7. Heininger A, Yusuf R, Lawrence RJ, Draheim RR (2016) Identification of transmembrane helix 1 (TM1) surfaces important for EnvZ dimerisation and signal output. Biochim Biophys Acta 1858:1868–1875. https://doi.org/10.1016/j.bbamem.2016.05.002
8. Yusuf R, Yau JJY, Baker-Valla GW et al (2025) EnvZ signal output correlates with capacity for transmembrane water encapsulation. bioRxiv 2025.03.25.645265. https://doi.org/10.1101/2025.03.25.645265
9. Jumper J, Evans R, Pritzel A et al (2021) Highly accurate protein structure prediction with AlphaFold. Nature 596:583–589. https://doi.org/10.1038/s41586-021-03819-2
10. Evans R, O'Neill M, Pritzel A et al (2021) Protein complex prediction with AlphaFold-Multimer. bioRxiv 2021.10.04.463034. https://doi.org/10.1101/2021.10.04.463034
11. Pogozheva ID, Armstrong GA, Kong L et al (2022) Comparative molecular dynamics simulation studies of realistic eukaryotic, prokaryotic, and archaeal membranes. J Chem Inf

Model 62:1036–1051. https://doi.org/10.1021/acs.jcim.1c01514

12. Li Y, Liu J, Gumbart JC (2021) Preparing membrane proteins for simulation using CHARMM-GUI. In: Schmidt-Krey I, Gumbart JC (eds) Structure and function of membrane proteins. Springer, New York, pp 237–251
13. Jo S, Kim T, Im W (2007) Automated builder and database of protein/membrane complexes for molecular dynamics simulations. PLoS One 2:e880
14. Wu EL, Cheng X, Jo S et al (2014) CHARMM-GUI membrane builder toward realistic biological membrane simulations. J Comput Chem 35:1997–2004. https://doi.org/10.1002/jcc.23702
15. Feng S, Park S, Choi YK, Im W (2023) CHARMM-GUI membrane builder: past, current, and future developments and applications. J Chem Theory Comput 19:2161–2185. https://doi.org/10.1021/acs.jctc.2c01246
16. Jo S, Kim T, Iyer VG, Im W (2008) CHARMM-GUI: a web-based graphical user interface for CHARMM. J Comput Chem 29:1859–1865. https://doi.org/10.1002/jcc.20945
17. Phillips JC, Hardy DJ, Maia JDC et al (2020) Scalable molecular dynamics on CPU and GPU architectures with NAMD. J Chem Phys 153:044130. https://doi.org/10.1063/5.0014475
18. Michaud-Agrawal N, Denning EJ, Woolf TB, Beckstein O (2011) MDAnalysis: a toolkit for the analysis of molecular dynamics simulations. J Comput Chem 32:2319–2327. https://doi.org/10.1002/jcc.21787
19. Humphrey W, Dalke A, Schulten K (1996) VMD: visual molecular dynamics. J Mol Graph 14:33–38. https://doi.org/10.1016/0263-7855(96)00018-5
20. Abramson J, Adler J, Dunger J et al (2024) Accurate structure prediction of biomolecular interactions with AlphaFold 3. Nature 630:493–500. https://doi.org/10.1038/s41586-024-07487-w

Chapter 14

Protein–Protein Interaction Site Prediction via EquiPPIS and Its Application in Studying Viral Replication

Rahmatullah Roche, Sumit Tarafder, Parkesh Suseendran, Xiaofeng Wang, and Debswapna Bhattacharya

Abstract

Understanding the binding sites within proteins is vital for determining their functional properties including the host–virus interactions. Therefore, to meet the pressing demand for rapid identification of these binding sites, fast computational prediction techniques have become essential. The recently developed method EquiPPIS enables accurate and scalable prediction of generalized, partner-independent protein–protein interaction sites for a wide range of proteins and thus can deepen our understanding of viral replication. The predicted interaction sites within these proteins can guide us how to disrupt these viral replications, offering rapid antiviral strategies that are applicable to multiple (+)RNA viruses. Here, we present the detailed procedures of protein–protein interaction site prediction by EquiPPIS and its application in studying viral replication.

Key words Protein–protein interactions, Protein–protein interaction site prediction, Viral replication, Deep learning

1 Introduction

The recent predictive modeling utilizing deep learning techniques enables the prediction of protein structures and their interactions in the absence of experimental structural information. AlphaFold2 [1] has shown groundbreaking performance in predicting accurate three-dimensional (3D) structure of proteins from the corresponding amino acid sequences. In addition to AlphaFold2, multiple methods including RoseTTAFold [2], ESMFold [3], and OmegaFold [4] have been emerged for accurate protein structure prediction. To facilitate the complicated and computationally expensive structural prediction process accessible to a wide range of scientific community, ColabFold [5] is developed integrating multiple protein structure prediction methods and by utilizing a fast and effective multiple sequence alignment-based method

Shahid M. Khan and Florencio Pazos (eds.), *Protein Evolution: Methods and Protocols*, Methods in Molecular Biology, vol. 2979,
https://doi.org/10.1007/978-1-0716-4828-5_14,

MMSeq2 [6]. Therefore, utilizing the ColabFold pipeline is an uncomplicated way to predict fast and accurate protein 3D structures.

Determining protein–protein interactions (PPIs) is crucial for our understanding of numerous biological processes in living organisms [7–13]. Multiple methods have been developed over the last decades for the computational prediction of partner-independent protein–protein interaction interfaces and their interaction sites. Among those methods, sequence-based prediction methods [14–20] predict the protein–protein interaction (PPI) sites from their amino acid sequences. The predictions from these methods are usually less accurate as those do not use structural information from proteins. On the other hand, the structure-aware methods [21–25] are designed to predict the PPI site with higher accuracy by utilizing the protein structural information in addition to sequence information. However, the structure-aware methods rely on experimentally determined protein monomers and extract structural information from their bound conformations, limiting the practicality of the PPI site prediction approach. The experimental structural information unbound conformation of protein can be used with a tradeoff of lower prediction accuracy. Even the known unbound protein structures are sparse, hindering the promise of large-scale PPI site predictions. Given the recent progress in protein structure prediction by AlphaFold, is it possible to utilize predicted structures replacing the experimentally determined structures to accurately predict PPI site for any target protein? EquiPPIS [26], a recent deep learning-based method, is suitable for predicting the PPI sites effectively utilizing predicted protein structures. EquiPPIS employs a symmetry-aware graph neural network [27] that accounts for equivariance in terms of reflections, rotations, and translations. The equivariance in terms of reflection is particularly fitting when dealing with predicted protein structures, which are noisy to some extent, and can contain molecular chirality in them. Using a predicted protein structure as an input, the EquiPPIS pipeline first converts it into an undirected protein residue–residue graph. Then, the graph along with additional sequence- and structure-based input node and edge features is passed into a series of E(3)-equivariant convolutions [27]. The Cα coordinates are directly employed by the network to ensure equivariance in the graph learning. Finally, by considering PPI site prediction as a node classification task, the pipeline predicts the probability of each residue being in a PPI site. The promisingly accurate computational prediction of protein structures and PPI sites can significantly reduce the experimental efforts for determining the possible viral interaction location in proteins.

Here, we present the detailed procedures we follow to predict the residue-level protein–protein interactions sites within brome mosaic virus (BMV) and reticulons (RTNs). First, we predict the protein 3D structures from their sequences by AlphaFold2 using ColabFold [5] pipeline. Then, we employ the protein structure predictions along with multiple sequence-based features generated by freely available tools and databases to predict the residue-level PPI sites using EquiPPIS. Afterward, the predicted PPI sites along with the residue-level self-assessment confidence scores in the predicted structures direct laboratory experiments in viral disease managements for the two target proteins.

2 Materials

1. Protein Data Bank (PDB) database [28] for obtaining protein amino acid sequence located at https://www.rcsb.org/.
2. PSIBLAST program [29] obtained from https://ftp.ncbi.nlm.nih.gov/blast/executables/blast+/LATEST/ and Uniref90 sequence database [30] obtained from https://ftp.uniprot.org/pub/databases/uniprot/uniref/uniref90/uniref90.fasta.gz for generating position-specific scoring metrics (PSSM).
3. ColabFold [5] obtained from https://github.com/sokrypton/ColabFold for protein structure prediction from the sequence.
4. DSSP [31] obtained from https://swift.cmbi.umcn.nl/gv/dssp/.
5. ESM-2 [3] obtained from https://github.com/facebookresearch/esm for generating protein language model (pLM)-based features.
6. EquiPPIS [26] obtained from https://github.com/Bhattacharya-Lab/EquiPPIS for predicting PPI sites.
7. Computational system with graphics processing unit (GPU) and capable of "conda" installation.

3 Methods

Predicting PPI Sites from Amino Acid Sequences

Follow the procedure to predict PPI sites using EquiPPIS [26]:

3.1 Clone and then install EquiPPIS from this GitHub location https://github.com/Bhattacharya-Lab/EquiPPIS

3.1.1 Run following commands

```
'conda env create -f EquiPPIS_environment.yml'
'conda activate EquiPPIS'
```

3.1.2 After successful installation, to see usage instruction run

```
'python EquiPPIS/EquiPPIS.py -h'
```

3.2 Input feature generation

3.2.1. To generate PSSM features from sequences of BMV1a and Yeast_Rtn1 proteins, install PSIBLAST at some '</path_to_psiblast/>' and extract the UniRef90 database at some '</path_to_uniref/>'. Place the sequences (FASTA format) files `'BMV1a.fasta'` (which can be obtained from https://www.ncbi.nlm.nih.gov/protein/CAA26228.1?report=fasta&log$=seqview&format=text) and `'Yeast_Rtn1.fasta'` (which can be obtained from https://rest.uniprot.org/uniprotkb/Q04947.fasta) inside `'EquiPPIS/Preprocessing/input/'`. Run `'psiblast'` program as follows:

```
'</path_to_psiblast/>/ncbi-blast-2.13.0+/bin/psiblast
-db </path_to_uniref/>/uniref90/uniref90.fasta -num_itera-
tions 3 -num_alignments 1 -query EquiPPIS/Preprocessing/in-
put/BMV1a.fasta -out EquiPPIS/Preprocessing/input/BMV1a.bla
-out_ascii_pssm EquiPPIS/Preprocessing/input/BMV1a.pssm
-num_threads 32'
```

```
'</path_to_psiblast/>/ncbi-blast-2.13.0+/bin/psiblast
-db </path_to_uniref/>/uniref90/uniref90.fasta -num_itera-
tions 3 -num_alignments 1 -query EquiPPIS/Preprocessing/in-
put/Yeast_Rtn1.fasta -out EquiPPIS/Preprocessing/input/
Yeast_Rtn1.bla -out_ascii_pssm EquiPPIS/Preprocessing/input/
Yeast_Rtn1.pssm -num_threads 32'
```

3.2.2. To generate predicted structures for BMV1a and Yeast Rtn1 proteins from their sequences, locally install ColabFold at some '</path_to_colabfold/>' from and run following command (*see* **Notes 1** and **2**):

```
python </path_to_colabfold/>/ColabFold/colabfold/
batch.py --num-models 1 --num-recycle 1 --recompile-padding
1 EquiPPIS/Preprocessing/input/BMV1a.fasta BMV1a/
```

```
python </path_to_colabfold/>/ColabFold/colabfold/
batch.py --num-models 1 --num-recycle 1 --recompile-padding
1 EquiPPIS/Preprocessing/input/Yeast_Rtn1.fasta Yeast_Rtn1/
```

Now, copy the predicted rank1 models as `EquiPPIS/Preprocessing/input/BMV1a.pdb` and `EquiPPIS/Preprocessing/input/Yeast_Rtn1.pdb`

3.2.3. To generate DSSP files from the predicted structures, first download precompiled DSSP from here https://swift.cmbi.umcn.nl/gv/dssp/DSSP_5.html and run following commands:

```
./dsspcmbi -i EquiPPIS/Preprocessing/input/Yeast_Rtn1.pdb -o EquiPPIS/Preprocessing/input/Yeast_Rtn1.dssp
```

```
./dsspcmbi -i EquiPPIS/Preprocessing/input/BMV1a.pdb -o EquiPPIS/Preprocessing/input/BMV1a.dssp
```

3.2.4. To generate pretrained protein language model (pLM)-based features, download and install ESM-2 at some '</path_to_esm/>' from here https://github.com/facebookresearch/esm and copy this feature generation script `'EquiPPIS/Preprocessing/supporting_scripts/ESM 2_features.py'` inside '</path_to_esm/>esm/esm/'. Then, run following commands:

```
python /path_to_esm/>esm/esm/esm/esm2_to_33featnpy.py -i EquiPPIS/Preprocessing/input/BMV1a.fasta -o EquiPPIS/Preprocessing/input/BMV1a.esm2_33
```

```
python /path_to_esm/>esm/esm/esm/esm2_to_33-featnpy.py -i EquiPPIS/Preprocessing/input/Yeast_Rtn1.fasta -o EquiPPIS/Preprocessing/input/Yeast_Rtn1.esm2_33
```

3.2.5. Generate distance maps `'BMV1a.dist'` and `'Yeast_Rtn1.dist'` from the predicted structures using any publicly available scripts such as https://github.com/multicom-toolbox/CONFOLD2/blob/master/confold-v2.0/suppl-scripts/pdb2rr.pl, and place the distance maps inside `'EquiPPIS/distmaps/'`. The first two columns indicate the residue pairs, the second two columns indicate the lower and upper bound distances (0 and 14 angstroms), and the final columns indicate the real-valued distances between the corresponding residue pairs.

3.3 Preprocess input features:

Put the target names `'BMV1a'` and `'Yeast_Rtn1'` inside `'input.list'`. Then, run following instructions:

3.3.1 Go to preprocessing directory:

```
cd EquiPPIS/Preprocessing/
```

3.3.2 Generate structure-based features:

```
python gen_aa_structural_features.py -t input.list
```

3.3.3 Generate PSSM-based features:

```
python genpssmto20feat.py -i input -o tmp/ -t input.list
```

3.3.4 Generate node features:

```
python gen_preprocessed_node_features.py -t input.list
```

3.4. Run EquiPPIS to predict protein–protein interaction sites for BMV1a and Yeast Rtn1 by running following commands (*see* **Notes 3–5**):

Go to current directory:

```
cd –
```

Make output directory:

```
mkdir output/
```

Run EquiPPIS for protein–protein interaction site prediction:

```
python EquiPPIS.py --model_state_dict Trained_model/
EquiPPIS_model/E-110-256.pt --indir Preprocessing/ --outdir
output/
```

Now, the residue-wise protein–protein interaction site predictions are generated as follows:

```
'output/BMV1a.out' and 'output/Yeast_Rtn1.out'
```

Implications in Studying Viral Replication

Brome mosaic virus (BMV) is a positive-strand RNA [(+)RNA] virus that typically infects monocot plants such as barley, rice, and wheat [32]. Viruses in the class of (+)RNA viruses include many important pathogens of humans, animals, and plants, such as severe acute respiratory syndrome coronavirus 2 (SARS-CoV-2), hepatitis C virus, foot and mouth disease virus, and cucumber mosaic virus, among others. Despite having very different host range and morphology, all (+)RNA viruses share similar replication strategies: They all have negative-strand RNA as replication template, form membrane-bound viral replication complexes at specific organelle membranes, and recruit host proteins to complete their life cycles, among others.

BMV serves as a model for examining (+)RNA viruses in gene expression, viral replication, and recombination, due to its small genome, robust replication, and ability to replicate in yeast [33]. BMV encodes two replication proteins, 1a and $2a^{pol}$. While $2a^{pol}$ contains a polymerase domain and serves as a replicase [34, 35], 1a is the major player in building the membrane-invaginated viral replication complexes (VRCs) at the perinuclear endoplasmic reticulum (ER) membrane. During VRC formation, several host proteins, including reticulons (RTNs) and sucrose non-fermenting 7 (Snf7), are recruited into VRCs [36, 37]. RTNs belong to the reticulon homology proteins (RHPs), and there are three members, *RTN1*, *RTN2*, and *YOP1*, in yeast [38]. RHPs are membrane-shaping proteins that function to generate and maintain curvature in the ER membrane [38]. BMV 1a interacts with and recruits RHP proteins into VRCs. Deleting all three *RHP* genes abolishes the formation of BMV VRCs [36]. Similarly, human RTN3 and RTN3.1A are recruited by replication protein 2C of enterovirus 71 [39] and NS4A of West Nile virus [40] to VRCs. It was recently shown that RTNLB2 from *Nicotiana benthamiana* was recruited to VRCs of beet black scorch virus and required for viral replication [41]. All these data indicate a general requirement of RTNs in the formation of VRCs of multiple (+)RNA viruses. A better understanding of the interactions between viral replication protein with RHPs, such as how to disrupt the interaction, can provide antiviral strategies that are applicable to multiple (+)RNA viruses. In our first attempt targeting interactions 1a-RHPs, we will focus on the 1a and RTN1 interaction (*see* Fig. 1).

Structurally, 1a contains a N-proximal RNA capping (CAP) domain and a C-terminal NTPase/helicase-like (HEL) domain [42, 43], linked by a short proline-rich, disordered region. The interaction regions between 1a and 2a [35, 44] as well as interactions between CAP and CAP domain [45], HEL and HEL [45], and CAP and HEL [44] domains have been reported. However, no specific region or residuals in 1a or interacting host proteins have been identified.

4 Notes

1. The predicted structures can be also predicted by recent AlphaFold3 [46] web server (https://alphafoldserver.com/).

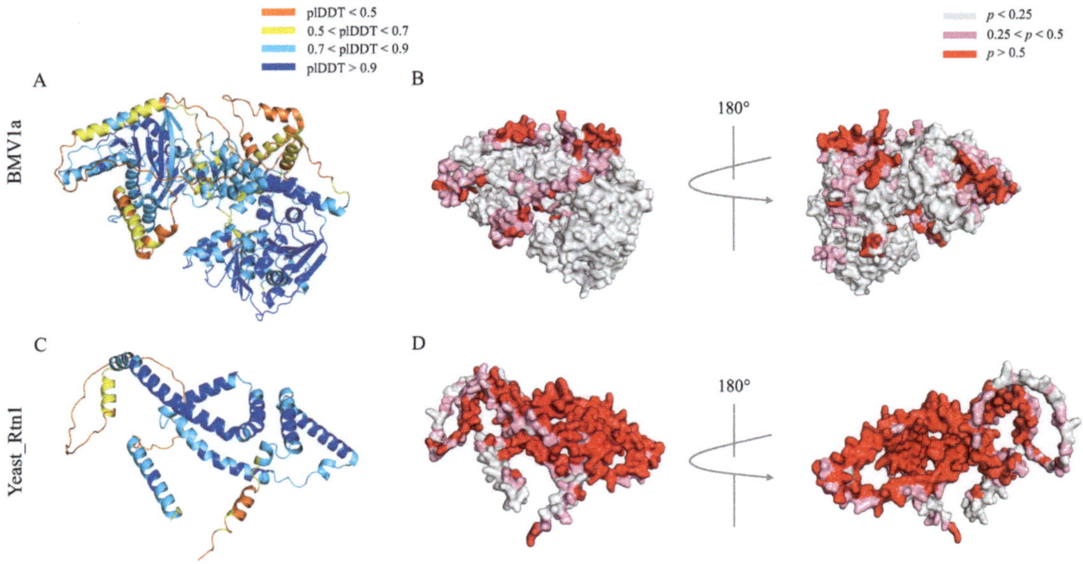

Fig. 1 (**a**) Predicted structure of BMV1a. (**b**) Predicted interaction sites in BMV1a. (**c**) Predicted structure of Yeast_Rtn1. (**d**) Predicted interaction sites in Yeast_Rtn1. The predicted structural models are shown in cartoon representations, colored based on the self-assessment confidence scores defined by predicted local distance difference test (pLDDT) scores. The predicted interaction sites are shown in surface representations, colored based on the prediction probability (*p*) of EquiPPIS

2. The quality of the predicted protein structure can affect the PPI site prediction. The reliability of the predicted protein structure can be estimated by looking at the self-assessment scores. In particular, the eleventh column in the predicted PDB file (same position of the b-factor column of the experimental PDB file), contains the residue-level self-assessment confidence scores, ranging from 0 to 100. The higher the self-assessment confidence scores, the higher the chance that the part of the structure is predicted accurately.
3. The output of the EquiPPIS consists of a list of numbers representing the residue-level generalized (irrespective of its interacting partner) protein–protein interaction site prediction probabilities, ranging from 0 to 1. For instance, the fifth number in the file indicates the PPI site prediction probability of the fifth residue. Higher prediction probabilities indicate a greater likelihood of a residue being part of the PPI site.
4. The combination of the self-estimation confidence score of the predicted structure and the EquiPPIS outputs is used to determine whether a residue is potentially part of a PPI site. Specifically, if both the self-estimation confidence score and the PPI prediction for a particular residue are high, the residue is considered for testing as a PPI site.

5. The databases and software packages utilized here were fetched and installed in or before 2022. Updated versions can be used.

Acknowledgments

This work was partially supported by the National Institute of General Medical Sciences (R35GM138146 to D.B.) and the National Science Foundation (DBI2208679 to D.B.). The authors gratefully acknowledge high-performance computing resources and technical support from the National AI Research Resource Pilot award (NAIRR240093 to D.B.).

References

1. Jumper J, Evans R, Pritzel A, Green T, Figurnov M, Ronneberger O, Tunyasuvunakool K, Bates R, Žídek A, Potapenko A (2021) Highly accurate protein structure prediction with AlphaFold. Nature 596: 583–589
2. Baek M, DiMaio F, Anishchenko I, Dauparas J, Ovchinnikov S, Lee GR, Wang J, Cong Q, Kinch LN, Schaeffer RD et al (2021) Accurate prediction of protein structures and interactions using a three-track neural network. Science 373:871–876
3. Lin Z, Akin H, Rao R, Hie B, Zhu Z, Lu W, Smetanin N, Verkuil R, Kabeli O, Shmueli Y (2023) Evolutionary-scale prediction of atomic-level protein structure with a language model. Science 379:1123–1130
4. Wu R, Ding F, Wang R, Shen R, Zhang X, Luo S, Su C, Wu Z, Xie Q, Berger B (2022) High-resolution de novo structure prediction from primary sequence. BioRxiv, 2022.2007.2021.500999
5. Mirdita M, Schütze K, Moriwaki Y, Heo L, Ovchinnikov S, Steinegger M (2022) ColabFold: making protein folding accessible to all. Nat Methods 19:679–682
6. Steinegger M, Söding J (2017) MMseqs2 enables sensitive protein sequence searching for the analysis of massive data sets. Nat Biotechnol 35:1026–1028
7. Jones S, Thornton JM (1996) Principles of protein-protein interactions. Proc Natl Acad Sci 93:13–20
8. Sharan R, Suthram S, Kelley RM, Kuhn T, McCuine S, Uetz P, Sittler T, Karp RM, Ideker T (2005) Conserved patterns of protein interaction in multiple species. Proc Natl Acad Sci 102:1974–1979
9. Zhang J, Kurgan L (2018) Review and comparative assessment of sequence-based predictors of protein-binding residues. Brief Bioinform 19:821–837
10. Kuzmanov U, Emili A (2013) Protein-protein interaction networks: probing disease mechanisms using model systems. Genome Med 5:1–12
11. Acuner Ozbabacan SE, Engin HB, Gursoy A, Keskin O (2011) Transient protein–protein interactions. Protein Eng Des Sel 24:635–648
12. Chène P (2006) Drugs targeting protein–protein interactions. ChemMedChem 1:400–411
13. De Las Rivas J, Fontanillo C (2010) Protein–protein interactions essentials: key concepts to building and analyzing interactome networks. PLoS Comput Biol 6:e1000807
14. Murakami Y, Mizuguchi K (2010) Applying the Naïve Bayes classifier with kernel density estimation to the prediction of protein–protein interaction sites. Bioinformatics 26:1841–1848
15. Wang B, Mei C, Wang Y, Zhou Y, Cheng M-T, Zheng C-H, Wang L, Zhang J, Chen P, Xiong Y (2019) Imbalance data processing strategy for protein interaction sites prediction. IEEE/ACM Trans Comput Biol Bioinform 18:985–994
16. Deng A, Zhang H, Wang W, Zhang J, Fan D, Chen P, Wang B (2020) Developing computational model to predict protein-protein interaction sites based on the XGBoost algorithm. Int J Mol Sci 21:2274
17. Zhang J, Kurgan L (2019) SCRIBER: accurate and partner type-specific prediction of protein-binding residues from proteins sequences. Bioinformatics 35:i343–i353
18. Hu S-S, Chen P, Wang B, Li J (2017) Protein binding hot spots prediction from sequence

only by a new ensemble learning method. Amino Acids 49:1773–1785

19. Zhang B, Li J, Quan L, Chen Y, Lü Q (2019) Sequence-based prediction of protein-protein interaction sites by simplified long short-term memory network. Neurocomputing 357:86–100
20. Li Y, Golding GB, Ilie L (2021) DELPHI: accurate deep ensemble model for protein interaction sites prediction. Bioinformatics 37: 896–904
21. Zeng M, Zhang F, Wu F-X, Li Y, Wang J, Li M (2020) Protein–protein interaction site prediction through combining local and global features with deep neural networks. Bioinformatics 36:1114–1120
22. Gainza P, Sverrisson F, Monti F, Rodola E, Boscaini D, Bronstein MM, Correia BE (2020) Deciphering interaction fingerprints from protein molecular surfaces using geometric deep learning. Nat Methods 17:184–192
23. Porollo A, Meller J (2007) Prediction-based fingerprints of protein–protein interactions. Proteins 66:630–645
24. Mahbub S, Bayzid MS (2022) EGRET: edge aggregated graph attention networks and transfer learning improve protein–protein interaction site prediction. Brief Bioinform 23:bbab578
25. Yuan Q, Chen J, Zhao H, Zhou Y, Yang Y (2022) Structure-aware protein–protein interaction site prediction using deep graph convolutional network. Bioinformatics 38:125–132
26. Roche R, Moussad B, Shuvo MH., Bhattacharya D (2022) E(3) equivariant graph neural networks for robust and accurate protein–protein interaction site prediction. bioRxiv, 2022.2012.2014.520476
27. Satorras VCG, Hoogeboom E, Welling M (2021) In: Marina M, Tong Z (eds) Proceedings of the 38th international conference on machine learning PMLR, proceedings of machine learning research, vol 139, pp 9323–9332
28. Bank PD (1971) Protein data bank. Nat New Biol 233(10):1038
29. Altschul SF, Madden TL, Schäffer AA, Zhang J, Zhang Z, Miller W, Lipman DJ (1997) Gapped BLAST and PSI-BLAST: a new generation of protein database search programs. Nucleic Acids Res 25:3389–3402
30. Suzek BE, Wang Y, Huang H, McGarvey PB, Wu CH, UniPort Consortium (2015) UniRef clusters: a comprehensive and scalable alternative for improving sequence similarity searches. Bioinformatics 31:926–932
31. Kabsch W, Sander C (1983) Dictionary of protein secondary structure: pattern recognition of hydrogen-bonded and geometrical features. Biopolymers 22:2577–2637
32. He G, Zhang Z, Sathanantham P, Diaz A, Wang X (2021) Brome mosaic virus (Bromoviridae). In: Encyclopedia of virology, vol 252. Elsevier, Amsterdam
33. Janda M, Ahlquist P (1993) RNA-dependent replication, transcription, and persistence of brome mosaic virus RNA replicons in S. cerevisiae. Cell 72:961–970
34. Kao CC, Ahlquist P (1992) Identification of the domains required for direct interaction of the helicase-like and polymerase-like RNA replication proteins of brome mosaic virus. J Virol 66:7293–7302
35. Chen J, Ahlquist P (2000) Brome mosaic virus polymerase-like protein 2a is directed to the endoplasmic reticulum by helicase-like viral protein 1a. J Virol 74:4310–4318
36. Diaz A, Wang X, Ahlquist P (2010) Membrane-shaping host reticulon proteins play crucial roles in viral RNA replication compartment formation and function. Proc Natl Acad Sci 107:16291–16296
37. Diaz A, Zhang J, Ollwerther A, Wang X, Ahlquist P (2015) Host ESCRT proteins are required for bromovirus RNA replication compartment assembly and function. PLoS Pathog 11:e1004742
38. Voeltz GK, Prinz WA, Shibata Y, Rist JM, Rapoport TA (2006) A class of membrane proteins shaping the tubular endoplasmic reticulum. Cell 124:573–586
39. Tang W-F, Yang S-Y, Wu B-W, Jheng J-R, Chen Y-L, Shih C-H, Lin K-H, Lai H-C, Tang P, Horng J-T (2007) Reticulon 3 binds the 2C protein of enterovirus 71 and is required for viral replication. J Biol Chem 282:5888–5898
40. Aktepe TE, Liebscher S, Prier JE, Simmons CP, Mackenzie JM (2017) The host protein reticulon 3.1 A is utilized by flaviviruses to facilitate membrane remodelling. Cell Rep 21:1639–1654
41. Zhang Q, Wen Z, Zhang X, She J, Wang X, Gao Z, Wang R, Zhao X, Su Z, Li Z (2023) RETICULON-LIKE PROTEIN B2 is a proviral factor co-opted for the biogenesis of viral replication organelles in plants. Plant Cell 35: 3127–3151
42. Kong F, Sivakumaran K, Kao C (1999) The N-terminal half of the brome mosaic virus 1a protein has RNA capping-associated activities: specificity for GTP and S-adenosylmethionine. Virology 259:200–210

43. Wang X, Lee W-M, Watanabe T, Schwartz M, Janda M, Ahlquist P (2005) Brome mosaic virus 1a nucleoside triphosphatase/helicase domain plays crucial roles in recruiting RNA replication templates. J Virol 79:13747–13758
44. O'Reilly EK, Paul JD, Kao CC (1997) Analysis of the interaction of viral RNA replication proteins by using the yeast two-hybrid assay. J Virol 71:7526–7532
45. Diaz A, Gallei A, Ahlquist P (2012) Bromovirus RNA replication compartment formation requires concerted action of 1a's self-interacting RNA capping and helicase domains. J Virol 86:821–834
46. Abramson J, Adler J, Dunger J, Evans R, Green T, Pritzel A, Ronneberger O, Willmore L, Ballard AJ, Bambrick J et al (2024) Accurate structure prediction of biomolecular interactions with AlphaFold 3. Nature 630:493–500

Chapter 15

Modeling Protein–Protein Complexes by Combining pyDock and AlphaFold

Luis Ángel Rodríguez-Lumbreras, Víctor Monteagudo, and Juan Fernández-Recio

Abstract

The lack of experimental structures for the majority of protein–protein complexes has motivated the development of a variety of strategies for the structural modeling of protein complexes, such as computational docking, in active development for the last decades, and the more recent artificial intelligence (AI)-based ground-breaking methodologies. Among the existing computational docking methods, Python docking (pyDock) has shown competitive predictive rates and high robustness over the years. However, the field has dramatically changed with the appearance of artificial intelligence (AI)-based methods, like AlphaFold. While structure prediction of individual proteins is virtually solved by this program, the focus is now on how to improve the prediction of challenging cases like antibody–antigen complexes, multiprotein complexes, weak interactions, or highly flexible interacting proteins. Successful strategies are based on the generation of more diverse sets of models and the integration with other "classical" approaches that facilitate the identification of the correct models. Here, we will show in practical terms how to combine the structural modeling capabilities of AlphaFold with the energy-based scoring function in pyDock to improve structural predictions in challenging protein–protein complexes.

Key words Protein–protein interactions, Protein structure prediction, Computational docking, Artificial intelligence, pyDock, AlphaFold, Binding energy, Model confidence

1 Introduction

The structural determination of all possible protein–protein interactions in humans and other organisms represents a major challenge in structural biology, and would largely contribute to our understanding of protein functional mechanisms, with multiple biomedical and biotechnological applications. The lack of experimental structures for the majority of protein–protein complexes is motivating the development of a variety of strategies for structural modeling of protein complexes. Computational docking is the major approach that has been used for several decades for the prediction of the structure of protein complexes [1]. Among the

Shahid M. Khan and Florencio Pazos (eds.), *Protein Evolution: Methods and Protocols*, Methods in Molecular Biology, vol. 2979, https://doi.org/10.1007/978-1-0716-4828-5_15,

variety of docking methods that have been reported, one of the most successful ones is Python docking (pyDock) [2], which uses an energy-based function to score docking poses generated by different sampling methods like FTDock 2.0 [3] or ZDOCK 2.1 [4]. The method is also available as a web server, pyDockWEB, which includes a custom parallel FTDock version [5].

The scoring function of pyDock is a combination of Coulomb electrostatics, Lennard-Jones van der Waals (VDW) potential (limited to 10% to cope with potential clashes from docking), and desolvation energy based on atomic solvation parameters (ASP), which were originally optimized for a small set of cases, and have proven to be robust afterward in their application to new benchmark cases and to different types of interactions. The method has been successfully validated in internal tests and in blind conditions in critical assessment of prediction of interactions (CAPRI) (http://www.ebi.ac.uk/msd-srv/capri/capri.html) and critical assessment of protein structure prediction (CASP) (http://predictioncenter.org/) community-wide assessment experiments. The program pyDock has a modular architecture with a variety of functionalities for the structural and energetic characterization of protein interactions. One of the most relevant modules is pyDockNIP, based on a statistical analysis of interface residues in the top 100 models generated by pyDock [6], which can be applied to identify interface hot spot residues [7] in cases where the structure of the complex is not available. Another important module is pyDockRST, which can be used to impose distance restraints from experimental data based on bioinformatics predictions, mutational experiments, nuclear magnetic resonance (NMR), cross-linking, etc. [8]. The pyDock protocol has been adapted to model complexes formed between proteins and other biomolecules, such as protein–RNA [9] or protein–DNA, with pyDockDNA module [10]. The method has been applied to many different interesting biological problems and cases of biomedical and biotechnological interest [11–14]. More details about installation and execution have been reported elsewhere [15].

On the other side, structural modeling of proteins and their interactions has dramatically changed with the appearance of artificial intelligence (AI)-based methods, like AlphaFold, which proved to be better than any other existing method in blind conditions, as tested in CASP13 [16]. But the revolution came 2 years later in CASP14 with the improved version AlphaFold2 (AF2) [17], which not only was better than any other previous method, but also provided accurate models for all kinds of difficult proteins in blind conditions, at a quality equivalent to those of experimentally determined structures. However, when applied to heterodimeric protein–protein complexes, AlphaFold2 showed limited success [18]. Then a new version specific for the prediction of protein–protein complexes, AlphaFold2-Multimer, was released [19]. This version was tested in blind conditions in CASP15, providing improved predictive rates with respect to "classical" docking methods, but still showing difficulties in challenging cases like antibody–antigen complexes, multiprotein complexes, weak interactions, or highly flexible interacting proteins. The best results were obtained by groups that tried to increase variability in the models generated by AlphaFold, using massive sampling, with and without templates, increased number of recycles, dropout layers during inference, random selection of multiple sequence alignments (MSAs), or combination with docking and scoring functions [20, 21]. More recently, a new version AlphaFold3 (AF3) has been released, available only through a web server at the time of writing, with a limit of 100 structures generated per user per day, but later, the source code and model weights were released for academic use. This version has been specifically improved for the prediction of protein–protein complexes, but it still fails in a number of challenging cases [22]. Thus, research is currently focused on how to improve the predictions in these cases where AlphaFold fails, with successful strategies based on the generation of a higher diversity of models, and its combination with other more "classical" approaches that facilitate the identification of the correct models.

In this chapter, we will show in practical terms how to combine the structural modeling capabilities of AlphaFold with the energy-based scoring function in pyDock to improve structural predictions in challenging protein–protein complexes.

2 Materials

2.1 Input

To generate a three-dimensional (3D) model for a protein using AlphaFold, we only need a text file with the polypeptide sequence in FASTA format, that is, beginning with a ">" followed by an identifier (header) and then the sequence. A single FASTA file with several protein sequences under different headers will make AlphaFold to generate different models for each of the provided sequences. For modeling the structure of a heteromeric protein–protein dimer in AlphaFold, we need a FASTA file with a single header and the sequences of the interacting proteins separated by a colon (":"). See, as an example, the input FASTA file for a complex formed by an engineered variable heavy domain of heavy chain (VHH) antibody and ribonuclease A (RNase A) (3D structure available at Protein Data Bank [PDB] with code 4POU):

```
>4POU
  QVQLVESGGGLVQAGGSLRLSCAASGYPHPYLHMGWFRQAPGKEREGVAAMDSGGGG
TLYADSVKGRFTISRDKGKNTVYLQMDSLKPEDTATYYCAAGGYQLRDR
TYGHWGQGTQVTVSS:
  KETAAAKFERQHMDSSTSAASSSNYCNQMMKSRNLTKDRCKPVNTFVHESLADV
```

```
QAVCSQKNVACKNGQTNCYQSYSTMSITDCRETGSSKYPNCAYKTTQANKHIIVACEGN
PYVPVHFDASV
```

For a homomeric protein–protein dimer, the polypeptide sequence is repeated, separated by a colon (":"). In case of oligomers of a higher degree of polymerization (trimers, tetramers, etc.), the number of copies of the polypeptide sequence will increase according to the oligomerization degree.

The 3D coordinates of the protein–protein complex models generated by AlphaFold will be used for scoring.

2.2 AlphaFold

There are different AlphaFold versions available to generate 3D structural models of protein–protein complexes.

2.2.1 AlphaFold2-Multimer

AlphaFold2-Multimer is an extension of AlphaFold2 [17] that was specifically trained to predict the 3D structure of multichain proteins [19]. The AF2-Multimer package and instructions are available through DeepMind's GitHub at this link: `https://github.com/google-deepmind/alphafold`. Due to its complex installation process, a Docker container compatible with compute unified device architecture (CUDA) is preferred. CUDA is a technology by NVIDIA that allows software to run faster by using the power of NVIDIA graphics processing units (GPUs) for parallel processing. An NVIDIA graphics card is recommended, and it should be compatible with CUDA v11.8 and have at least 8GB of video random access memory (VRAM). The size of the possible protein complex that can be generated depends on the available VRAM. While the central processing unit (CPU) can be used, it is extremely slow, and the software must be adapted to utilize the neural processing unit (NPU) of newer processors effectively.

Additionally, this version requires downloading the following databases:

- Big fantastic database (BFD).
- MGnify.
- PDB70.
- PDB (structures in the macromolecular crystallographic information file [mmCIF] format).
- PDB seqres—only for AlphaFold-Multimer.
- UniRef30 (formerly known as Uniclust30).
- UniProt—only for AlphaFold-Multimer.
- UniRef90.

Once decompressed, these databases require approximately 2.62 TB of storage, which can be a significant barrier to installation on workstations. Additionally, for faster access, it is recommended that they be stored on a solid-state drive (SSD), preferably using the nonvolatile memory express (NVMe) specification.

AlphaFold-Multimer has evolved significantly across its versions, each one bringing key improvements. Version 1 (v2.1.0, model weights: params_2022-01-19) was the initial attempt to adapt AlphaFold architecture for protein complexes but struggled with accurately handling heteromeric complexes and sometimes exhibited overconfidence in its predictions. Version 2 (v2.2.0, model weights: params_2022-03-02) addressed these issues by enhancing the model's ability to manage heteromeric and symmetric interactions, refining confidence metrics, and reducing structural clashes. It also improved performance and allowed for the generation of more structures by default, making it a more robust tool. Version 3 (v2.3.0, model weights: params_2022-12-06) introduced even more substantial architectural enhancements, including optimized attention mechanisms within the Triangle Multiplication module, which improved scalability for larger complexes and further boosted accuracy and confidence in predictions. This progression reflects a steady advancement in data utilization, architectural refinement, and predictive performance, driven by ongoing feedback and developments in protein structure research.

2.2.2 AlphaFold2 in ColabFold

ColabFold [23] is an open-source software to facilitate the use of AlphaFold2 and other AI-based modeling programs for researchers without large computational resources. It consists of a set of notebooks generated using Google Colaboratory (Google Colab), a free cloud-based platform that allows users to execute Python code in a web browser. Google Colab provides access to virtual machines equipped with NVIDIA GPUs, typically offering around 12GB–16GB of GPU VRAM. This makes it a valuable alternative for those who lack the computational power or storage capacity to run AlphaFold2 on a local workstation. However, it is important to note that there may be limitations, since these are free resources. The available computation time for larger proteins may be insufficient, and the instance could expire before the calculation is completed. Despite these limitations, ColabFold can still be a viable option, offering results that, while not as robust as those running DeepMind package locally or on a cluster, may still be sufficient for specific use cases.

The ColabFold notebooks are well organized on the `sokrypton` GitHub: `https://github.com/sokrypton/ColabFold`, where one can find helpful protein folding tools such as AlphaFold2 and AlphaFold2-Multimer, with options for generating MSAs using different programs like MMseqs2 and JackHMMER (with or without templates), as well as other folding tools like ESMFold [24], RoseTTaFold2 [25], and OmegaFold [26].

2.2.3 ColabFold on a Local Machine with LocalColabFold

The YoshitakaMo LocalColabFold GitHub (`https://github.com/YoshitakaMo/localcolabfold`) is an installation script that brings ColabFold functionality to the user in a local machine. It supports a wide range of operating systems, including Windows 10 or later (using Windows Subsystem for Linux 2), macOS, and Linux. LocalColabFold is best suited for advanced applications, such as batch processing of structure predictions for natural complexes, nonnatural proteins, or predictions with manually specified MSAs or templates. The advantages of LocalColabFold include a faster structure inference and relaxation if the personal computer (PC) has an NVIDIA GPU with CUDA drivers, and avoiding the need to prepare the large database required for native AlphaFold2.

It is recommended to install the latest graphics card driver, particularly the one provided by each distribution, especially when installing on Linux.

For instance, in Debian:

```
sudo apt-get install nvidia-driver-550
sudo reboot
```

In the case of a Red Hat/CentOS-based distribution, one needs to add the NVIDIA repository:

```
sudo yum install epel-release
sudo yum-config-manager --add-repo https://developer.download.nvidia.com/compute/cuda/repos/rhel8/x86_64/cuda-rhel8.repo
sudo yum install nvidia-driver-550
sudo reboot
```

Before running the installation script, it is advisable to follow the instructions on NVIDIA's website to install the CUDA Toolkit for each given distribution (see, e.g., the instructions for Ubuntu 22.04 LTS):

```
wget https://developer.download.nvidia.com/compute/cuda/repos/ubuntu2404/x86_64/cuda-keyring_1.1-1_all.deb
sudo dpkg -i cuda-keyring_1.1-1_all.deb
sudo apt-get -y install cuda-toolkit-12-6
```

Now, you can use the LocalColabFold GitHub script `install_colabbatch_linux.sh` to install LocalColabFold on your Linux machine.

```
wget https://raw.githubusercontent.com/YoshitakaMo/localcolabfold/main/install_colabbatch_linux.sh
bash install_colabbatch_linux.sh
```

This will take a few minutes, as it will create a Conda environment within the `localcolabfold` folder where all the necessary packages for running LocalColabFold will be installed. To be able to use `colabfold_batch` and `colabfold_relax`, export the path to the binaries by adding the following line to your `.bashrc` file:

```
echo 'export PATH="/path/to/your/localcolabfold/colabfold-conda/bin:$PATH"' >> ~/.bashrc
source ~/.bashrc
```

For more details about installation and use of LocalColabFold in computing clusters, *see* **Note 1**.

2.2.4 AlphaFold3

This version has been optimized for modeling protein–protein, protein–DNA, and protein–RNA complex interactions, increasing accuracy in antibody–antigen complexes, as well as in other protein–protein complexes. One of the key features of AlphaFold3 is its ability to model posttranslational modifications (PTMs), which are crucial for regulating protein function. AlphaFold3 is accessible through a user-friendly web server (`https://alphafoldserver.com`), allowing researchers to quickly obtain predictions without needing to interact with the command line or install it locally [22].

In addition, the source code and model weights were released under the CC BY-NC-SA 4.0 license for academic purposes only. Local execution of AlphaFold3 enables the generation of a larger number of models and the inclusion of additional ligands. Before installing AlphaFold3 locally, the model weights need to be requested through a Google form, as indicated in this link: `https://github.com/google-deepmind/alphafold3`.

Alternatively, you can follow our installation guide for AlphaFold3 using a Conda environment in this link: `https://github.com/pyDock/AlphaFold3-Conda-Install`, which outlines the complete process, from setting up Miniconda and creating the environment to installing dependencies and configuring the model, without the need for additional system-level packages.

2.3 pyDock Scoring

The pyDock 3.0 package is available at the program web page (`https://life.bsc.es/pid/pydock/`). This software has a complete protein–protein docking pipeline through many different modules, although we will use here only the functionality of scoring calculations. A new version pyDock4 with new functionalities will be soon available (upcoming publication).

To get pyDock3, you need to apply for a license (for academic use, you will receive a link to the pyDock3 distribution file by email; for commercial use, you need to contact the authors).

Uncompress and untar the pyDock distribution file to extract the `pyDock3` directory. Change permissions of the `pyDock3/data` directory:

```
chmod go+rx data
```

The `pyDock3` directory can be moved to any location of your choice. If you move it to `/usr/local/software/` directory, you can define the `PYDOCK` variable in your `.bashrc` file:

```
export PYDOCK=/usr/local/software/pyDock3/
```

so that the executable of pyDock can be called in a more convenient way:

```
$PYDOCK/pyDock3
```

See more installation and execution details in **Note 2** and in a recent publication [15].

2.4 SCWRL

The program SCWRL (`http://dunbrack.fccc.edu/lab/scwrl`) [27] is used by pyDock to rebuild incomplete side chains of the input PDB files. The version SCWRL3 can be used automatically within pyDock3, but since it is outdated, you need to obtain the installation file `scwrl3_lin.tar.gz` from their authors or through the SBGrid Consortium (https://sbgrid.org/software/titles/scwrl3). Uncompressing this file will extract its contents to a new directory called `scwrl3_lin`. Within this directory, run:

```
./setup
```

This will create an SCWRL3 binary (`scwrl3`) in such directory.

The upcoming pyDock4 distribution (unpublished) will be compatible with the updated version SCWRL4.

2.5 Greasy

Greasy is a tool designed to simplify the deployment of embarrassingly parallel simulations across various environments. It efficiently executes a list of tasks in parallel, scheduling them based on available resources. This makes it an ideal solution for scenarios where a serial program needs to run numerous instances with different parameters. Greasy bundles these individual runs together and maximizes resource usage by running as many tasks in parallel as possible. As tasks are completed, Greasy automatically initiates the following tasks in the queue.

To obtain Greasy, first you may need to install make and wget programs in Debian distributions:

```
sudo apt-get update
```

```
sudo apt-get install build-essential
sudo apt-get install wget
```

In Red Hat/CentOS distributions, you also need to install those programs:

```
sudo yum groupinstall "Development Tools"
sudo yum install wget
```

Once this software is installed, go to the installation directory, download, uncompress, configure, compile, and install the source code with the following instructions:

```
sudo mkdir -p /opt/GREASY_2.2
sudo chown -R user:group /opt/greasy/
cd /opt/greasy/
 wget https://github.com/BSC-Support-Team/GREASY/archive/
refs/tags/2.1.2.1.tar.gz
tar -xvf 2.1.2.1.tar.gz
cd GREASY-2.1.2.1
make clean
./configure
prefix=/opt/greasy/GREASY_2.2
make
make install
```

The `GREASY` directory can be moved to any location of your choice. If you move it to `/home/user/software/` directory, you can define the `GREASY` variable in your `.bashrc` file:

```
export GREASY=/home/user/software/GREASY_2.2/
```

so that the executable of Greasy can be called more conveniently:

```
$GREASY/greasy
```

2.6 CCharPPI Web Server

The pyDock scoring functions are also available at the computational characterization of protein–protein interactions (CCharPPI) web server (`https://life.bsc.es/pid/ccharppi`) [28]. The user will need to upload the 3D coordinates of a protein–protein complex structure, and the server will return the values for up to 108 descriptors calculated on the input protein–protein interface, including pyDock total binding energy (PYDOCK_TOT), pyDock van der Waals (VDW), pyDock electrostatics (ELE), and pyDock desolvation (DESOLV).

There are three different options to submit the input data: (i) two separate PDB files for each of the two proteins forming a given complex, (ii) the PDB code of the complex, and (iii) a compressed file with the coordinates for up to 100 models. See more details on the CCharPPI web page.

3 Methods

In order to overcome current limitations of AlphaFold in modeling the 3D structure of challenging complexes like antibody–antigen interactions or those involving highly dynamic proteins, we have devised a procedure that combines the extraordinary capabilities of AlphaFold to generate 3D models for proteins and complexes, and the efficient energy-based protein–protein scoring functions in pyDock. This procedure is being evaluated in community-wide assessment experiments like CASP or CAPRI and can be described as follows. First, we will use several versions of AlphaFold under different conditions to generate a large variety of models for a given protein–protein complex. Then, these models will be scored by a combination of AlphaFold model confidence (AF-MC) and pyDock scoring functions (*see* Fig. 1). Below are the details of the specific steps involved in this modeling strategy, which will be illustrated by an example, the complex formed by an engineered VHH antibody and RNase A (3D structure available at PDB with code 4POU).

3.1 Generating 3D Models for Protein–Protein Complexes with AlphaFold

3.1.1 AlphaFold2-Multimer

To use AlphaFold2-Multimer (AF2-Multimer) within LocalColabFold, first, a folder is created where we will keep all the needed files to model the structure of our example case (PDB 4POU):

```
export CASE=4POU
mkdir -p ${CASE}/AF2/
```

Then, the FASTA sequences of the interacting proteins separated by colon (":") are copied into a single FASTA file (*see* Subheading 2.1), using your favorite text edit program (vi, nano, or gedit):

```
vi ${CASE}/AF2/${CASE}.fasta
```

Next, the three AF2-Multimer versions are executed one by one using a loop:

```
for h in v1 v2 v3; do
for i in */AF2/*fasta; do
 colabfold_batch 'pwd'/$i 'pwd'/${i/.fasta}.colabfold.$h --
save-recycles --model-type alphafold2_multimer_${h} --re-
```

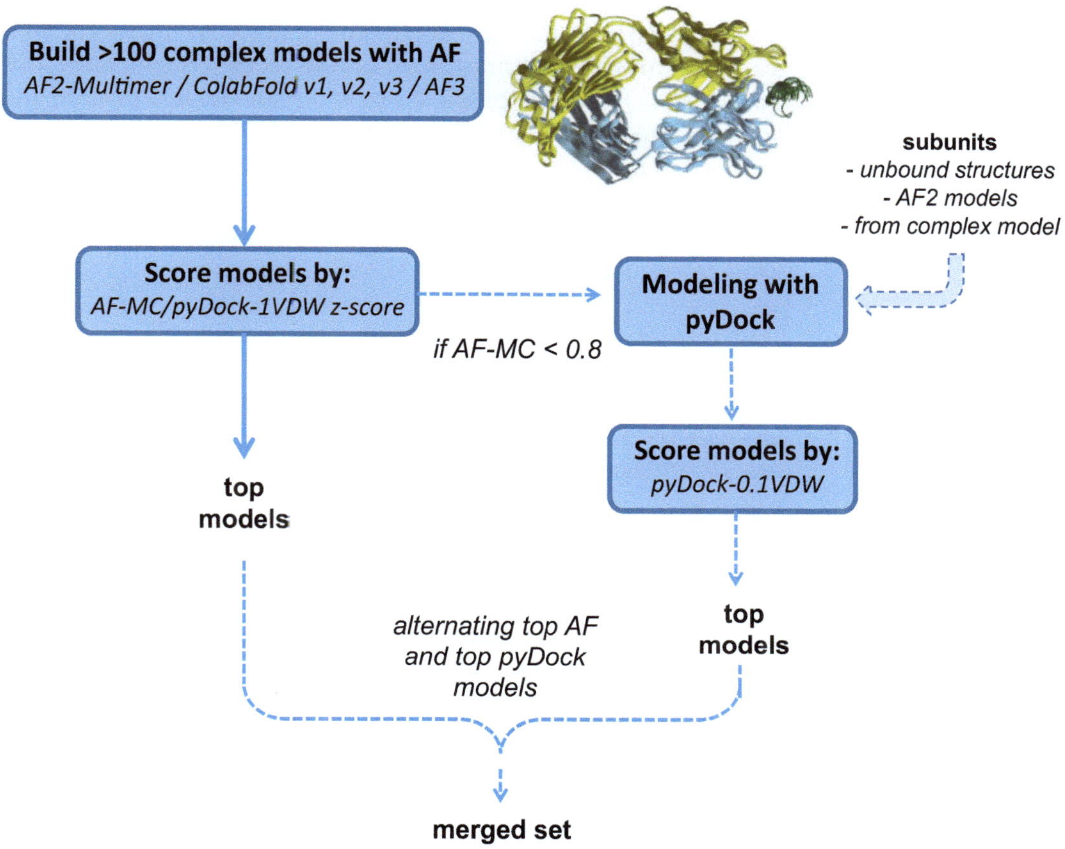

Fig. 1 Scheme of the strategy to integrate AlphaFold (AF) and Python docking (pyDock) modeling of protein–protein complexes. A diversity of models are generated by different versions of AlphaFold and sorted by a combined function composed of AlphaFold model confidence (AF-MC) and pyDock-1VDW (100% van der Waals [VDW]) z-scores to provide a final set of predictions. However, if there are no models with high AF-MC scores (>0.8), new docking orientations will be generated by pyDock and sorted by default pyDock scoring. In that case, both sets will be merged by alternating the top docking models sorted by pyDock scoring with the top AF models sorted by the combined AF-MC and pyDock-1VDW (100% VDW) z-scores

```
cycle-early-stop-tolerance 0 --rank multimer --use-dropout --
num-recycle 20 --amber --use-gpu-relax
 done
 done
```

where `--save-recycles` option is enabled to save the recycles (intermediate models), the `--num-recycle` parameter is set to 20, and `--recycle-early-stop-tolerance` is set to 0, which in combination make the model generation stop after reaching 20 recycles, and `--use-dropout` is activated. Additionally, the `--rank` is set to multimer, and the `--model-type` is specified as alphafold2_multimer_[v1, v2, v3]. The option `--amber` and `--use-gpu-relax` is enabled to perform relaxation on the GPU.

Additionally, when the process is complete, the `log.txt` file is generated, which informs about the quality of the structures according to AF2-Multimer based on the predicted template modeling (pTM), interface predicted template modeling (ipTM), and predicted local distance difference test (pLLDT) scores (explained in Subheading 3.2).

Only the final models are relaxed by default, but since we are going to use all the AlphaFold2-generated models, including the recycles, we need to relax them as well.

```
for f in `find -name '*_*.r*[0-9].pdb'`;do
echo "colabfold_relax --max-iterations 2000 --tolerance 2.39
--stiffness 10.0 --max-outer-iterations 3 --use-gpu $f ${f/
unrelaxed/relaxed}" ;
done > ${CASE}.colabfold.relax.greasy
```

Then, we use Greasy to send the models to the graphics card. In some cases, especially with version 1, the calculations may take a long time or even fail due to a high number of clashes or because the GPU has been overloaded (running out of VRAM) by sending too many models simultaneously. Therefore, we recommend sending a maximum of 4 or 2 models at a time to the GPU, depending on the VRAM available. In this example, we set the environment variable GREASY_NWORKERS to 4.

```
export GREASY_NWORKERS=4
$GREASY/greasy ${CASE}.colabfold.relax.greasy
```

3.1.2 AlphaFold3 Server

We can also model a protein–protein complex using the recently released AlphaFold3 server (https://alphafoldserver.com). To use the web, we have to log in with a Google account and accept the terms of service. The web provides an interface where we can select the type of molecule and number of copies, and write the sequences of the molecules. We can also add PTMs and other modifications by pressing the three points at the right of the block of the sequence. The available modifications as well as other details can be found in the FAQ menu at the top left of the web. If more than one molecule needs to be modeled, we can add them by pressing "add entity." After creating the job, we can save it as a draft by pressing save job or run it by clicking continue and preview the job. By pressing the latter, a new menu will be displayed to check if all the inputs are correct before running it. Finally, by clicking "*confirm and submit job*," the modeling process will start. Alternatively, we can create a JavaScript Object Notation (*JSON*) file to submit one or more jobs as drafts by clicking "*Upload JSON*." The formatting of the file is detailed in the `README.md` of the Google DeepMind AlphaFold GitHub in the server folder (https://github.com/google-deepmind/alphafold/blob/main/server/README.md). Each job will usually run for a few minutes, but depending on the job size, the day, or the submission time, the job can take from a few minutes to 1 or 2 hours.

When the job finishes, we can see the results directly on the web or download them as a zipped file, which will contain: the models in CIF format, a `job_request.json` file, and `summary.json` and `full_data.json` files for each of the five models. The CIF file is similar to a PDB; it can be visualized with programs like ChimeraX or PyMOL and can include a number indicating the rank assigned by AF3 according to the parameters gathered in the `summary_-confidence.json`. The `job_request.json` file gathers the information of the job provided by the user, including the seed assigned automatically or manually, and can be used to replicate the job. The summary *JSON* file gathers the information to rank the five models generated by the job, including the ipTM and the pTM values. The `full_data.json` provides additional information of the individual atoms and residues.

3.2 AlphaFold Model Confidence Score

To assess the accuracy of the models, the `log.txt` file from AF2-Multimer provides different quality scores: a per-residue measure of local accuracy, the predicted local distance difference test (pLDDT), and two global accuracy measures, the predicted template modeling (pTM) score and the interface predicted template modeling (ipTM) score, which are derived from the template modeling (TM) score [29] and evaluate how similar the generated model is to the templates used in the MSA [19]. The values for these scores range from 0 to 1, where the higher the values, the better the models are expected to be.

The authors of AF2-Multimer suggest using another score metric, the model confidence (hereinafter referred to as AF-MC), which is a combination of pTM and ipTM:

$$\text{Model confidence} = 0.8 \cdot \text{ipTM} + 0.2 \cdot \text{pTM}$$

The values for this score range from 0 to 1, where the models with the highest scores are expected to be the most accurate ones. AlphaFold2-Multimer uses this model confidence score to rank their models.

The way to get the ipTM and pTM values for a given model differs between versions. In the case of AlphaFold2, these values can be extracted from the log.txt file, a plain text file that indicates all the steps performed by AlphaFold2, by using a Python script. In the case of AF3, the ipTM and pTM values can be extracted from the `summary_confidence.json` file, which gathers the information to rank the five models generated by the job. While the internal scoring of models in AF3 differs from that of AF2-Multimer, the calculation method of ipTM and pTM remains the same. The protocols to extract these values from the mentioned files can be found at pyDock GitHub (https://github.com/pyDock/AF_pyDock).

3.3 Scoring Protein–Protein Complex Models by pyDock

3.3.1 Computing pyDock Score for a Single Complex

We can compute the pyDock energy-based score for a given complex structure (either experimentally determined or modeled) with the pyDock *bindEy* module. For the sake of clarity, we will show here how to calculate such pyDock docking energy on one of the models generated by AF2-Multimer for the example complex formed by VHH antibody and RNase A [30].

First, we need to create a text file called `DOCKNAME.ini`, containing all the needed information for the scoring calculations, where `DOCKNAME` is the name used as the execution command argument and the base name for the output files. In our case, the file will be named `4POU_relaxed_rank_001_alphafold2_multimer_v3_model_3_seed_000.ini` with the following format:

```
[receptor]
pdb = 4POU_relaxed_rank_001_alphafold2_multimer_v3_model_3_-
seed_000.pdb
mol = A
newmol = A

[ligand]
pdb = 4POU_relaxed_rank_001_alphafold2_multimer_v3_model_3_-
seed_000.pdb
mol = B
newmol = B
```

The `pdb` field defines the coordinate file of the complex (`4POU_relaxed_rank_001_alphafold2_multimer_v3_model_3_seed_000.pdb`), which can be generated using LocalColabFold (*see* Subheading 2.2.3). The `mol` field is the original chain ID (one character) of the protein chain/s for the receptor and ligand proteins forming the complex interface that will be scored (here, `A` and `B`, respectively). The `newmol` will be the name of the chain ID of the interacting proteins in the output files, although this has no effect on the scoring calculations (*see* **Note 3**). The decision of which protein is defined as the receptor or ligand is not relevant for scoring calculations (in docking calculations, usually the largest protein is the receptor and the other one, the ligand).

Now, we just need to run the *bindEy* module as follows:

```
$PYDOCK/pyDock3 4POU_relaxed_rank_001_alphafold2_multi-
mer_v3_model_3_seed_000 bindEy
```

First, this command will generate the coordinate files correctly parsed for pyDock, from the receptor and ligand PDB files indicated in the *DOCKNAME*.ini parameter file. In our example, it will create new parsed files for receptor and ligand proteins, 4POU_relaxed_rank_001_alphafold2_multimer_v3_model_3_seed_000_rec.pdb and 4POU_relaxed_rank_001_alphafold2_multimer_v3_model_3_seed_000_lig.pdb, respectively. Some PDBs may have incomplete side chains due to missing atoms in the structures, which make them not valid for pyDock scoring calculations. These side chains can be rebuilt with external programs, like SCWRL, which can be run automatically within pyDock if its location is indicated in the pydock.conf file (*see* **Note 4**).

Then, the command will calculate the binding energy for the complex using the coordinates of the parsed receptor and ligand molecules, and a new table will be created (4POU_relaxed_rank_001_alphafold2_multimer_v3_model_3_seed_000.ene in our example), with the detail of the different energy terms calculated for the interface defined in the 4POU_relaxed_rank_001_alphafold2_multimer_v3_model_3_seed_000.ini file. See below an example:

```
Conf Ele Desolv VDW Total RMSD RANK
- -
- -
- -
- -
- -
- -
- -
- -
- -
- -
- -
- -
- -
------------------------------------------------------------
------------------------------------------------------------
 1 -12.770 -3.990 -42.313 -20.991 20.335 1
```

Conf	Conformation number (not relevant here; in the case of docking calculations, this number will be different for each docking pose)
Ele	Electrostatic energy term
Desolv	Desolvation energy term
VDW	van der Waals energy term
Total	Total binding energy (Ele + Desolv +0.1*VDW)
RMSD	

Ligand root mean square deviation (RMSD) with respect to the reference indicated in the *.ini file (*see* Subheading 3.5)

RANK: Rank according to its total binding energy (not relevant here; in the case of docking calculations, this number will be different for each docking pose)

The value indicated by the "Total" column is the energy-based score that pyDock assigns to all docking models by default and includes electrostatics, desolvation, and a limited van der Waals contribution (10%). This factor that reduces the contribution of van der Waals is optimized for the scoring docking models due to the existence of interatomic clashes in the interfaces created by rigid-body docking from the unbound conformations of the interacting proteins. However, in the case of the models generated by AlphaFold, the geometries are more accurate, and the number of clashing atoms is much more reduced. In these cases, we could score the models by using 100% of the van der Waals term. We will call this new score *pyDock-1VDW*. Hereinafter, we will use

$$pyDock\text{-}0.1\,VDW = \text{Ele} + \text{Desolv} + 0.1^{*}\text{VDW}$$

$$pyDock\text{-}1\,VDW = \text{Ele} + \text{Desolv} + \text{VDW}$$

3.3.2 Computing pyDock Scores for a Set of Complexes

We need to create a bash script to generate the `*.ini` files for all the models produced by AlphaFold2 for the 4POU example. This script will also generate a text file containing all the *bindEy* commands. Using the embarrassingly parallel tool GREASY, we will calculate these energy-based scores in parallel.

```
cd 4POU/AF2/
```

Using your favorite editor (vim, nano, or gedit), copy and paste this script and save it as `ini_creator_bindEy.sh`:

```
#!/bin/bash
CHAINS_REC_LIG_VALUES=("A B")
for h in *cola*v*/ fold*/; do
echo "$h"
cd "$h" || exit


for j in *[0-9].pdb; do
echo "$j"
for value in "${CHAINS_REC_LIG_VALUES[@]}"; do
REC=$(echo "$value" | cut -d" " -f1)
LIG=$(echo "$value" | cut -d" " -f2)
CH=${LIG/,}
ini_file="${j/.pdb}_LIG_${CH}.ini"
```

```
cat <<EOF > "$ini_file"
[receptor]
pdb = $j
mol = $REC
newmol = $REC
[ligand]
pdb = $j
mol = $LIG
newmol = $LIG
EOF
echo "cd ${h}; timeout 5m pydock4 $ini_file bindEy; cd -" >>
Greasy_BindE_mul_ligs.txt
done
done
cd - || exit
done
```

Grant execution permissions to the newly created script, run it, and then immediately execute Greasy to start the parallel calculations.

```
chmod +x ini_creator_bindE.sh
./ini_creator_bindE.sh
```

This process will produce a corresponding `*.ini` and `*.ene` file for each model PDB. The `*.ene` files, containing the energetic scoring from pyDock, and the `log.txt` from AlphaFold2, which includes the ipTM and pTM scores, will be merged into a pandas DataFrame to calculate the combined model confidence and pyDock score (see the code at pyDock GitHub: https://github.com/pyDock/AF_pyDock).

3.4 Combined Model Confidence and pyDock Score

In many cases, using pyDock scores and AlphaFold model confidence (AF-MC) values separately may not select the best structure among a set of models generated by AF. Both scores have different bases, and they can complement each other, so their combination can provide a more accurate identification of the correct models. Since they cannot be directly combined, and in order to avoid the necessity of parameter optimization, one straightforward option to standardize the two functions is to combine them using z-scores, which in general enables the combination of different scoring functions whose values are inherently different in nature and range, as follows:

$$Z = \frac{X - \mu}{\sigma}$$

where Z represents the z-score value for a given data point X, μ is the mean, and σ the standard deviation of the studied dataset or that of a reference one. By applying the z-score to scoring functions like AlphaFold model confidence (AF-MC) and pyDock-1VDW (100% VDW) energies, we can derive $Z_{AF\text{-}MC}$ (z-score of model confidence) and $Z_{pyDock\text{-}1VDW}$ (z-score of pyDock total energy with 100% van der Waals). After this standardization, the z-scores retain the same sign interpretation as their raw counterparts: High (positive) $Z_{AF\text{-}MC}$ values indicate reliable models, and low (negative) $Z_{pyDock\text{-}1VDW}$ values suggest better models in terms of energy.

These standardized scores can now be combined to create a new function:

$$Z_{AF/pyDock-1VDW} = Z_{AF-MC} - Z_{pyDock-1VDW}$$

This combination ensures that both terms have the same sign, facilitating their utility and interpretation. Thus, according to this score, the best models will be those with the highest (positive) $Z_{AF/pyDock\text{-}1VDW}$ values. A similar approach can be applied using pyDock default total energy (with 0.1 van der Waals) to form $Z_{AF/pyDock\text{-}0.1VDW}$. A script to extract these values for a given model can be found at pyDock GitHub (https://github.com/pyDock/AF_pyDock).

3.5 Evaluating the Quality of the Models with Respect to a Reference Complex Structure

In cases where a reference structure for a modeled complex is available, for instance, when running a modeling protocol for testing purposes, we can compare the resulting models with such reference complex structure. There are different quality measures that are used by the protein–protein interaction modeling community. In CAPRI, results have been traditionally classified in acceptable, medium, and high quality based on three criteria: ligand RMSD (*L_rmsd*), interface RMSD (*I_rmsd*), and fraction of native contacts (*Fnat*) (*see* Table 1) [31]. More recently, the DockQ score, a combination of these three metrics, can provide a single value that can also be used to classify the models according to different thresholds (*see* Table 1) [32].

Table 1
Model quality classification according to CAPRI criteria and equivalent categories based on DockQ score

Classification	Fnat	and	(L_rmsd	or	I_rmsd)	DockQ
High	≥0.5		≤1.0		≤1.0	≥0.80
Medium	≥0.3		≤5.0		≤2.0	≥0.49
Acceptable	≥0.1		≤10.0		≤4.0	≥0.23
Incorrect	<0.1					<0.23

The CAPRI metrics can be computed by pyDock. For instance, the value of *ligand RMSD* (*see L_rmsd* in Table 1), that is, the RMSD between the positions of the ligand in the reference and modeled complexes, after superimposing the receptor chains of both complexes, can be automatically computed by pyDock *bindEy* module if a suitable reference complex is indicated and properly set up in the `DOCKNAME.ini` file.

In our example, we can use as reference the file 4pou.pdb (PDB entry 4POU, downloaded from www.pdb.org) and define the following `*.ini` file:

```
[receptor]
pdb = 4POU_relaxed_rank_001_alphafold2_multimer_v3_model_3_-
seed_000.pdb
mol = A
newmol = A

[ligand]
pdb = 4POU_relaxed_rank_001_alphafold2_multimer_v3_model_3_-
seed_000.pdb
mol = B
newmol = B

[reference]
pdb = 4pou.pdb
recmol = B
ligmol = A
newrecmol = A
newligmol = B
```

The `newrecmol` field must indicate the same chain ID as `newmol` for the receptor, and the `newligmol` must be the same as `newmol` for the ligand. In this way, when running the *setup* module, the reference file `*_ref.pdb` will be created. And, when running the *bindEy* module, the ligand RMSD value for the model will be calculated and shown in the `*.ene` file, in the *RMSD* column. If we have an energy table without RMSD values from a previous execution, and we do not want to execute *bindEy* again, we can run pyDock *RMSD* module to calculate and save RMSD values in a separate file `*.RMSD` (which has to be manually added to `*.ene`). We can also use the pyDock module *capriRMSD* to calculate and save all CAPRI metrics in a separate file called `*.RMSD.capri` (which has to be manually added to `*.ene`).

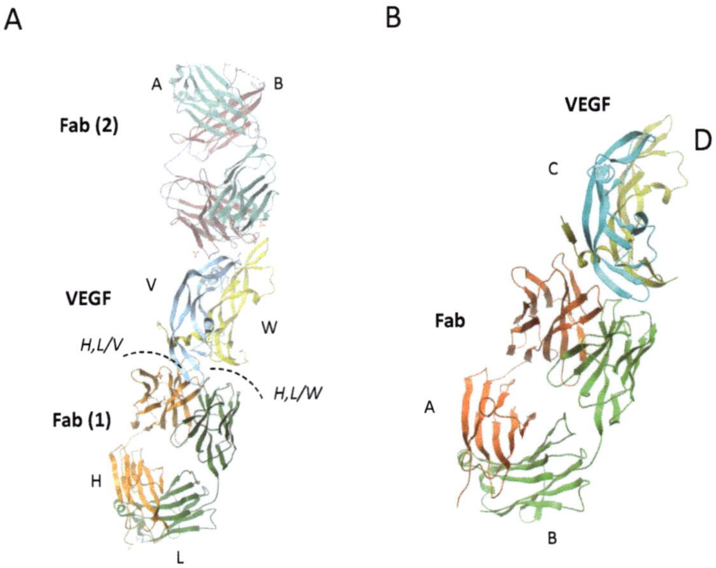

Fig. 2 Assessment of model quality in a complex with a symmetric ligand dimer. (**a**) X-ray structure of antigen-binding (Fab) fragment bound to Vascular endothelial growth factor (VEGF) dimer (Protein Data Bank [PDB] 2FJG) with two Fab copies in the asymmetric cell. (**b**) Model generated by AlphaFold2 (AF2) (ColabFold version 3) and ranked 4 by pyDock-1VDW. Chain IDs are shown as found in the PDB and in the model

3.5.1 Evaluating Model Quality when Receptor or Ligand Has Symmetric Copies of the Same Chain

When analyzing the quality of the resulting models with respect to the reference X-ray structure, we should consider all distinct interfaces found in the structure. In the case of receptor or ligand formed by symmetric copies of the same chain, to calculate the RMSD and other quality measures, we should first find the equivalence between the receptor (or ligand) chains in the model and those in the reference, which is not always defined by the chain IDs, since the chain names chosen in the reference and in the model are usually arbitrary. We can illustrate this with an example, the complex formed by G6 Fab, a phage-derived antigen-binding (Fab) fragment, and vascular endothelial growth factor (VEGF), which is included in a recently reported structural prediction benchmark set formed by challenging antibody–antigen complexes [33]. The 3D structure of this complex (PDB 2FJG) is formed by two copies of the Fab molecule (each composed of heavy and light chains, named H and L in the first Fab copy, and B and A in the second Fab copy, respectively) and two VEGF chains forming a symmetric dimer (chain IDs V and W) (*see* Fig. 2a). Due to C2 cyclic symmetry of VEGF dimer, the two interfaces formed by each of the Fab copies with the VEGF dimer are virtually the same, so we can disregard the second Fab copy. However, each VEGF chain shows two different interfaces with Fab, one with respect to each Fab copy, that is, the interface of VEGF chain V and the first Fab copy (which is equivalent to the interface of VEGF chain W with the second Fab copy) is different from the interface of VEGF chain W and the first Fab copy (which in turn is the same as that of chain V and the second Fab copy). This has to be considered when selecting the correct VEGF chains to compare the models with respect to the reference structure. For instance, in Fig. 2b, we show one of the models generated by AF2 ColabFold version 3, which is ranked 4 by pyDock-1VDW (100% VDW). In this model, Fab heavy and light chains are named A and B, respectively, and VEGF chains are C and D. To calculate the ligand RMSD, after superimposing chains A and B of the Fab model onto those of the reference structure (H and L), we have to compute the RMSD of VEGF chains C and D in the model with respect to VEGF chains V and W in the reference. In Fig. 2, we can visually check that chain C in the model is equivalent to chain V in the reference, and that chain D in the model is equivalent to chain W in the reference. But for automatic calculation, since a given VEGF chain in the model could be equivalent to any of the two VEGF chains in the reference, we should calculate RMSD twice: one assuming that chains C and D in the model are equivalent to chains V and W in the reference, respectively (in our case, this value is 11.6 Å), and another one assuming that chains C and D in the model are equivalent to chains W and V in the reference, respectively (in our case, this value is 36.3 Å). The smallest of these computed RMSD values is the one that correctly represents the ligand RMSD for this model applied on the truly equivalent chains. In a similar manner, we need to calculate all the parameters that form the DockQ function for each resulting model.

4 Case Studies

4.1 Scoring with pyDock Improves AF2-Multimer Predicting Performance: VHH Bound to RNase A

The first case study is the complex formed by an engineered VHH antibody and RNase A, which is included in a recently reported structural prediction benchmark set formed by challenging antibody–antigen complexes [33]. The 3D structure of this complex is available in the PDB with code 4POU, so we can check the results of the modeling strategy. We applied AF2-Multimer (ColabFold version 3) to model the interaction. We started a standard AF2 execution (one seed) with five model trajectories, saving all the recycled models (*see* Subheading 3.1.1). This produced 21 unrelaxed models (intermediate models from recycle r0 to r20) plus the relaxed final model for each of the five model trajectories, which yielded 110 different conformations in total. We note that here only the final model was relaxed (as default), although it is advisable to relax all intermediate recycles if possible. Then, we scored these 110 models using pyDock binding energy (*see* Subheading 3.3.2), AF-MC (*see* Subheading 3.2), and the combined AF-MC/pyDock-

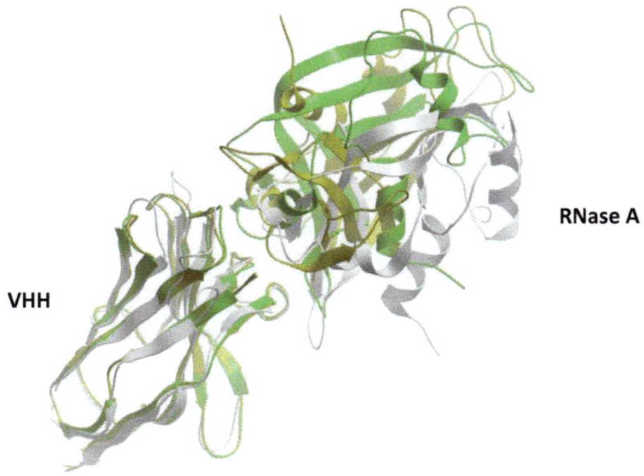

Fig. 3 Models generated by AF2-Multimer (version 3) for the example case: variable heavy domain of heavy chain antibody/ribonuclease A (VHH/RNase A). Incorrect model generated by AF2 as rank 1 (based on AF-MC) is shown in yellow. Acceptable model ranked 4 by pyDock-1VDW is shown in green (this model ranked 10 by AF-MC/pyDock-1VDW z-score, but it was ranked 65 by AF-MC). For comparison, X-ray structure is shown in white (PDB 4POU)

1VDW z-score (*see* Subheading 3.4). Finally, we ranked the models according to each score and computed quality measures (L_rmsd, I_rmsd, Fnat, and DockQ) using the original 4POU PDB as the reference (*see* Subheading 3.5).

After comparison with the reference structure, we found that the model provided by AF2-Multimer as the one with the best model confidence score is not acceptable (DockQ 0.07) (*see* Fig. 3). The best-ranked acceptable model according to AF-MC is found with rank 28 (DockQ 0.28). On the other side, when we use pyDock-1VDW (100% VDW) to score the models generated by AF2-Multimer, we found an acceptable model with rank 4 (DockQ 0.24) (*see* Fig. 3). When we score the models by the combined AF-MC/pyDock-1VDW z-score, this acceptable model is still ranked within the top 10 (but it was ranked 65 by AF-MC alone). Interestingly, this model is ranked 2 by default pyDock (10% VDW). This is a good example in which the scoring of pyDock can improve the selection of models generated by AF2-Multimer.

4.2 Scoring with pyDock Improves AF3 Predicting Performance: Fab Bound to Cytochrome c

This case is the complex formed by Fab fragment of E8 antibody and horse cytochrome c (cyt c), which is included in the above-mentioned antibody–antigen benchmark set [33]. The 3D structure of this complex is available at PDB with code 1WEJ, so we can check the results of the modeling strategy. We applied AF3 to model the interaction between an antibody molecule (formed by heavy and light chains) and the monomeric cytochrome c. First, we

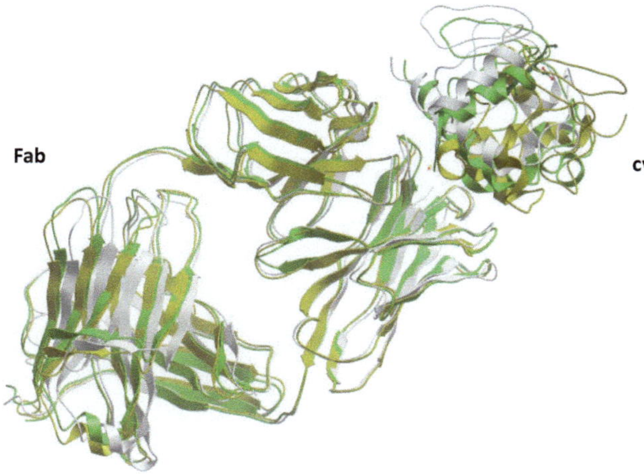

Fig. 4 Models generated by AlphaFold3 (AF3) for the example case: Fab/cytochrome c (cyt c). Incorrect model generated by AF3 as rank 1 (based on AF-MC) is shown in yellow. Acceptable model ranked 1 by pyDock-1VDW is shown in green (this model ranked 11 by AF-MC and by AF-MC/pyDock-1VDW z-score). For comparison, X-ray structure is shown in white (PDB 1WEJ)

created a *JSON* file including the input sequences for 20 jobs, which gave us 100 different models in total. Second, we uploaded the *JSON* file to the AF3 server and ran the 20 jobs individually. Finally, we downloaded the zipped files with the results and processed them. The protocol can be found at pyDock GitHub (https://github.com/pyDock/AF_pyDock). We calculated the same scores and quality parameters as in the previous case study (*see* Subheading 4.1).

We found that the model provided by AF3 as rank 1 (based on AF-MC score) is not acceptable (DockQ 0.06), and there are no other acceptable models within the top 10 predictions. Actually, the best-ranked acceptable model according to AF-MC is found with rank 12 (DockQ 0.79). Interestingly, this model is ranked 1 by pyDock-1VDW (100% VDW) (*see* Fig. 4). This is a nice example in which the scoring of pyDock is essential to improve the selection of models generated by AF3.

4.3 Docking with pyDock Can Help Where AF3 Fails: Fab Bound to Der p 1 Allergen

Another interesting case study is the complex formed by Fab fragment of monoclonal 4C1 antibody and *Dermatophagoides pteronyssinus* allergen 1 (Der p 1), which is included in the above-mentioned antibody–antigen benchmark set [33]. The 3D structure of this complex is available in the PDB with code 5VPG (supersedes 3RVW), so we can check the results of our modeling strategy. We applied AF3 to model the interaction between an antibody molecule (formed by heavy and light chains) and the monomeric Der p 1 protein. First, we created a *JSON* file including

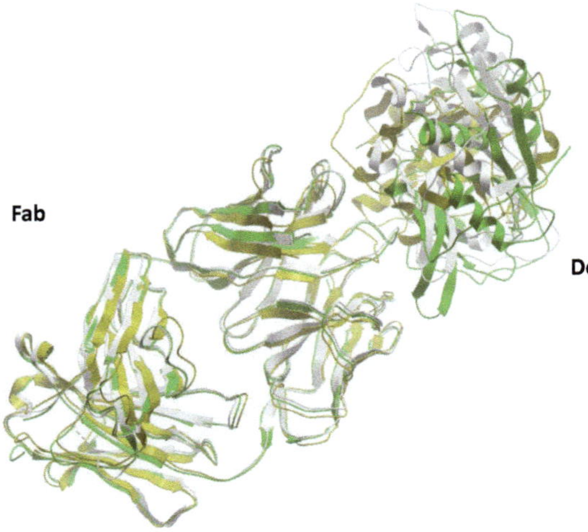

Fig. 5 Models generated by AF3 and pyDock for the example case: Fab/*Dermatophagoides pteronyssinus* allergen 1 (Der p 1). Incorrect model generated by AF3 as rank 1 (based on AF-MC) is shown in yellow (actually, no acceptable model was generated by AF3 in the entire set). Medium-quality model generated by pyDock docking and ranked 57 by default pyDock scoring (pyDock-0.1VDW) is shown in green. For comparison, X-ray structure is shown in white (PDB 5VPG)

the input sequences for 20 jobs, which gave us 100 different models in total. Second, we uploaded the *JSON* file to the AF3 server and ran the 20 jobs individually. Finally, we downloaded the zipped files with the results and processed them. The protocol can be found at pyDock GitHub (https://github.com/pyDock/AF_pyDock). We calculated the same scores and quality parameters as in the previous case studies.

The model provided by AF3 as rank 1 (based on the AF-MC score) is far from being acceptable (DockQ 0.07). Moreover, we found that there is no other acceptable model generated within the 100 generated models (best model is ranked 95, DockQ 0.15). In fact, none of the models generated by AF3 for this case showed model confidence above 0.8, which is an indication that AF models are not fully reliable. In blind conditions, we normally use this threshold to decide when to generate models by docking that can complement the low-reliability models generated by AF (*see* Fig. 1). Thus, we applied our standard pyDock protocol [15] to the X-ray unbound structures, and we obtained a medium-quality model ranked 57 by default pyDock scoring, with DockQ 0.44 (*see* Fig. 5). This is a nice example in which the failure of AF3 to generate suitable models for a protein–protein complex can be complemented by classical docking.

5 Notes

1. LocalColabFold can also be useful for installation on computing clusters, as this allows you to maintain an experimental version (where you can make changes) in your home directory. However, some clusters have security measures that prevent internet access. To install LocalColabFold in such environments, you must replicate the cluster directory structure on a workstation and follow the previous installation instructions. Afterward, you can copy the installation to the cluster. The steps are as follows:

```
sudo mkdir -p /path/existing/on/the/cluster
sudo chown -R user:group /path/existing/on/the/cluster
cd /path/existing/on/the/cluster
wget https://raw.githubusercontent.com/YoshitakaMo/localco-
labfold/main/install_colabbatch_linux.sh
bash install_colabbatch_linux.sh
scp -r localcolabfold/ user@cluster.com: /path/existing/on/
the/cluster
ssh user@cluster.com
echo 'export PATH="/path/existing/on/the/cluster/localcolab-
fold/colabfold-conda/bin:$PATH"' >> ~/.bashrc
source ~/.bashrc
```

Since the machine is not connected to the internet, the server query to generate the MSAs must be done on your local PC with the `--msa-only` parameter:

```
for h in v1 v2 v3; do
for i in ${CASE}/AF2/*fasta; do
colabfold_batch 'pwd'/$i 'pwd'/${i/.fasta}.colabfold.$h --
msa-only --save-recycles --model-type alphafold2_multimer_
${h} --recycle-early-stop-tolerance 0 --rank multimer --use-
dropout --num-recycle 20 --amber --use-gpu-relax
done
done
```

Afterward, the generated folders are copied to the cluster and executed:

```
rsync -av --relative $[CASE}/AF2/* user@cluster.com:/destina-
tion
```

Additionally, by default, when LocalColabFold is installed, only version 3 of the AlphaFold2-Multimer and AlphaFold2-Monomer models are downloaded. To automatically download the rest of the models before uploading, you need to run LocalColabFold locally for each version. Alternatively, you can download the weights for all three versions from storage.googleapis.com, using the URLs directly in your browser, or using `curl` or `wget` commands to get them, and then upload them to the cluster using `scp` and decompress them:

```
wget https://storage.googleapis.com/alphafold/alphafold_params_2021-07-14.tar . #Version 1
wget https://storage.googleapis.com/alphafold/alphafold_params_2022-03-02.tar . #Version 2
wget https://storage.googleapis.com/alphafold/alphafold_params_2022-12-06.tar . #Version 3
scp alphafold_params_*.tar user@cluster.com: /path/existing/on/the/cluster/localcolabfold/colabfold/params/
ssh user@cluster.com
cd /path/existing/on/the/cluster/localcolabfold/colabfold/params/
for i in alphafold_params_* ;do tar -xvf $i ;done
```

2. We recommend the execution of pyDock in a 32-bit operating system (OS), since the binary has been compiled for Linux 32-bit. However, it is possible to run pyDock in a 64-bit OS, after installing the necessary compatibility packages.

   In Debian-like distributions, you need to install the zlib1g 32-bit package by following these commands:

```
sudo dpkg --add-architecture i386
sudo apt-get update
sudo apt-get install zlib1g:i386
```

   In Red Hat/CentOS distributions, you need to install glibc and zlib 32-bit package by the following command:

```
sudo yum install glibc.i686 zlib.i686
```

3. The PDB names in the *DOCKNAME*`.ini` file must correspond to the exact names of the PDB files you are using (1C5K.pdb, 1c5k.pdb, pdb1c5k.ent.Z, etc.). If the chain ID in a PDB file is empty, use "-" or "" in the `mol` field to select that chain. If a PDB file contains several copies of the same protein, select only the desired chain by indicating its ID in the `mol` field. If a protein to dock contains several chains (for example, L and H chains for antibodies) that are relevant for docking, you may indicate their chain IDs in the `mol` field, separated by a comma. The `newmol` field is relevant only when running docking (not for the scoring calculations). It can be used to rename a protein chain in the docking output file, assigning it a name different

from `mol`, but it can also be left unchanged. The `newmol` chain IDs must be different for the receptor and the ligand molecules (so that their chains can be distinguished in the docking models).

4. For automatic use of SCWRL program within pyDock, after installing it locally (*see* Subheading 2), you should indicate the full path of the SCWRL binary by modifying the corresponding lines in the `$PYDOCK/pyDock3/etc./pydock.conf` file, as follows:

```
(...)
SCWRL=/<your-installation-directory>/scwrl3_lin/scwrl3
(...)
```

Acknowledgments

This work was partially supported by grant number PID2022-143215OB-I00, funded by MCIN/AEI/10.13039/501100011033/ FEDER, UE. V.M. was supported by "Programa Investigo. Mecanismo de Recuperación y Resiliencia. La Rioja" program, funded by Next Generation EU Plan de Recuperación, Transformación y Resiliencia.

References

1. Porter KA, Desta I, Kozakov D, Vajda S (2019) What method to use for protein-protein docking? Curr Opin Struct Biol 55:1–7. https://doi.org/10.1016/j.sbi.2018.12.010
2. Cheng TM-K, Blundell TL, Fernandez-Recio J (2007) pyDock: electrostatics and desolvation for effective scoring of rigid-body protein–protein docking. Proteins 68(2):503–515. https://doi.org/10.1002/prot.21419
3. Gabb HA, Jackson RM, Sternberg MJE (1997) Modelling protein docking using shape complementarity, electrostatics and biochemical information. J Mol Biol 272(1):106–120. https://doi.org/10.1006/jmbi.1997.1203
4. Chen R, Weng Z (2003) A novel shape complementarity scoring function for protein-protein docking. Proteins 51(3):397–408. https://doi.org/10.1002/prot.10334
5. Jiménez-García B, Pons C, Fernández-Recio J (2013) pyDockWEB: a web server for rigid-body protein–protein docking using electrostatics and desolvation scoring. Bioinformatics 29(13):1698–1699. https://doi.org/10.1093/bioinformatics/btt262
6. Fernández-Recio J, Totrov M, Abagyan R (2004) Identification of protein–protein interaction sites from docking energy landscapes. J Mol Biol 335(3):843–865. https://doi.org/10.1016/j.jmb.2003.10.069
7. Grosdidier S, Fernández-Recio J (2008) Identification of hot-spot residues in protein-protein interactions by computational docking. BMC Bioinformatics 9:447–447. https://doi.org/10.1186/1471-2105-9-447
8. Chelliah V, Blundell TL, Fernández-Recio J (2006) Efficient restraints for protein–protein docking by comparison of observed amino acid substitution patterns with those predicted from local environment. J Mol Biol 357(5):1669–1682. https://doi.org/10.1016/j.jmb.2006.01.001
9. Pérez-Cano L, Romero-Durana M, Fernández-Recio J (2017) Structural and energy determinants in protein-RNA docking.

Methods 118–119:163–170. https://doi.org/10.1016/j.ymeth.2016.11.001

10. Rodríguez-Lumbreras LA et al (2022) pyDockDNA: a new web server for energy-based protein-DNA docking and scoring. Front Mol Biosci 9. https://doi.org/10.3389/fmolb.2022.988996
11. Lopez-Robles C, Scaramuzza S, Astorga-Simon EN, Ishida M, Williamson CD, Baños-Mateos S, Gil-Carton D, Romero-Durana M, Vidaurrazaga A, Fernandez-Recio J, Rojas AL, Bonifacino JS, Castaño-Díez D, Hierro A (2023) Architecture of the ESCPE-1 membrane coat. Nat Struct Mol Biol 30(7): 958–969. https://doi.org/10.1038/s41594-023-01014-7
12. Venezian J, Bar-Yosef H, Ben-Arie Zilberman H et al (2024) Diverging co-translational protein complex assembly pathways are governed by interface energy distribution. Nat Commun 15:2638. https://doi.org/10.1038/s41467-024-46881-w
13. Alegre-Martí A et al (2023) A hotspot for post-translational modifications on the androgen receptor dimer interface drives pathology and anti-androgen resistance. Sci Adv 9:eade2175. https://doi.org/10.1126/sciadv.ade2175
14. Jiménez-Panizo A, Alegre-Martí A, Tettey TT, Fettweis G, Abella M, Antón R, Johnson TA, Kim S, Schiltz RL, Núñez-Barrios I, Font-Díaz J, Caelles C, Valledor AF, Pérez P, Rojas AM, Fernández-Recio J, Presman DM, Hager GL, Fuentes-Prior P, Estébanez-Perpiñá E (2022) The multivalency of the glucocorticoid receptor ligand-binding domain explains its manifold physiological activities. Nucleic Acids Res 50(22):13063–13082. https://doi.org/10.1093/nar/gkac1119
15. Rosell M, Rodríguez-Lumbreras LA, Fernández-Recio J (2020) Modeling of protein complexes and molecular assemblies with pyDock. In: Kihara D (ed) Protein structure prediction, Methods in molecular biology, vol 2165. Humana, New York. https://doi.org/10.1007/978-1-0716-0708-4_10
16. Green T, Qin C, Žídek A, Nelson AWR, Bridgland A, Penedones H, Petersen S, Simonyan K, Crossan S, Kohli P, Jones DT, Silver D, Kavukcuoglu K, Hassabis D (2020) Improved protein structure prediction using potentials from deep learning. Nature 577: 706–710. https://doi.org/10.1038/s41586-019-1923-7
17. Jumper J, Evans R, Pritzel A et al (2021) Highly accurate protein structure prediction with AlphaFold. Nature 596:583–589. https://doi.org/10.1038/s41586-021-03819-2
18. Bryant P, Pozzati G, Elofsson A (2022) Improved prediction of protein-protein interactions using AlphaFold2. Nat Commun 13: 1265. https://doi.org/10.1038/s41467-022-28865-w
19. Evans R, O'Neill M, Pritzel A, Antropova N, Senior A, Green T, Žídek A, Bates R, Blackwell S, Yim J, Ronneberger O, Bodenstein S, Zielinski M, Bridgland A, Potapenko A, Cowie A, Tunyasuvunakool K, Jain R, Clancy E, Kohli P, Jumper J, Hassabis D (2021) Protein complex prediction with AlphaFold-Multimer. bioRxiv. https://doi.org/10.1101/2021.10.04.463034
20. Wallner B (2023) Improved multimer prediction using massive sampling with AlphaFold in CASP15. Proteins 91:1734–1746. https://doi.org/10.1002/prot.26562
21. Olechnovič K, Valančauskas L, Dapkūnas J, Venclovas Č (2023) Prediction of protein assemblies by structure sampling followed by interface-focused scoring. Proteins 91:1724–1733. https://doi.org/10.1002/prot.26569
22. Abramson J, Adler J, Dunger J et al (2024) Accurate structure prediction of biomolecular interactions with AlphaFold 3. Nature 630: 493–500. https://doi.org/10.1038/s41586-024-07487-w
23. Mirdita M, Schütze K, Moriwaki Y et al (2022) ColabFold: making protein folding accessible to all. Nat Methods 19:679–682. https://doi.org/10.1038/s41592-022-01488-1
24. Lin Z et al (2023) Evolutionary-scale prediction of atomic-level protein structure with a language model. Science 379:1123–1130. https://doi.org/10.1126/science.ade2574
25. Baek M, Anishchenko I, Humphreys IR, Cong Q, Baker D, DiMaio F (2023) Efficient and accurate prediction of protein structure using RoseTTAFold2. bioRxiv. https://doi.org/10.1101/2023.05.24.542179
26. Wu R, Ding F, Wang R, Shen R, Zhang X, Luo S, Su C, Wu Z, Xie Q, Berger B, Ma J, Peng J (2022) High-resolution de novo structure prediction from primary sequence. bioRxiv. https://doi.org/10.1101/2022.07.21.500999
27. Canutescu AA, Shelenkov AA, Dunbrack RL Jr (2003) A graph-theory algorithm for rapid protein side-chain prediction. Protein Sci 12(9):2001–2014. https://doi.org/10.1110/ps.03154503
28. Moal IH, Jiménez-García B, Fernández-Recio J (2015) CCharPPI web server: computational characterization of protein-protein interactions from structure. Bioinf (Oxf) 31(1):123–125.

https://doi.org/10.1093/bioinformatics/btu594

29. Zhang Y, Skolnick J (2004) Scoring function for automated assessment of protein structure template quality. Proteins 57:702–710. https://doi.org/10.1002/prot.20264
30. Fanning SW, Walter R, Horn JR (2014) Structural basis of an engineered dual-specific antibody: conformational diversity leads to a hypervariable loop metal-binding site. Protein Eng Des Sel 27(10):391–397. https://doi.org/10.1093/protein/gzu033
31. Mendez R, Leplae R, De Maria L, Wodak SJ (2003) Assessment of blind predictions of protein-protein interactions: current status of docking methods. Proteins 52(1):51–67. https://doi.org/10.1002/prot.10393
32. Basu S, Wallner B (2016) DockQ: a quality measure for protein-protein docking models. PLoS One 11(8):e0161879. https://doi.org/10.1371/journal.pone.0161879
33. Guest JD, Vreven T, Zhou J et al (2021) An expanded benchmark for antibody-antigen docking and affinity prediction reveals insights into antibody recognition determinants. Structure 29(6):606–621.e5. https://doi.org/10.1016/j.str.2021.01.005

Chapter 16

Binding Affinity Prediction of Membrane Protein–Protein Complexes Using MPA-Pred

Fathima Ridha and M. Michael Gromiha

Abstract

Membrane protein–protein interactions, essential for various cellular functions, are primarily governed by their binding affinities. Although numerous computational tools are available for predicting the binding affinity of globular protein–protein complexes, no specific tool exists for membrane protein–protein complexes. Experimental approaches to determine these affinities are costly and time-consuming, limiting large-scale applications. In this chapter, we describe MPA-Pred, a novel machine-learning-based method developed to quantitatively predict the binding affinity of membrane protein–protein complexes. We have derived both structure- and sequence-based features, as well as classified membrane proteins based on their type and function, resulting in improved performance. Extensive evaluations, including training, cross-validation, and independent test sets, demonstrate that MPA-Pred outperforms existing methods in the prediction task. The method is implemented as a user-friendly web server, freely available at https://web.iitm.ac.in/bioinfo2/MPA-Pred/, and is built using HTML and Python, supporting recent versions of major browsers such as Chrome, Firefox, and Safari. The method serves as a valuable tool for large-scale predictions of novel membrane protein complex affinities and can aid in improving drug design strategies. Here, we overview the methodology, demonstrate the utility of our method through two case studies, and provide guidance on how to run the tool and interpret the results.

Key words Binding affinity, Membrane proteins, Function, Protein–protein interaction, Machine learning, Sequence-based features, Structure-based features, Web server

1 Introduction

Membrane proteins (MPs) are an essential group of proteins making up one-third of the human proteome [1]. Their significance lies in their diverse functions as molecular transporters, signal receptors, ion channels, and enzymes, rendering them primary targets for medicinal drugs [2]. Membrane proteins mostly function as complexes [3], and the strength of their binding (binding affinity), quantified by binding free energy (ΔG), is critical for their proper functioning [4]. However, due to their complex structural

Shahid M. Khan and Florencio Pazos (eds.), *Protein Evolution: Methods and Protocols*, Methods in Molecular Biology, vol. 2979, https://doi.org/10.1007/978-1-0716-4828-5_16,

characteristics, the binding affinity of MPs has been less extensively studied compared to globular proteins.

Experimentally, the binding affinity of protein–protein complexes is determined by different methods such as surface plasmon resonance (SPR), isothermal titration calorimetry (ITC), bio-layer interferometry (BLI), and fluorescence-based methods. Consequently, a significant amount of experimental affinity data is available in the literature. Recently, Ridha et al. (2023) developed thse first database, membrane protein complex binding affinity database (MPAD) [5], which contains ~5400 experimental affinity data of membrane protein–protein complexes and their mutants along with other features, with an easy-to-use interface.

On the other hand, resource-intensive and time-consuming experimental techniques make large-scale affinity prediction studies unfeasible, necessitating the indispensability of accurate computational approaches to determine the affinity of new MP complexes. Recent years have seen the emergence of various computational approaches developed for binding affinity prediction, ranging from rigorous free energy perturbation methods [6] to reliable machine learning- and deep learning-based models using structure- and sequence-based features [7–11]. However, all the methods are developed for protein–protein complexes in general and there is no specific method available in the literature for predicting the binding affinity of membrane protein–protein complexes.

To address this gap, we developed the first machine learning-based method for predicting the affinity of novel membrane protein complexes using several structure- and sequence-based features [12]. In terms of predictive performance, the method showed an average correlation (r) of 0.83 and mean absolute error (MAE) of 0.91 kcal/mol upon evaluation using the jackknife test, on a set of 114 complexes. Classification of complexes based on membrane protein type and function improved the performance of the method. Further, we found that aromatic and charged residues at the interface are important for predicting the binding affinity. Upon validation using an external blind test set, the method consistently predicted binding affinities within an MAE of 0.87 kcal/mol. We also developed a user-friendly web server, MPA-Pred, that can help with large-scale analysis.

This chapter provides a comprehensive overview of the architecture of MPA-Pred, highlighting the methodology behind its development and the prediction performance achieved by our method. We also introduce the web server, showcasing its user-friendly interface and providing step-by-step guidance on how to utilize the tool effectively and interpret the results. To demonstrate the reliability and practical utility of MPA-Pred, we present two case studies, where the predicted binding affinities closely align with experimental values, underscoring its potential applications in future research and large-scale analysis.

2 Materials and Methods

2.1 Dataset

For this study, we curated binding affinity data for membrane protein–protein complexes from our in-house MPAD [5] database. After applying stringent filtering criteria, we selected a training dataset comprising 114 membrane protein–protein complexes. These complexes were further categorized based on function (enzyme, transporter, receptor, and miscellaneous) and type of membrane protein (single pass and multi-pass). The dataset is available at https://web.iitm.ac.in/bioinfo2/MPA-Pred/dataset.html [12].

2.2 Algorithm

To uncover the factors influencing binding affinity in MPs, we computed various structure- and sequence-based features for each complex. Feature preprocessing was followed by an exhaustive systematic search for all three-feature combinations, and the best-performing combination was utilized for forward feature selection where the number of features per combination is incremented by 1, to include additional features. This process was repeated until no further improvement in Pearson's correlation (r) or reduction in mean absolute error (MAE) was observed. We developed multiple linear regression models for each class of complexes using the scikit-learn package in Python. The performance of the model was evaluated using Pearson's correlation and MAE, with the jackknife test or the leave-one-out cross-validation (LOOCV) employed to assess robustness. For validation, we used an independent test dataset comprising complexes with both known three-dimensional (3D) structures and modeled structures. Figure 1 illustrates the overall workflow of the MPA-Pred methodology, and further

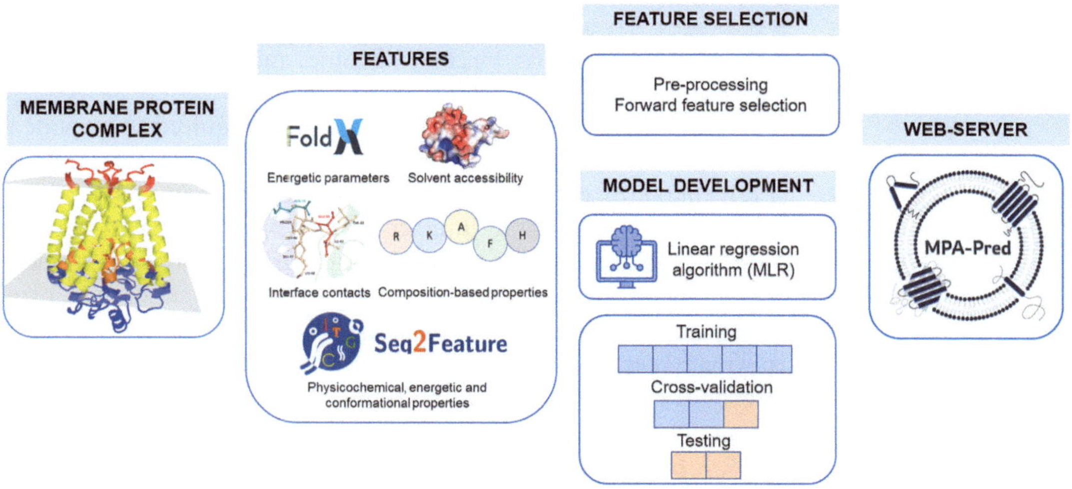

Fig. 1 Overview of the MPA-Pred methodology workflow

details on the features and methodology are provided in our earlier article [12].

2.3 Predictive Performance

The predictive quality of our method was assessed in detail in the original manuscript [12]. Our method for predicting the binding affinity of membrane protein–protein complexes demonstrated robust performance, achieving an overall correlation of 0.90 with an MAE of 0.72 kcal/mol on training and a correlation of 0.83 with an MAE of 0.91 kcal/mol in jackknife testing. Notably, the method performed exceptionally well for the multi-pass membrane protein class, with correlations of 0.94 and 0.89 for training and jackknife tests, respectively, and the lowest MAE of 0.44 and 0.59 kcal/mol. For single-pass proteins, the transporter class showed the best performance, achieving a correlation of 0.81 and an MAE of 0.90 kcal/mol in the jackknife test. When validated against a separate test dataset, our method predicted binding affinity with an MAE of 0.87 kcal/mol. A comparison of our method on the blind test dataset revealed that the performance of the present method is better than existing methods available in the literature for predicting the affinity of protein–protein complexes. The superior performance of our method can be attributed to (i) hybrid model, which integrates both structure- and sequence-based features, (ii) classification approach based on structural and functional characteristics, and (iii) specifically designed for membrane protein-protein complexes.

2.4 Web Server

We have developed a user-friendly web server, MPA-Pred, which is freely accessible at https://web.iitm.ac.in/bioinfo2/MPA-Pred/ for predicting the binding affinity of membrane protein–protein complexes (*see* Fig. 1). The frontend is built using HyperText Markup Language (HTML), Cascading Style Sheet (CSS), JavaScript, and Bootstrap, while the backend functionality uses Python 3 packages. To use the web server, the end user needs a modern web browser (tested on Firefox, Chrome, and Safari) and an Internet connection.

To submit a job for prediction, users are required to provide a wild-type structure, either by uploading a Protein Data Bank (PDB) file or by entering a valid PDB accession code. Additionally, users need to specify the interacting protein chain identifiers (*see* **Note 1**). They must also select the membrane protein class, indicating whether it is multi-pass or single pass. For single-pass proteins, the functional class must be chosen from the options: enzyme, receptor, transporter, or miscellaneous. The output page displays the predicted value of binding affinity ΔG and the dissociation constant (K_d) computed using the equation, $\Delta G = -RT \ln (1/K_d)$. For guidance on submitting jobs and navigating the platform, a detailed tutorial page is accessible at https://web.iitm.ac.in/bioinfo2/MPA-Pred/tutorial.html.

3 Case Study

3.1 Case 1: Potassium Channel/Charybdotoxin Complex

The KcsA potassium channel in complex with the high-affinity antagonist charybdotoxin provides critical insights into the mechanism of channel inhibition. Structural data reveal that charybdotoxin binds to the closed conformation of KcsA by making precise contact with the extracellular surface of the channel. This interaction effectively blocks the ion-conducting pore, preventing ion passage through the channel [13]. The KcsA–charybdotoxin complex has been deposited in the Protein Data Bank (PDB ID: 2A9H). This complex has an experimental binding affinity of −8.24 kcal/mol [13]. In Fig. 2, we outline the step-by-step procedure for utilizing MPA-Pred to predict the binding affinity of this complex.

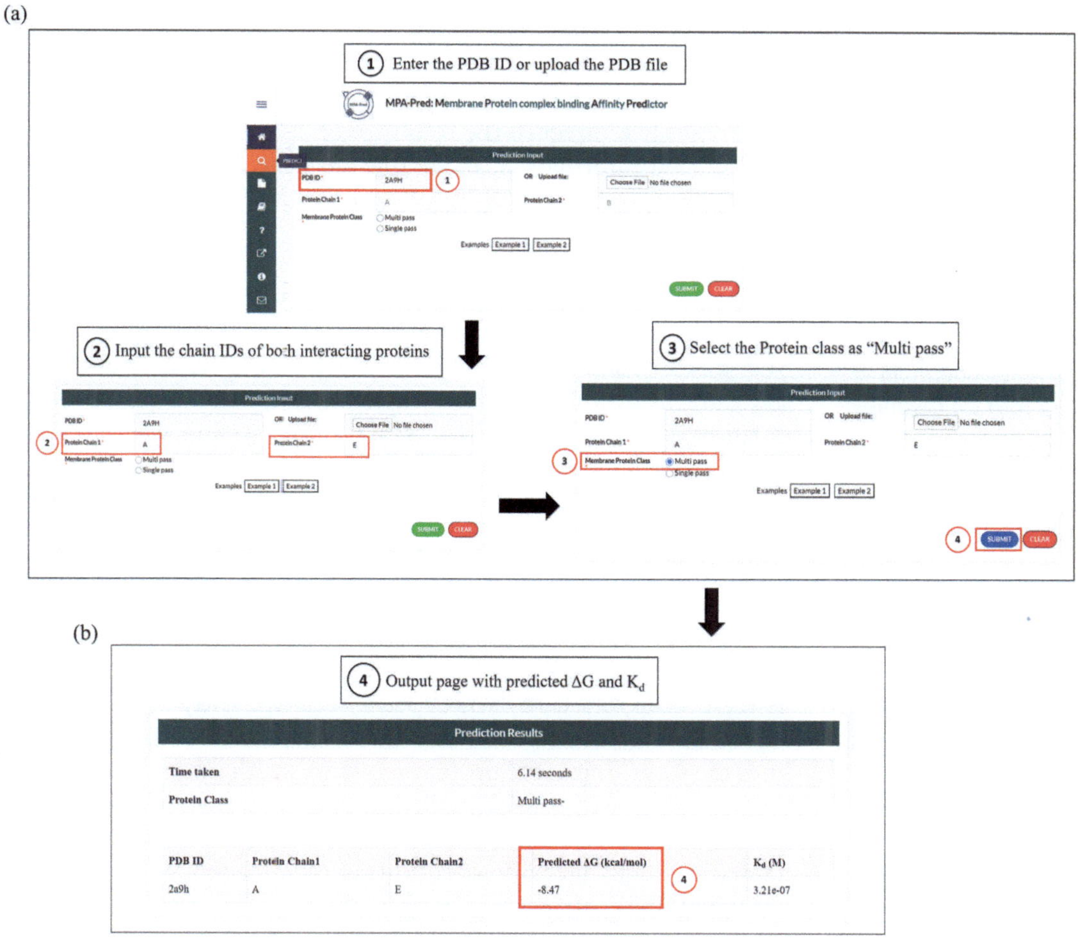

Fig. 2 Example of predicting the binding affinity of KcsA–charybdotoxin complex using MPA-Pred. (**a**) Input query and (**b**) output result page

1. The "Predict" page can be accessed from the sidebar menu, offering users two options to input the complex structure. If the PDB ID of the complex is known, it can be entered directly into the designated input field; for example, enter "2A9H" in this case. Alternatively, users with a preprocessed PDB file, such as one with a renumbered structure or specific chains selected, can upload their file in PDB format using the "Upload File" input field.
2. Provide the chain IDs for both interacting proteins in the respective input fields.
3. Under the "Membrane Protein Class," two options are available: single pass and Multi-pass. Single-pass membrane proteins traverse the membrane only once, whereas multi-pass membrane proteins span the membrane more than once. This classification can be determined from the UniProt [14] "Sub-cellular Location" annotation. Since the KcsA potassium channel is classified as a multi-pass membrane protein, select the "Multi-pass" radio button in the prediction form.
4. Click the "Submit" button to start the prediction.
5. Once the prediction is complete, the web browser will automatically redirect to the results page. The output table displays detailed information, including the PDB ID, Chain 1, Chain 2, Predicted ΔG (kcal/mol), and Kd (M).

3.2 Case 2: Staphylococcal Enterotoxin Type A (SEA)-Glycoprotein 130 (gp130) Complex

Staphylococcal enterotoxin type A (SEA), a superantigen from the Staphylococcus bacteria, has been shown to bind to the cytokine signaling receptor glycoprotein 130 (gp130) and forms SEA-gp130 complex [15]. This interaction may lead to abnormal signaling, contributing to the emetic response and immune dysregulation observed in staphylococcal food poisoning. However, the complex structure has not been resolved in the PDB. Here, we provide a detailed example of predicting the binding affinity of the SEA-gp130 complex using the modeled structure.

3.2.1 Preparing the Wild-Type Complex Structure

If the structure of the complex is not experimentally solved, it can be modeled using structure prediction tools. For small-scale predictions, you can use the ColabFold [16] (https://colab.research.google.com/github/sokrypton/ColabFold/blob/main/AlphaFold2.ipynb) for easy and interactive execution. Fig. 3 outlines the steps: Input the protein sequences with inter-protein chain breaks (e.g., MK...SK:PI...SR) for complex modeling. Click "Runtime" → "Run all" to execute the pipeline. Finally, a "jobname.result.zip" file with all outputs is downloaded, which includes predicted PDB structures (ranked by pLDDT and pTMscore), model quality plots, and MSA coverage plots, which can be used to select the best model. For large-scale predictions, the user can

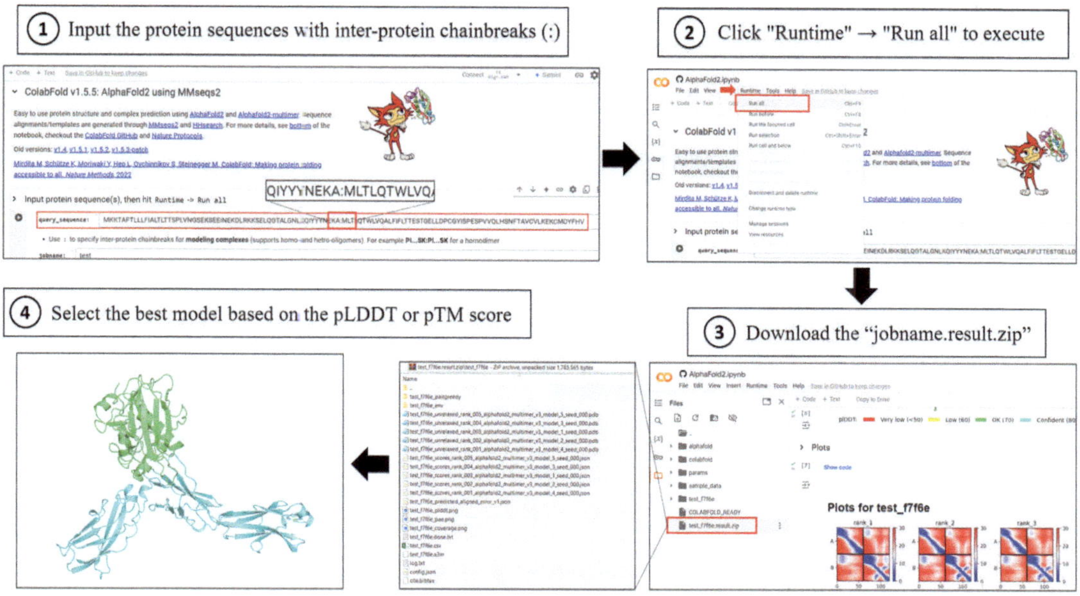

Fig. 3 Detailed steps for modeling the SEA-gp130 complex structure using ColabFold

download and install the AlphaFold-Multimer [17] model (https://github.com/google-deepmind/alphafold) or the local version of ColabFold.

3.2.2 Predicting the Binding Affinity of SEA-gp130 Complex

1. In the "Predict" page, the user is required to provide the wild-type complex structure. Hence, upload the modeled complex file in PDB format.
2. Enter the chain IDs of both the interacting proteins in the respective input boxes.
3. Select "Single pass" under the "Membrane protein class" options, as the protein is a single-pass membrane protein.
4. The user should select "Receptor" as the functional class provided in the dropdown for the prediction. The functions of the interacting proteins are primarily classified as enzymes, transporters, receptors, or miscellaneous. Functional information for receptors, transporters, and enzymes can be obtained from UniProt [14], Transporter Classification Database (TCDB) [18], and BRENDA [19], respectively. If the function is unknown, select "miscellaneous."
5. Click the "Submit" button to start the prediction.
6. After the prediction is completed, the user's web browser is automatically redirected to the page with the results (Fig. 4). The first box on the result page shows the time taken for the prediction (*see* **Note 2**) and the membrane protein class. The second table provides details including the PDB ID, Chain 1, Chain 2, Predicted ΔG (kcal/mol), and Kd (M).

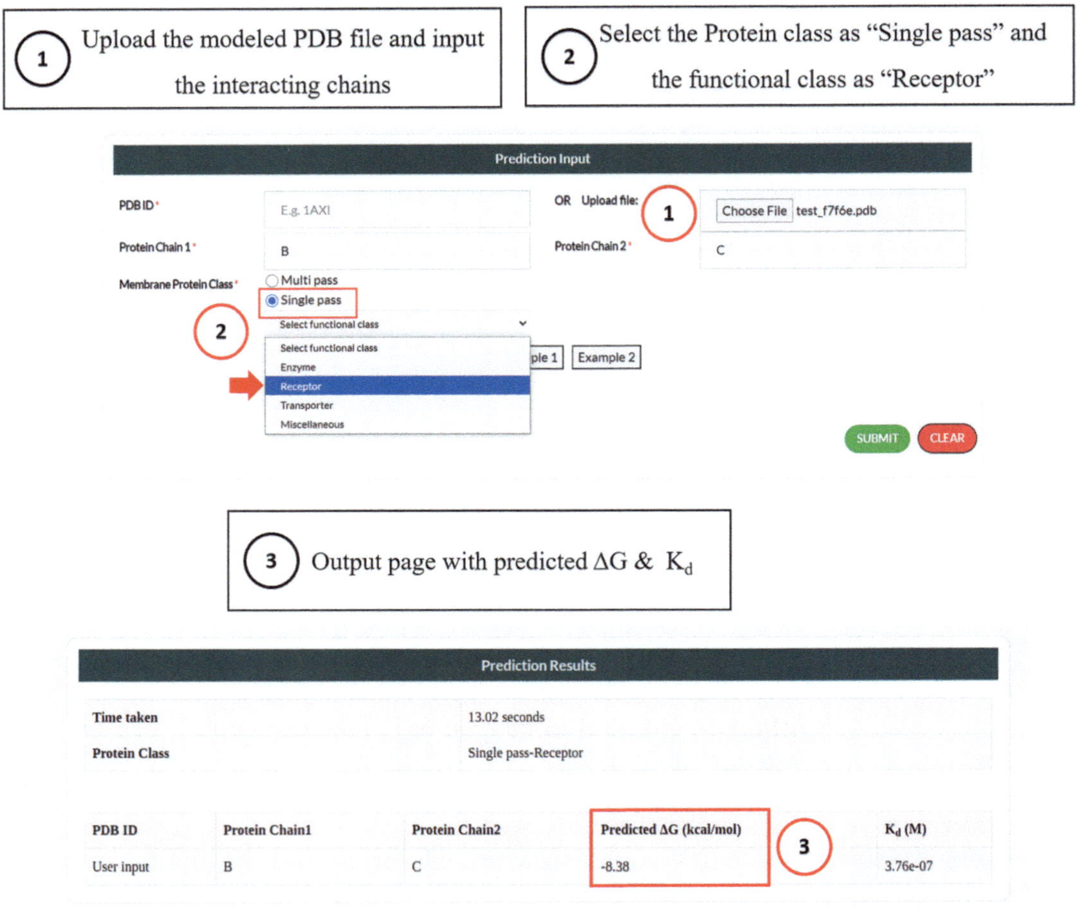

Fig. 4 Input and output pages of the MPA-Pred web server for predicting the binding affinity of the SEA-gp130 complex using the modeled structure. Steps 1 and 2 (highlighted in red) represent the input query, and Step 3 shows the prediction output

Overall, we conclude from both case studies that our web server generates accurate predictions closely aligning with the experimental value, offering valuable insights for large-scale analysis of membrane protein interactions, which can accelerate drug discovery and therapeutic design.

4 Notes

1. If your input is not being accepted by the server, ensure that all required fields marked with * are filled, the PDB ID and chain identifiers are correct and available in the PDB database, and the uploaded PDB file adheres to the proper PDB format.
2. For the MPA-Pred server, the prediction typically completes quickly, though the waiting time can vary depending on the

size of the membrane protein complex and the server's job queue. On average, the waiting time is around 2–3 min.

Acknowledgments

We thank the Department of Biotechnology and Indian Institute of Technology Madras for computational facilities. Authors also thank the members of the Protein Bioinformatics Laboratory for their valuable discussions and help at various stages of the work. The work is partially supported by the Department of Biotechnology (DBT), India, to MMG (BT/PR40237/BTIS/137/79/2023).

References

1. Almén MS, Nordström KJ, Fredriksson R, Schiöth HB (2009) Mapping the human membrane proteome: a majority of the human membrane proteins can be classified according to function and evolutionary origin. BMC Biol 7:50
2. Almeida JG, Preto AJ, Koukos PI, Bonvin A, Moreira IS (2017) Membrane proteins structures: a review on computational modeling tools. Biochim Biophys Acta Biomembr 1859: 2021–2039
3. Daley DO (2008) The assembly of membrane proteins into complexes. Curr Opin Struct Biol 18:420–424
4. Nikam R, Ridha F, Jemimah S, Yugandhar K, Gromiha MM (2024) Binding affinity changes upon mutation in protein–protein complexes. In: Gromiha MM (ed) Protein mutations: consequences on structure, functions, and diseases. World Scientific, Singapore, pp 105–122
5. Ridha F, Kulandaisamy A, Michael Gromiha M (2023) MPAD: a database for binding affinity of membrane protein-protein complexes and their mutants. J Mol Biol 435:167870
6. Siebenmorgen T, Zacharias M (2020) Computational prediction of protein–protein binding affinities. WIREs Comput Mol Sci 10:e1448
7. Vangone A, Bonvin AMJJ (2017) PRODIGY: a contact-based predictor of binding affinity in protein-protein complexes. Bio Protoc 7: e2124
8. Romero-Molina S, Ruiz-Blanco YB, Mieres-Perez J, Harms M, Münch J, Ehrmann M, Sanchez-Garcia E (2022) PPI-affinity: a web tool for the prediction and optimization of protein-peptide and protein-protein binding affinity. J Proteome Res 21:1829–1841
9. Yugandhar K, Gromiha MM (2014) Protein-protein binding affinity prediction from amino acid sequence. Bioinformatics 30:3583–3589
10. Abbasi WA, Yaseen A, Hassan FU, Andleeb S, Minhas FUAA (2020) Island: in-silico proteins binding affinity prediction using sequence information. Biodata Min 13:20
11. Nikam R, Yugandhar K, Gromiha MM (2023) Deep learning-based method for predicting and classifying the binding affinity of protein-protein complexes. Biochim Biophys Acta Proteins Proteom 1871:140948
12. Ridha F, Gromiha MM (2024) MPA-Pred: a machine learning approach for predicting the binding affinity of membrane protein-protein complexes. Proteins 92:499–508
13. Yu L, Sun C, Song D et al (2005) Nuclear magnetic resonance structural studies of a potassium channel-charybdotoxin complex. Biochemistry 44:15834–15841
14. UniProt Consortium (2021) UniProt: the universal protein knowledgebase in 2021. Nucleic Acids Res 49:D480–D489
15. Uzunçayır S, Vera-Rodriguez A, Regenthal P, Åbacka H, Emanuelsson C, Bahl CD, Lindkvist-Petersson K (2022) Analyses of the complex formation of staphylococcal enterotoxin a and the human gp130 cytokine receptor. FEBS Lett 596:910–923
16. Mirdita M, Schütze K, Moriwaki Y, Heo L, Ovchinnikov S, Steinegger M (2022) ColabFold: making protein folding accessible to all. Nat Methods 19:679–682
17. Bryant P, Pozzati G, Elofsson A (2022) Improved prediction of protein-protein interactions using AlphaFold2. Nat Commun 13: 1265
18. Saier MH, Reddy VS, Moreno-Hagelsieb G, Hendargo KJ, Zhang Y, Iddamsetty V,

Lam K, Tian N, Russum S, Wang J, Medrano-Soto A (2021) The transporter classification database (TCDB): 2021 update. Nucleic Acids Res 49:D461–D467

19. Chang A, Jeske L, Ulbrich S, Hofmann J, Koblitz J, Schomburg I, Neumann-Schaal M, Jahn D, Schomburg D (2021) BRENDA, the ELIXIR core data resource in 2021: new developments and updates. Nucleic Acids Res 49: D498–D508

Part IV

Integrated 1D-Sequence/3D-Structure Pipelines

Chapter 17

Computational Methods to Engineer Cas Proteins for Efficient Genome Editing

Muhammad Qaiser Fatmi, Aneeqa Nadeem, Mehraj Abbasov, and Muhammad Sajjad

Abstract

The clustered regularly interspaced short palindromic repeats/CRISPR-associated protein (CRISPR/Cas) system has revolutionized genome editing through programmable, sequence-specific deoxyribonucleic acid (DNA) targeting. Yet, its broader application remains limited by off-target effects and context-dependent efficiency. To address these challenges, we present an integrated computational protocol with easy-to-do steps for researchers to guide the rational design of CRISPR/Cas variants with improved stability and specificity. The integrated workflow begins with coevolutionary coupling analysis to identify conserved and covarying residues critical for function. These residues are then evaluated for energetically favorable substitutions through mutant stability prediction, followed by network centrality analysis to evaluate the impact of mutations on intramolecular communication pathways, preserving key allosteric interactions. Finally, molecular dynamics (MD) simulations validate the structural integrity and dynamic behavior of the selected variants. Network analysis and molecular dynamics (MD) simulations are applied iteratively, allowing insights from MD to refine network-based evaluations and vice versa. This multiscale strategy offers a streamlined and systematic approach for engineering optimized Cas proteins for genome editing applications.

Key words CRISPR/Cas9, Genome editing, Coevolutionary coupling, Mutant stability prediction, Network centrality, Molecular dynamics simulation, Protein engineering

1 Introduction

Genome editing technology modifies deoxyribonucleic acid (DNA) sequences at targeted genomic locations by creating new DNA sequences through site-specific DNA breaks. The site-specific DNA breaks are repaired by innate DNA repair mechanisms [1]. The four programmable nucleases, namely zinc finger nucleases (ZFN), meganucleases, transcription activator-like effector nucleases (TALEN), and clustered regularly interspaced short palindromic repeats/CRISPR-associated protein system (CRISPR/Cas), are currently available for genome editing. The

Shahid M. Khan and Florencio Pazos (eds.), *Protein Evolution: Methods and Protocols*, Methods in Molecular Biology, vol. 2979, https://doi.org/10.1007/978-1-0716-4828-5_17,

CRISPR/Cas dominated other genome editing systems because it is simple, cheaper, and highly efficient. The widely used Cas9 endonuclease contains two nuclease domains, resolvase C (RuvC) and Histidine–Asparagine–Histidine (HNH), along with a protospacer adjacent motif (PAM)-interacting (PI) domain. The RuvC and HNH domains function together to cleave double-stranded DNA, but this cleavage is dependent on the recognition of a protospacer adjacent motif (PAM) sequence near the target site. Once a double-stranded DNA break (DSB) is generated, the cell employs endogenous repair mechanisms, primarily non-homologous end joining (NHEJ) or homology-directed repair (HDR), to repair the break. While NHEJ is error-prone and often leads to insertions or deletions (indels), HDR requires a homologous DNA template, which can be exogenous single-stranded DNA (ssDNA) or double-stranded DNA (dsDNA), a sister chromatid, a homologous chromosome, or closely related sequences [2]. Cas9's ability to target specific genome segments is driven by two key factors: sequence complementarity between the single guide ribonucleic acid (sgRNA) and the target DNA and PAM recognition, which ensures target specificity (*see* Fig. 1). The CRISPR/Cas9 system requires an sgRNA, which is an engineered fusion of CRISPR RNA (crRNA) and trans-activating CRISPR RNA (tracrRNA). In an optimized system, the crRNA is a synthetic ~20-nucleotide sequence complementary to the target DNA, while the tracrRNA scaffold facilitates Cas9 binding and activity. Since Cas9 has two nuclease domains, it can cleave both sense and antisense strands of the DNA. However, a PAM sequence is essential for target site recognition, meaning that not every DNA sequence can serve as a CRISPR/Cas target site. The PAM acts as a binding signal for the Cas enzyme, and without it, Cas9 cannot cleave the DNA. Thus, PAM recognition is a critical requirement for all CRISPR/Cas-based genome editing strategies [3].

Recent advancements in computational biology have shown great promise for overcoming these challenges. By integrating machine learning (ML), molecular dynamics simulations, and structure-guided design, researchers can now engineer Cas variants with enhanced specificity, expanded PAM compatibility, and reduced off-target effects. Class II CRISPR systems, utilizing Cas9 and Cas12a, dominate genome editing applications due to their simplicity. However, Class I systems, which use multiprotein effector complexes like Cascade–Cas3, offer unique advantages such as long-range DNA deletions and reduced off-target activity in complex genomes. Computational tools like AlphaFold 3 [4] and Rosetta [5] can accelerate the engineering of these systems by predicting protein–DNA interactions and enabling in silico mutagenesis to refine nuclease activity. Furthermore, deep learning frameworks trained on multiomics datasets now guide the design of synthetic Cas chimeras, such as xCas9 and SpRY-Cas9, which recognize noncanonical PAM sequences while maintaining high precision [6].

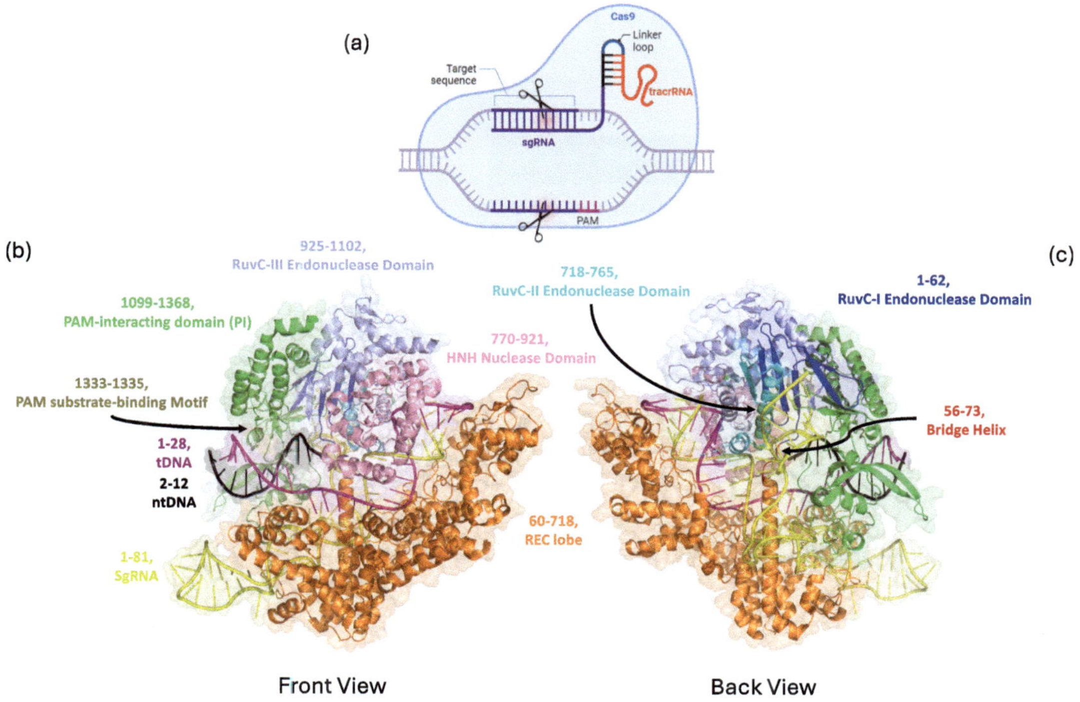

Fig. 1 (**a**) Clustered regularly interspaced short palindromic repeats/CRISPR-associated protein 9 (CRISPR/Cas9) genetic scissors. To edit a genome using the CRISPR/Cas9 genetic scissor, a single guide ribonucleic acid (sgRNA) is artificially constructed, which matches the deoxyribonucleic acid (DNA) sequence where the cut is to be made. The scissor protein, Cas9, forms a complex with sgRNA and directs the scissor to the precise targeted location in the genome where the cut is desired to be made. (**b**, **c**) Front and back views of the three-dimensional (3D) structure of the scissor protein Cas9 (Protein Data Bank [PDB] entry 4UN3)

Computational pipelines have been used to develop some Cas variants optimized for polyploid crops wherein off-target effects are too high [7]. Similarly, in mammalian systems, artificial intelligence (AI)-driven platforms predict chromatin accessibility and epigenetic barriers to improve editing efficiency in hard-to-transfect cell types [8]. This protocol chapter integrates computational methodologies for CRISPR/Cas engineering (*see* Fig. 2), highlighting strategies to tailor these molecular scissors for applications ranging from designing climate-resilient crops to therapeutic genome corrections in humans.

2 Materials

2.1 Data Source

1. Protein structure data can be retrieved from Protein Data Bank (PDB) (https://www.rcsb.org).
2. Sequence data can be retrieved from UniProt (https://www.uniprot.org/uniprotkb).

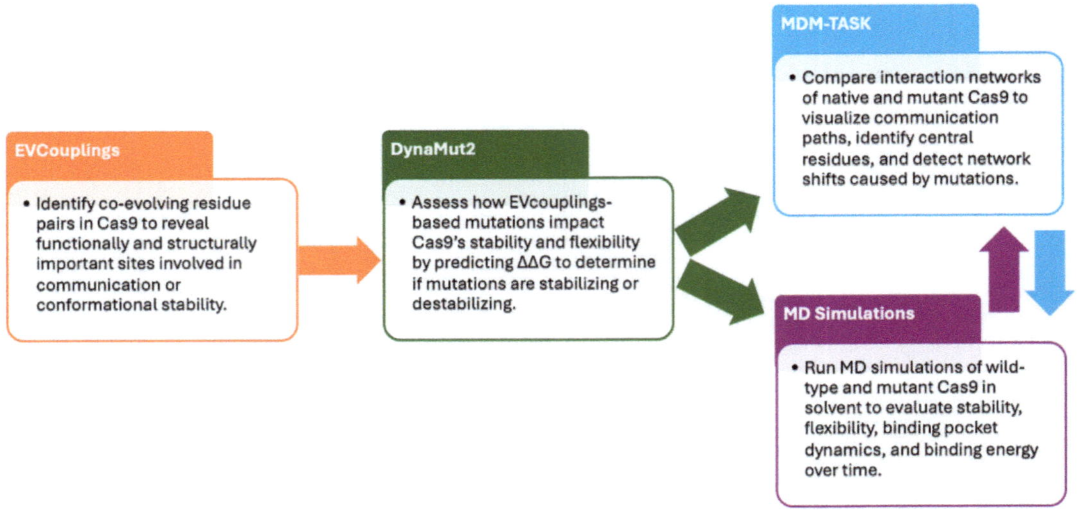

Fig. 2 Integrated computational workflow for rational engineering of Cas9

2.2 Web Servers

1. Evolutionary couplings (EVcouplings) web server (https://v2.evcouplings.org).
2. DynaMut2 web server (https://biosig.lab.uq.edu.au/dynamut2).
3. MDM-TASK web server (https://mdmtaskweb.rubi.ru.ac.za).
4. Chemistry at HARvard Macromolecular Mechanics-Graphical User Interface (CHARMM-GUI) (https://www.charmm-gui.org).

2.3 Software and Tools

1. We assume that the following tools and software are already installed on your system:
   (a) Python Molecular Viewer (PyMOL) version 2.3.0 (https://www.pymol.org/).
   (b) GROningen MAchine for Chemical Simulations (GROMACS) version 2020.4 (https://www.gromacs.org).
   (c) AmberTools version 2020 (https://ambermd.org/AmberTools.php).
   (d) Visual Molecular Dynamics (VMD) version 1.9.4 (https://www.ks.uiuc.edu/Research/vmd).
   (e) Chimera version 1.14 (https://www.cgl.ucsf.edu/chimera/download.html).

3 Methods

3.1 Predicting Residue Coevolution of a Cas9 Using EVcouplings Web Server

This section provides a step-by-step guide for predicting coevolutionary relationships between amino acid residues in Cas9 using EVcouplings [9]. These coevolving residues are mutated together across evolution, indicating that they are structurally or functionally connected. The identified coevolving residues guide the engineering of Cas9 to enhance activity, increase specificity, and minimize off-target effects by prioritizing mutations that are evolutionarily supported.

3.1.1 Accessing the EVcouplings Web Server

1. Open a web browser and navigate to the EVcouplings server at https://v2.evcouplings.org and register yourself (*see* **Note 1**).
2. Once the page is open, click on the option to "Predict your protein."

3.1.2 Input Sequence Selection

1. Obtain the amino acid sequence of Cas9.
   (a) You can retrieve it from UniProt (e.g., *Streptococcus pyogenes* Cas9: Q99ZW2) or from the PDB database (e.g., 4UN3).
2. Copy the Cas9 sequence in FASTA format.
3. Paste the sequence into the input box on the EVcouplings submission page or simply enter UniProt ID "Q99ZW2."
4. Cas9 consists of 1368 amino acids, which exceeds the web server's sequence length limit of 700. To proceed, use the slider to select a specific region. For this protocol, we have chosen the PAM-interacting domain (PI) with a sequence range of 1099–1368.

3.1.3 Configuring EVcouplings Parameters

1. Next, click on "Show Advanced Settings."
2. Go to Alignments > Homology search and do the following settings:
   (a) Alignment threshold type > bitscore (length-normalized).
   (b) Bitscore sequence inclusion thresholds > keep intermediate threshold values as 0.10, 0.30, 0.50, and 0.70. A lower threshold value incorporates a greater number of distantly related sequences, enhancing evolutionary diversity but potentially introducing alignment noise and non-homologous sequences. In contrast, a higher threshold value selects only closely related sequences, ensuring strong similarity to the query and producing a cleaner, high-quality alignment, though it may limit the coevolutionary signal due to reduced sequence variation.

   (c) Search iterations >5.
   (d) Sequence database > UniRef90.

3. Go to Alignments > Sequence and position filters:
   (a) Position filter >70%: Removes alignment columns with more than 30% gaps, ensuring higher reliability and reducing alignment noise.
   (b) Sequence fragment filter >50%: Retains only sequences covering at least 50% of the target length, discarding partial or fragmented sequences.
   (c) Removing similar sequences >90%: Eliminates sequences with more than 90% identity to reduce redundancy and prevent bias from overrepresented species.
   (d) Down weighting similar sequences >80%: Assigns lower weights to sequences with over 80% identity, ensuring a balanced contribution of diverse sequences to the analysis.
4. Go to Evolutionary couplings > Statistical inference method > Pseudo-likelihood maximization: Uses a probabilistic model to approximate the likelihood of a sequence based on the observed multiple sequence alignment (MSA), considering each residue pair independently while maximizing the likelihood of the entire alignment. This approach is more accurate than the mean-field approximation in detecting direct interactions; however, it is computationally expensive.
5. Go to Results evaluation:
6. Contact distance cutoff >5 Å: Residues are considered in contact if their interatomic distance is within 5 Å, which helps define residue–residue interactions in structural analysis.
7. PDB structure search method > Conservative: A strict method is used to find homologous structures, ensuring only highly reliable matches with strong sequence similarity are selected.
8. Bitscore threshold for finding known homologous three-dimensional (3D) structures >0.50: A minimum bitscore of 0.50 is required to identify homologous structures in the PDB, ensuring that only significant matches with strong sequence similarity are considered.

3.1.4 Submitting the Job

1. Enter the job name as PAM-interacting domain (PI).
2. Enter your email address.
3. Click "Submit" and wait for processing (this may take several hours).
4. You will receive an email with a link to your results once processing is completed.

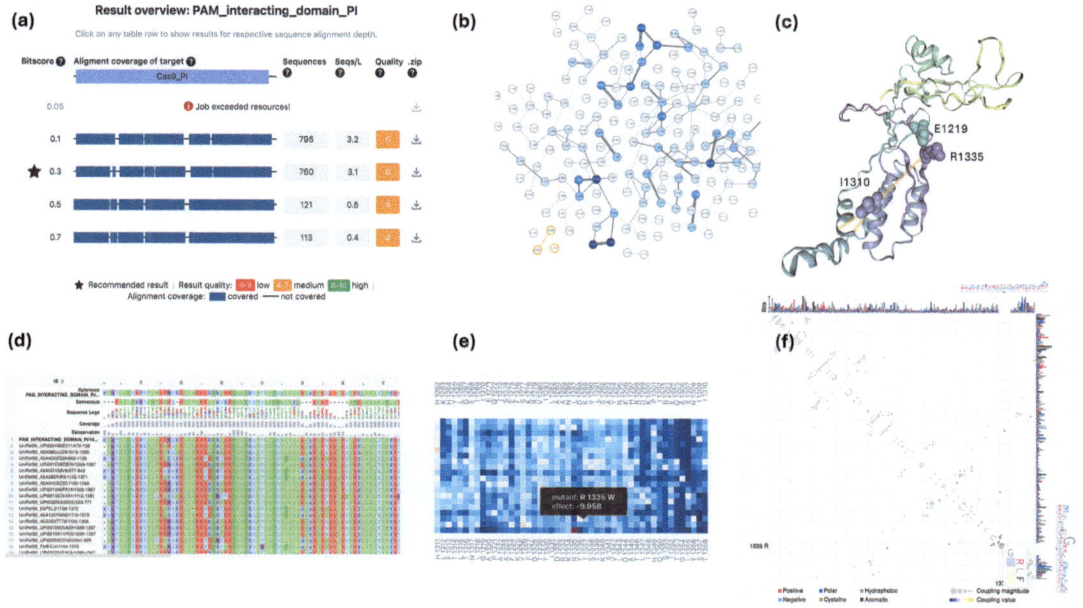

Fig. 3 An example of interactive and comprehensive EVcouplings analysis of Cas9 including (**a**) an overview of the results, (**b**) identification of evolutionarily coupled residue pairs, (**c**) 3D structural representation of coevolving residues, (**d**) multiple sequence alignment of homologous Cas9 sequences, (**e**) predicted mutation effect strengths, and (**f**) a probabilistic coupling model visualizing the magnitude of residue–residue couplings

3.1.5 EVcouplings Output

1. Once the job is completed, click the provided link in your email to access the results overview (*see* Fig. 3). In our case, a bitscore value of 0.3 produced the best results, yielding a Quality (Q) score of 6 and identifying 760 homologous sequences. You can explore an interactive session with the following details:
2. Evolutionary Coupling Maps—a visual representation of coevolving residue pairs.
3. Network of Strongly Coupled Sites—higher scores indicate strong evolutionary coupling, suggesting functional or structural significance.
4. Mutation Analysis—insights into potential functional impacts of mutations.
5. MSA Alignment Visualization—highlights conserved and variable regions within the multiple sequence alignment.
6. Probability Models—can be visualized on https://evzoom.org for further analysis.
7. Results can also be downloaded in .zip format directly from the results overview page. The zip archive contains the following results:

8. Evolutionary couplings.
    (a) Couplings file (all pairs, with distance information).
    (b) Couplings file (long-range pairs, with distance information).
    (c) Contact map plots.
    (d) Strongly coupled residues.
    (e) EVzoom JSON file (can be visualized on https://evzoom.org).
    (f) EVcouplings model parameters (binary file, for use with EVcouplings Python package).
9. Sequence alignment.
    (a) Sequence alignment (in A2M format, only matching and deleted positions relative to the target sequence are kept, while insertions are removed for a cleaner alignment).
    (b) Target sequence region used for computing alignment.
    (c) Positional amino acid frequencies/conservation.
10. Mutation effects.
    (a) Predicted single substitution effects.
    (b) Single substitution matrix plots.
11. Experimental 3D structures (for result evaluation).
    (a) Homologous 3D structures remapped to target numbering.
    (b) Homologous 3D structures remapped to target numbering and residue types (backbone only).
    (c) Summary table of identified homologous 3D structures.

3.1.6 Identifying Key Mutations for Cas9 Engineering

1. Use the results to identify key residues for mutation that could improve Cas9 activity, specificity, or PAM compatibility.
2. Navigate to the settings menu and adjust various parameters interactively to observe their effects.
3. Focus on residues within the PAM-interacting domain (PI), such as Arg1335, for further analysis.
    (a) The R1335-I1310 pair has a moderate probability (0.49) of evolutionary coupling, with a weak coupling score (−0.05) and a large distance (17.6 Å), suggesting no direct interaction. This indicates minimal coevolutionary constraints, making it unlikely to significantly impact Cas9 function. However, potential allosteric effects may require further investigation.
    (b) The evolutionary coupling map identifies R1335-E1219 as a comparatively strongly coevolving pair with high

probability (0.92) and a strong coupling score (2.39), indicating functional significance. Their close distance (2.5 Å) suggests direct interactions, potentially stabilizing the PAM-interacting domain (PI) in Cas9. This interaction may influence PAM recognition, making it a key target for mutational studies to enhance Cas9 specificity.

3.2 Mutational Analysis and Stability Prediction Using DynaMut2

In this section, we will use the DynaMut2 web server [10] that predicts the impact of point mutations on protein stability and flexibility by integrating normal mode analysis and machine learning (random forest model). DynaMut2 can also be locally installed for automated high-throughput predictions.

3.2.1 Accessing the DynaMut2 Web Server

1. Open a web browser and navigate to the DynaMut2 web server https://biosig.lab.uq.edu.au/dynamut2/.
2. DynaMut2 offers three options:
   (a) Single Mutation.
   (b) Multiple Mutations.
   (c) Normal Mode Analysis (NMA).
3. For this protocol, click on "Single Mutation" to proceed.

3.2.2 Upload the Protein Structure and Define Mutations

1. Single Mutation again offers two options:
   (a) Single Mutation (enter details for one mutation).
   (b) Mutation List (provide multiple mutations in a list format).
2. For this protocol, select the "Mutation List" option.
3. Enter the PDB accession code "4UN3" to directly fetch the 3D coordinates of the Cas9 protein from the PDB database.
4. Alternatively, you can upload the PDB file from your local machine using the "Upload" option.
5. Now, create a text file (e.g., *mutations.txt*) with the mutation details in the following format (*see* **Note 2**):

   B R1333A.

   B R1335E
6. Optionally enter your email address and click "Submit" to proceed.

3.2.3 Interpret the Results

1. Once the calculation is complete, the results can be analyzed by clicking the "Details" option.
2. Check the values of predicted stability change: $\Delta\Delta G^{\mathrm{Stability}} = \Delta G_{\mathrm{mut}} - \Delta G_{\mathrm{wt}}$.

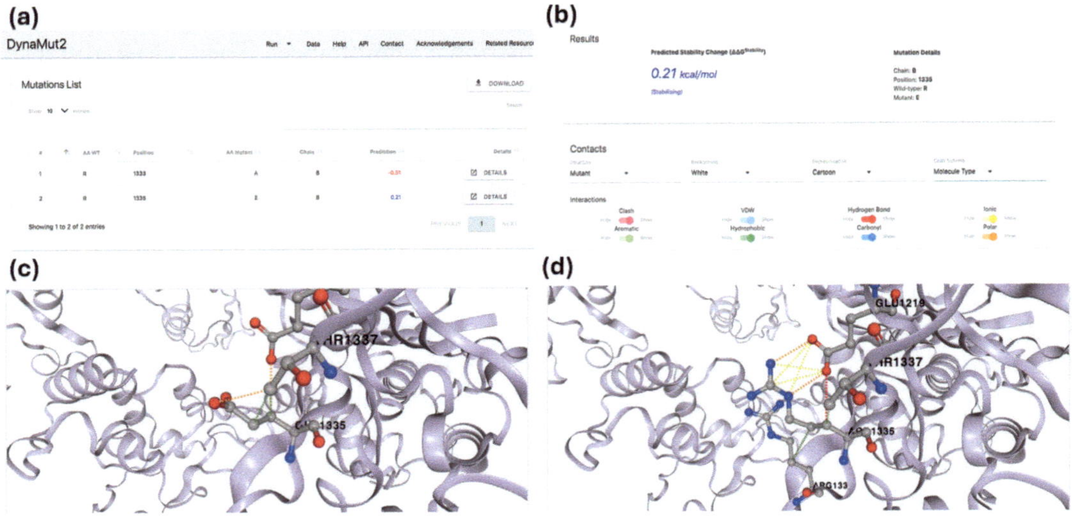

Fig. 4 An example of interactive and comprehensive DynaMut2 analysis of Cas9 showcasing (**a**) an overview of the results, (**b**) predicted ΔΔG for stability changes and a control panel for 3D visualization of various residue interactions, for (**c**) wild-type Cas9, and (**d**) R1335E mutation

(a) Positive ΔΔG (>0 kcal/mol) → Stabilizing mutation.

(b) Negative ΔΔG (<0 kcal/mol) → Destabilizing mutation.

3. The results show that the R1333A mutation destabilizes the Cas9 protein by −0.44 kcal/mol, whereas the R1335E mutation stabilizes it by 0.21 kcal/mol. However, their functional impact remains uncertain.
4. Residue interaction network analysis and molecular dynamics (MD) simulations of these mutated Cas9 variants can be performed to assess the binding affinity and functional consequences of these mutations.
5. The results can be interactively viewed online or downloaded as a PyMOL session for further analysis (*see* Fig. 4 and **Note 3**).

3.3 Residue Interaction Network (RIN) Analysis Using MDM-TASK Web Server

This section provides a structured workflow for residue interaction network (RIN) analysis using MDM-TASK [11], aiding in the identification of critical residues affecting protein stability and function. RIN analysis requires a single .pdb file as input. Additionally, trajectory files from molecular dynamics (MD) simulations can be analyzed through dynamic residue network (DRN) analysis. Therefore, MDM-TASK can be effectively used both before and after MD simulations to reveal critical residue interaction networks involved in protein function.

3.3.1 Access the MDM-TASK Web Server

1. Open a web browser and navigate to the MDM-TASK web server https://mdmtaskweb.rubi.ru.ac.za/ and register.

3.3.2 Upload the Input Structure

1. Navigate to Dynamic Residue Network (DRN) Analysis > Residue Interaction Network (RIN).
2. Optionally, enter Cas9 as the job name.
3. Upload the locally modified protein-only structure in .pdb format using the Choose File option (refer to **Note 4** for details on performing Dynamic Residue Network analysis and to **Note 5** for guidance on generating a single nucleotide polymorphism (SNP)-Weighted Residue Contact Map using a trajectory file).

3.3.3 Configure the Parameters

1. Make sure Betweenness Centrality (BC) is selected. This metric represents the fraction of times a node appears in the shortest paths connecting node pairs within the network graph. Other available options can also be selected as needed.
2. Set the cutoff distance to 6.7 Å.

3.3.4 Submit the Job

1. Click "Start" to submit the job.
2. Wait for the job to complete. Processing time depends on the structure size and network complexity.

3.3.5 Analyze the Results

1. Once processing is complete, results will be available on the web interface.
2. Visualization options include:
   (a) 2D Graph Representation: a graphical view of residue interactions.
   (b) 3D Visualization: Download and open the PDB with PyMOL, Chimera, or VMD for interactive exploration.
3. Key output includes network centrality scores to identify key residues involved in stability and function (*see* Fig. 5 and **Note 3**).

3.3.6 Download and Interpret the Data

1. Download results in comma-separated value (CSV) format for further analysis.
2. Compare the results of wild-type and mutated Cas9 to predict functionally important residues.
3. Perform molecular dynamics (MD) simulations using GROMACS to validate RIN insights.

3.4 Molecular Dynamics (MD) Simulation of a Cas9 Using CHARMM-GUI and GROMACS

This section provides a step-by-step guide to setting up and performing an MD simulation of a Cas9 protein using CHARMM-GUI [12] for system preparation and GROMACS for simulation execution. For the following protocol, we have chosen 4UN3.pdb, which represents the crystal structure of *Streptococcus pyogenes* Cas9 bound to a single guide RNA (sgRNA) and target DNA containing a canonical 5′-NGG-3′ PAM sequence. The structure demonstrates that the PAM motif is embedded within a base-

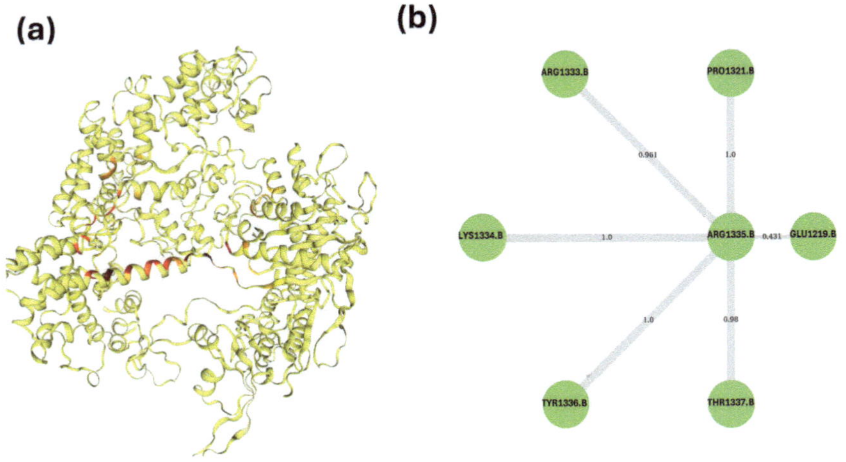

Fig. 5 An example of interactive and comprehensive MDM-TASK analysis of Cas9 showcasing (**a**) network centrality metrics from a residue interaction network (RIN). Red residues indicate highly central nodes with high betweenness values, acting as key hubs in structural communication, and (**b**) a 2D-weighted network graph using contact frequencies calculated at a cutoff radius of 6.7 Å around ARG1335 in Cas9. This approach is particularly useful for visualizing contact patterns at loci with high centrality identified through dynamic residue network (DRN) metrics

paired DNA duplex, where the GG dinucleotide on the non-complementary strand is recognized through major-groove interactions with conserved arginine residues in the carboxyl-terminal domain (CTD) of Cas9. Additionally, interactions involving the minor groove of the PAM duplex and the phosphodiester bond at the +1 position in the target DNA strand facilitate local DNA strand separation just upstream of the PAM sequence, a crucial step in Cas9-mediated target recognition and cleavage [13].

All calculations were performed on a Linux-x86_64 system running Ubuntu 20.04.6 LTS, equipped with an Intel Xeon E5–2678 v3 processor, 48 GB RAM, an NVIDIA GeForce GTX 1080 Ti GPU, and a Dell Precision Tower 7910 workstation.

3.4.1 Downloading the PDB Structure

1. Navigate to the Protein Data Bank (PDB) website: https://www.rcsb.org/.
2. Search for PDB ID: 4UN3 and download the PDB file.
3. The downloaded 4UN3.pdb structure contains:
   (a) Chain A: Single guide RNA (83 nt).
   (b) Chain B: Cas9 protein (1372 aa).
   (c) Chain C: Target DNA (28 nt).
   (d) Chain D: Non-target DNA (12 nt).
   (e) Mg^{2+} ions (8 ions) (although two catalytic Mg^{2+} ions are essential, we will retain all eight ions, as Mg^{2+} plays a crucial role in stabilizing the DNA structure).

3.4.2 Removing Water Molecules

1. Open 4UN3.pdb in a text editor (Notepad, Vim, Nano, etc.).
2. Remove all water molecules (HOH records).
3. Save the modified file as "01_4UN3_nowat.pdb."

3.4.3 Checking for Missing Residues

1. Open "01_4UN3_nowat.pdb" in a visualization tool such as VMD, PyMOL, or Chimera.
2. Identify missing loops in the Cas9 protein.
3. Before proceeding to molecular dynamics (MD) simulations, it is mandatory to fix the missing loops.

3.4.4 Predicting and Fixing Missing Loops Using AlphaFold

1. Access the FASTA sequence of 4UN3.pdb from the PDB website.
2. Copy the sequences of chains A, B, C, and D.
3. Navigate to the AlphaFold server: https://alphafoldserver.com/.
4. Paste the copied sequence into the input sequence box and submit the query.
5. Once results are generated, download the predicted structure.
6. Locate the output files (.cif format) and identify "*model_0.cif," which is the best-ranked model.

3.4.5 Aligning the AlphaFold Model with the Original Crystal Structure

1. Open PyMOL.
2. Load the following structures:
   (a) "model_0.cif" (AlphaFold model).
   (b) "01_4UN3_nowat.pdb" (crystal structure).
3. Align the AlphaFold model to the crystal structure:
   (a) In the PyMOL Object Control Panel, select AlphaFold model "A" > align > to molecule >4un3.
4. Observe the root mean square deviation (RMSD) value in the information panel above the command input area.
   (a) An RMSD <1 Å indicates an excellent match between the structures.
5. To save the aligned AlphaFold model as "02_model_4UN3.pdb":
   (a) Go to File > Export Molecule > Multi File > One File per Object > Save.
   (b) Choose PDB (*.pdb, *.pdb.gz) as the file format. Click "Save."

3.4.6 Restoring Chain Information, Adding Missing Mg^{2+} Ions, and Verification

1. Open "02_model_4UN3.pdb" in a text editor (Notepad, Vim, Nano, etc.).
2. Ensure that the termination (TER) record is present after each molecule (protein, RNA, and DNA).
3. If necessary, manually insert TER records after each macromolecule using a text editor.
4. Verify chain assignments in "02_model_4UN3.pdb" (notice that the chain assignment has changed):
   (a) Chain A: Cas9 (1372 aa).
   (b) Chain B: Target DNA (28 nt).
   (c) Chain C: Non-target DNA (12 nt).
   (d) Chain D: Single guide RNA (83 nt).
5. The Mg^{2+} ions are missing from this structure.
6. Copy the Mg^{2+} coordinates from "01_4UN3_nowat.pdb."
7. Paste the Mg^{2+} coordinates at the end of "02_model_4UN3.pdb" and save it as "03_model_4UN3_Mg.pdb."
8. Open "03_model_4UN3_Mg.pdb" in a visualization tool (VMD, PyMOL, or Chimera).
9. Ensure that no steric clashes exist between Mg^{2+} ions and biomolecules.

3.4.7 Accessing CHARMM-GUI

1. Navigate to https://www.charmm-gui.org/ and register/login.
2. From the left panel, go to Input Generator > Solution Builder.

3.4.8 Uploading the PDB File

1. Click "Choose File" and upload "03_model_4UN3_Mg.pdb."
2. Click "Next Step: Select Model/Chain."

3.4.9 Model and Chain Selection

1. In Model/Chain Selection Options, ensure that:
   (a) Protein, DNA, RNA, and Hetero (Mg^{2+} ions) are selected.
   (b) No engineered residues are present.
   (c) If engineered residues exist.
   (d) Convert them back to the original residue in the next step or modify them in "03_model_4UN3_Mg.pdb" and upload it again.
   (e) If they are required, provide parameter and topology files in the next step.
2. Click "Next Step: PDB Manipulation."

3.4.10 PDB Manipulation and Setting pH

1. Save the Job ID (in case of interruptions, retrieve it using "Job Retriever" from the left panel).
2. Set System pH to 7.4 (the physiological pH of most living systems). Click Apply.
3. Modify chains if needed:
   (a) Mutate residues if necessary.
   (b) Ensure Protonation State is checked (reflecting changes in pH).
   (c) If applicable, introduce phosphorylation, ubiquitination, or other post-translational modifications (PTMs).
4. As "03_model_4UN3_Mg.pdb" has no disulfide bonds, leave Disulfide Bonds unchecked.
5. Click "Next Step: Generate PDB."

3.4.11 Checking and Modifying Waterbox Size

1. In the CHARMM PDB preview (step1_pdbreader.pdb), verify that the structure is correct (it is advisable to keep viewing the structure after every step).
2. Under "Waterbox Size Options":
   (a) Change the Waterbox Type to Octahedral.
   (b) Set Edge Distance to 14.0 Å. As the DNA and RNA have highly charged phosphate backbones that require a well-hydrated environment to neutralize electrostatic repulsions. A box smaller than 14 Å may cause interactions between periodic images, leading to nonphysical conformational changes.
3. Click "Next Step: Solvate Molecule."

3.4.12 Adding Ions and Setting Ionic Concentration

1. Under "Add Ions," ensure "Include Ions" is checked.
2. Change "Ion Placing Method" from "Monte-Carlo" to "Distance-Based." The distance-based method ensures proper electrostatic neutralization by placing ions near the highly charged DNA/RNA phosphate backbone, stabilizing biomolecular interactions.
3. Select KCl as the "Basic Ion Type," because potassium (K^+) is the dominant intracellular ion, more biologically relevant than sodium (Na^+) for stabilizing nucleic acids and protein–nucleic acid interactions. Also, K^+ better mimics the physiological ionic environment, especially in the nucleus and cytoplasm, where DNA, RNA, and proteins interact.
4. Set KCl concentration to 0.15 M to match physiological conditions. In our case, the total ion counts for K^+ and Cl^- are ~267 and ~184, respectively, to ensure 0.15 M KCl concentration for our system (numbers may vary based on system size).

In addition to basic ion types, several buffer systems can also be added to mimic the experimental conditions.

5. Click Next Step: Solvate Molecule. This CHARMM process will take several minutes.

3.4.13 Setting Periodic Boundary Conditions (PBC)

1. Under Periodic Boundary Condition Options, select "Generate grid information for PME FFT automatically," which enables the automatic determination of grid parameters for particle mesh Ewald (PME) calculations in MD simulations. Since PME relies on a 3D grid to interpolate charge distributions in reciprocal space, this option ensures that the fast Fourier transform (FFT) grid dimensions are optimally chosen based on the simulation box size, improving computational efficiency and accuracy.
2. Click Next Step: Setup Periodic Boundary Condition.

3.4.14 Choosing Force Field and Simulation Engine

1. Under Force Field Options:
   (a) Change CHARMM36m to AMBER, as the CHARMM36m has been reported to make DNA more flexible, particularly in long simulations. Some studies have observed excessive DNA flexibility, leading to overbending and deviations from B-DNA conformations. Use FF19SB for protein, OL15 for DNA and OL3 for RNA. OL15 incorporates correct χ torsion angles to maintain B-DNA while OL3 improves RNA backbone and sugar puckering.
   (b) Ensure to select:
      (i) FF19SB for protein.
      (ii) OL15 for DNA (maintains B-DNA stability).
      (iii) OL3 for RNA (improves sugar puckering and backbone stability).
2. Under Input Generation Options, select GROMACS as the MD engine. Other options like Amber, OpenMM, or GENESIS are also available (*see* **Note 6**).

3.4.15 Equilibration and Dynamics Setup

1. Under "Equilibration Input Generation Options," select "NVT Ensemble" (stabilizes temperature before pressure equilibration).
2. Under Dynamics Input Generation Options:
   (a) Select NPT Ensemble.
   (b) Set Temperature to 310.15 K (normally, 310.15 K is set for MD simulation representing normal human body temperature or incubation temperature for in vitro mammalian cell culture).
3. Click Next Step: Generate Equilibrium and Dynamics Inputs. This CHARMM process will take several minutes.

3.4.16 Download and Extract Simulation Files

1. Download the simulation system by clicking on the "download.tgz" link provided by CHARMM-GUI and save the file to your local system or high-performance computing (HPC) cluster. The downloaded compressed file will be named "charmm-gui.tgz."
2. Open a terminal and navigate to the directory where the file is saved.
3. Extract the contents using the following command:

```
tar -xvzf charmm-gui.tgz
```

4. Navigate to the GROMACS directory:

```
cd charmm-gui > cd gromacs
```

5. This folder contains the following essential files:
   (a) index.ndx (defines index groups for protein, DNA, RNA, solvent, and ions).
   (b) README (step-by-step instructions for running GROMACS simulations).
   (c) step3_input.gro (stores atomic coordinates and system box dimensions).
   (d) step3_input.pdb (PDB file with atomic coordinates and connectivity).
   (e) step3_input.psf (CHARMM topology file with bonding and charge information).
   (f) step4.0_minimization.mdp (input parameter file for energy minimization).
   (g) step4.1_equilibration.mdp (input parameter file for NVT equilibration).
   (h) step5.0_production.mdp (input parameter file for long-term production MD).
   (i) topol.top (contains force field parameters, atomic charges, and system composition).
   (j) topper/ (folder containing topology-related files).
6. Open VMD and load the structure file using:

```
vmd step3_input.pdb
```

7. Our system contains ~222,214 atoms, including protein, DNA, RNA, ions, and solvent molecules.

3.4.17 Execution of Simulation Files

1. To execute the simulation, we will use custom input parameter files (*see* **Note 3**) instead of those generated by CHARMM-GUI. The following tailored files will be used, while the rest will remain unchanged:
   (a) em.mdp—energy minimization parameters.
   (b) nvt.mdp—NVT equilibration parameters.
   (c) npt.mdp—NPT equilibration parameters.
   (d) md.mdp—production MD simulation parameters.
2. Follow these steps to execute energy minimization, equilibrations, and production MD simulations using GROMACS 2020.4 compiled with CUDA on an NVIDIA GPU (*see* **Note 7**):

```
gmx grompp -f em.mdp -o em.tpr -c step3_input.gro -r step3_in-
put.gro -p topol.top -maxwarn 5
gmx mdrun -v -deffnm em

gmx grompp -f nvt.mdp -c em.gro -r step3_input.gro -p topol.top
-o nvt.tpr -maxwarn 5
gmx mdrun -nt 8 -v -deffnm nvt -nb gpu

gmx grompp -f npt.mdp -c nvt.gro -r nvt.gro -p topol.top -o
npt.tpr -maxwarn 5
gmx mdrun -nt 8 -v -deffnm npt -nb gpu

gmx grompp -f md.mdp -c npt.gro -r npt.gro -p topol.top -o md.
tpr -maxwarn 5
gmx mdrun -nt 8 -v -deffnm md -nb gpu
```

3.4.18 Final Output Files after Running MD Simulations in GROMACS

1. After successfully running all the energy minimization, equilibration, and production MD steps, GROMACS generates several important output files. Table 1 provides a brief explanation of the key output files.

3.4.19 Post-simulation Analysis

1. The following output files (*see* Table 2) are crucial for analyzing molecular interactions, stability, flexibility, and dynamics of the simulated system.
2. It is important to note that the md.xtc trajectory file requires periodic boundary condition (PBC) correction before analysis. Use the following GROMACS command to correct for PBC artifacts:

```
echo 1, 0 | gmx trjconv -s md.tpr -f md.xtc -o md_corr.xtc
-center -ur compact -pbc mol
```

Table 1
Key output files from MD simulation using GROMACS

File name	Description
md.xtc	(a) Compressed trajectory file containing atomic positions at each recorded timestep during the MD simulation (b) Used for post-simulation analysis
md.tpr	(a) Binary topology file containing all input parameters, coordinates, velocities, and topology information (b) Essential for restarting simulations and analyzing trajectories
md.log	(a) Log file storing all GROMACS execution details, including energy minimization, equilibration, and production run information
md.edr	(a) Energy file containing total energy, potential energy, kinetic energy, temperature, pressure, and other thermodynamic properties recorded throughout the simulation
md.gro	(a) Final structure file containing the last frame of the simulation (coordinates of all atoms) (b) Can be used for further analysis or visualization
md.cpt	(a) Checkpoint file that saves the simulation state (b) Used to restart simulations from where they stopped

Table 2
Summary of GROMACS output files used for post-simulation analyses

Analysis	Files used
Energy analysis (GROMACS energy module)	md.edr
Density and pressure calculations	md.edr
Visualization and analyses (RMSD, RMSF, $g_{(r)}$, hydrogen bonding, secondary structure analysis, clustering, PCA, DCCM, binding energy calculation, etc.)	md.xtc (needs PBC correction before analyses)

3. md_corr.xtc is free of periodic boundary artifacts and can be used for analyses such as root mean square deviation (RMSD), root mean square fluctuation (RMSF), hydrogen bonding, secondary structure analysis, clustering, principal component analysis (PCA), dynamics cross-correlation matrix (DCCM), binding energy calculation, etc. (*see* Fig. 6).
4. To visualize the PBC-corrected trajectory in VMD, follow these steps:

```
vmd step3_input.pdb md_corr.xtc
```

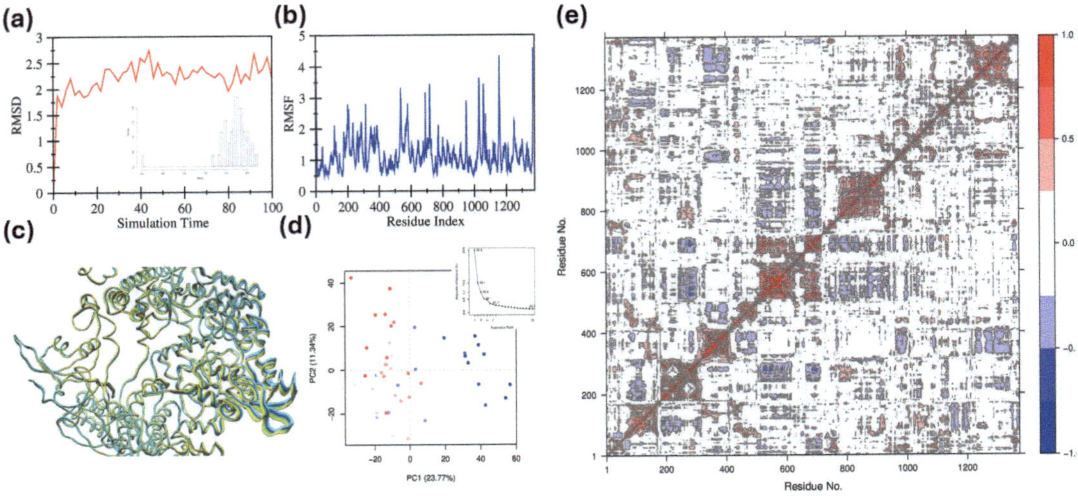

Fig. 6 An example of comprehensive post-simulation trajectory analysis of Cas9. (**a**) root mean square deviation (RMSD) plot, reflecting structural stability over time; a histogram of RMSD values can further reveal dominant conformational states, (**b**) root mean square fluctuation (RMSF), indicating residue-level flexibility, (**c**) 34 representative snapshots along principal component 1 (PC1), capturing major conformational shifts, (**d**) PC1 vs. principal component 2 (PC2) scatter plot and scree plot, illustrating the principal motions and conformational sampling, and (**e**) the dynamic cross-correlation matrix (DCCM), highlighting correlated and anticorrelated residue movements that suggest potential allosteric communication within Cas9

4 Notes

1. Alternatively, you can use the standalone version of EVcouplings by downloading it from https://github.com/debbiemarkslab/EVcouplings and following the provided installation instructions.
2. In the original PDB file, the Cas9 protein is represented by Chain B. Key residues, including Arg1333 and Arg1335 in the PAM-interacting (PI) domain, play a crucial role in recognizing the PAM sequence. Mutations in these residues, such as Arg1335Glu, can alter PAM specificity.
3. Input files and results can be downloaded from https://github.com/qaiserfatmi/Cas9_Files.
4. To analyze the MD trajectory file using MDM-TASK, begin by navigating to the DRN section and selecting Dynamic Residue Network. Optionally, assign a job name such as *Cas9_traj_wt_DRN* for easier identification. Under the Topology section, click Choose File and upload the appropriate .psf or .pdb file. Then, under Trajectory, click Choose File and upload the corresponding .xtc file. Make sure that only the protein structure is present in the uploaded files. In the DRN options, select BC as the network metric. Set the cutoff distance to 6.7 Å and define the Step size as 100. For DRN statistics, choose

Mean to calculate the average betweenness centrality across frames. Enable email notifications by selecting the email option to On. Finally, click Start to submit the job and initiate the DRN analysis.

5. To generate a weighted network graph for an MD trajectory using MDM-TASK, begin by navigating to the DRN section and selecting SNP-Weighted Residue Contact Map. Optionally, assign a job name such as *Cas9_traj_wt_SNP-WRCM* for easier tracking. Under the Topology section, click Choose File and upload the appropriate .psf or .pdb file. Next, under Trajectory, click Choose File and upload the corresponding .xtc/.dcd file. Ensure that only the protein structure is included in the uploaded files. In the Residue section, specify the three-letter amino acid code along with its residue ID (e.g., ARG1335). Please note that the residue ID may differ from the original PDB file. It is important to reconcile the trajectory file before assigning the residue ID here. In the Chain option, select chain ID to specify the relevant chain for analysis. Set the Contact cutoff distance to 6.7 Å and define the Step size as 100. To receive updates upon job completion, enable email notifications by selecting the email option to On. Finally, click Start to submit the job and initiate the DRN analysis.
6. Each molecular dynamics (MD) engine has its own advantages and limitations relative to GROMACS. AMBER offers highly accurate force fields and is well suited for biomolecular simulations, especially when using its native file formats and tools, though it is less efficient than GROMACS in terms of raw performance. OpenMM excels in GPU acceleration and Python-based flexibility, making it ideal for rapid prototyping and integration with machine learning workflows, but it may lack the breadth of built-in analysis tools that GROMACS provides. GENESIS is optimized for large-scale biomolecular simulations and supports advanced features like replica exchange and coarse-grained models, though it has a steeper learning curve and is less widely adopted. In contrast, GROMACS is renowned for its exceptional computational speed, scalability, and broad user community, but it may be less flexible for custom modifications compared to OpenMM or AMBER's scripting tools.
7. *-maxwarn* allows GROMACS to continue despite minor warnings.

   -nt 8 specifies eight CPU threads (adjust as needed).

   -nb gpu enables GPU acceleration for better performance.

References

1. Nambiar TS, Baudrier L, Billon P, Ciccia A (2022) CRISPR-based genome editing through the lens of DNA repair. Mol Cell 82(2):348–388. https://doi.org/10.1016/j.molcel.2021.12.026
2. Stinson BM, Loparo JJ (2021) Repair of DNA double-strand breaks by the nonhomologous end joining pathway. Annu Rev Biochem 90: 137–164. https://doi.org/10.1146/annurev-biochem-080320-110356
3. Gleditzsch D, Pausch P, Müller-Esparza H, Özcan A, Guo X, Bange G, Randau L (2019) PAM identification by CRISPR-Cas effector complexes: diversified mechanisms and structures. RNA Biol 16(4):504–517. https://doi.org/10.1080/15476286.2018.1504546
4. Jumper J, Evans R, Pritzel A, Green T, Figurnov M, Ronneberger O, Tunyasuvunakool K, Bates R, Žídek A, Potapenko A, Bridgland A, Meyer C, Kohl SAA, Ballard AJ, Cowie A, Romera-Paredes B, Nikolov S, Jain R, Adler J, Back T, Petersen S, Reiman D, Clancy E, Zielinski M, Steinegger M, Pacholska M, Berghammer T, Bodenstein S, Silver D, Vinyals O, Senior AW, Kavukcuoglu K, Kohli P, Hassabis D (2021) Highly accurate protein structure prediction with AlphaFold. Nature 596(7873):583–589. https://doi.org/10.1038/s41586-021-03819-2
5. Baek M, McHugh R, Anishchenko I, Jiang H, Baker D, DiMaio F (2024) Accurate prediction of protein–nucleic acid complexes using RoseTTAFoldNA. Nat Methods 21(1): 117–121. https://doi.org/10.1038/s41592-023-02086-5
6. Lee M (2023) Deep learning in CRISPR-Cas systems: a review of recent studies. Front Bioeng Biotechnol 11:1226182. https://doi.org/10.3389/fbioe.2023.1226182
7. Assa G, Kalter N, Rosenberg M, Beck A, Markovich O, Gontmakher T, Hendel A, Yakhini Z (2024) Quantifying allele-specific CRISPR editing activity with CRISPECTOR2.0. Nucleic Acids Res 52(16):e78. https://doi.org/10.1093/nar/gkae651
8. Dixit S, Kumar A, Srinivasan K, Vincent PMDR, Ramu Krishnan N (2024) Advancing genome editing with artificial intelligence: opportunities, challenges, and future directions. Front Bioeng Biotechnol 11:1335901. https://doi.org/10.3389/fbioe.2023.1335901
9. Fram B, Su Y, Truebridge I, Riesselman AJ, Ingraham JB, Passera A et al (2024) Simultaneous enhancement of multiple functional properties using evolution-informed protein design. Nat Commun 15(1):5141. https://doi.org/10.1038/s41467-024-49119-x
10. Rodrigues CHM, Pires DEV, Ascher DB (2021) DynaMut2: assessing changes in stability and flexibility upon single and multiple point missense mutations. Protein Sci 30(1): 60–69. https://doi.org/10.1002/pro.3942
11. Amamuddy OS, Glenister M, Tshabalala T, Bishop ÖT (2021) MDM-TASK-web: MD-TASK and MODE-TASK web server for analyzing protein dynamics. Comput Struct Biotechnol J 19:5059–5071. https://doi.org/10.1016/j.csbj.2021.08.043
12. Jo S, Kim T, Iyer VG, Im W (2008) CHARMM-GUI: a web-based graphical user interface for CHARMM. J Comput Chem 29(11):1859–1865. https://doi.org/10.1002/jcc.20945
13. Ran FA, Cong L, Yan WX, Scott DA, Gootenberg JS, Kriz AJ et al (2015) In vivo genome editing using Staphylococcus aureus Cas9. Nature 520(7546):186–191. https://doi.org/10.1038/nature14299

Chapter 18

Exploring the Functional Consequences of Protein-Coding Alternative Splice Variants at a Genome Scale: A Guide to Utilizing EXPANSION

Chakit Arora, Natalia De Oliveira Rosa, Marin Matic, Mariastella Cascone, Pasquale Miglionico, and Francesco Raimondi

Abstract

This chapter serves as a comprehensive user guide for EXPANSION (https://expansion.bioinfolab.sns.it/), an integrated web server dedicated to exploring the functional consequences of protein-coding alternative splice variants from genome-scale transcriptomics datasets. By amalgamating data from differentially expressed protein-coding transcripts, domain architecture, protein interaction networks, and gene enrichment analysis, EXPANSION provides researchers with a robust toolkit to dissect the impact of alternative splicing events on cellular functions. From navigating precalculated datasets to uploading custom data, visualizing transcript expression, scrutinizing splicing variations, and performing overrepresentation analysis, this chapter elucidates the key procedures involved in harnessing EXPANSION for a detailed exploration of protein alternative splicing from transcriptomics datasets in both healthy and disease conditions.

Key words Alternative splicing, Protein-coding transcripts, Cancer genomics, Protein interaction networks, Gene enrichment analysis, Protein domain architecture, Molecular biology, Cellular functions

1 Introduction

The huge amount of genetic information obtained from genome sequencing techniques requires methods to disentangle the intricate relationship between genetic alterations and cellular processes to unravel the complexities of disease pathogenesis and identify novel therapeutic targets. Alternative splicing, a fundamental mechanism in gene expression regulation, has garnered increasing attention for its role in generating protein diversity and modulating cellular functions [1]. The process of alternative splicing allows a

---

Chakit Arora, Natalia De Oliveira Rosa equally contributed to this work.

Shahid M. Khan and Florencio Pazos (eds.), *Protein Evolution: Methods and Protocols*, Methods in Molecular Biology, vol. 2979, https://doi.org/10.1007/978-1-0716-4828-5_18,

single gene to produce multiple mRNA transcripts, leading to the synthesis of distinct protein isoforms with diverse functions. This molecular diversity is crucial for cellular homeostasis, development, and response to environmental signals [1].

The advent of high-throughput sequencing technologies, particularly RNA sequencing (RNA-seq), has revolutionized the field of transcriptomics by enabling the comprehensive profiling of gene expression patterns and alternative splicing events across different biological conditions. Through RNA-seq data analysis, researchers can identify differentially expressed transcripts, including alternative splice variants, and investigate their functional implications in various biological processes. Alternative splicing is a tightly regulated process, influenced by a myriad of factors such as cis-acting elements, trans-acting factors, and epigenetic modifications [2]. Dysregulation of alternative splicing has been implicated in the pathogenesis of numerous diseases, including cancer, where aberrant splice variants can give rise to oncogenic proteins or disrupt tumor suppressor functions [3].

In the context of cancer, alternative splicing events can have profound effects on tumor biology, influencing key cellular processes such as proliferation, apoptosis, and metastasis. For instance, aberrant splicing of genes like CD44 and BCL2L1 has been linked to various cancer types, highlighting the importance of understanding the functional consequences of alternative splicing in oncogenesis [4, 5]. RNA-seq technology has significantly enhanced our ability to study alternative splicing and its impact on protein diversity and function [6]. By integrating transcriptomics data with protein functional annotations, researchers can gain valuable insights into the consequences of alternative splicing on protein structure, function, and interactions [7–10].

To address the growing need for advanced bioinformatics tools, we developed the EXPANSION web server as an integrated platform to explore the functional implications of protein-coding alternative splice variants [11]. EXPANSION provides a suite of features including differential expression visualization, splicing variation and posttranslational modification (PTM) information, multiple sequence alignment (MSA) viewing, and isoform-specific interaction networks. It also includes overrepresentation analysis (ORA) to evaluate functional categories associated with isoform-specific interactors. In addition to a user-friendly web interface, EXPANSION also offers a robust application programming interface (API) for easy programmatic access. The API provision allows seamless integration into automated workflows, such as with other preexisting bioinformatics tools, and thus enables large-scale, customizable studies.

EXPANSION facilitates the identification of functional splice transcripts that could represent novel biomarkers associated with patient survival and disease progression, via integration with patients' clinical data. This could lead to the development of prognostic indicators for early diagnosis and personalized treatment strategies. Additionally, EXPANSION could aid in the prioritization of neoantigens for immunotherapy by analyzing how alternative splicing generates certain unique protein fragments, which are highly expressed in specific conditions, and assessing their potential to be recognized by the immune system. This functionality could aid the identification of effective targets for drugs and other immunotherapies. Overall, EXPANSION has the potential to advance personalized medicine and deepen our understanding of alternative splicing in health and disease. The following sections of this chapter will delve into the functionalities of EXPANSION, guiding researchers through the process of utilizing this web server to unravel the impact of alternative splicing on protein function, cellular pathways, and disease mechanisms.

2 Materials

1. *EXPANSION Web Server:* A computer or mobile device with Internet access to navigate and utilize the EXPANSION web server for analyzing protein-coding alternative splice variants in cancer genomics.
2. *Query Transcript*: The "gene symbol" or "Ensembl (ENST) ID" for the transcript of interest.
3. *Custom Query Dataset (Optional, Requires User Upload):* Custom dataset obtained, for example, through differential expression (DE) analysis (such as from EBSeq [12]) performed on high-throughput transcript RNA sequencing transcript data from cross-condition comparison such as cancer samples versus normal tissues or within-condition comparisons such as z-scores for transcript expression in samples belonging to a specific tissue within a given biological condition. The dataset should be structured with transcript information in each row. The first column, labeled "*transcript*," should include the ENST IDs, followed by the column "*gene_symbol*," and then by two additional columns presenting the posterior probability of being differentially expressed (PPDE) values (or p-values) labeled as the "<tissue_name>" followed by log-fold changes (or z-scores) labeled as "logFC_ < tissue_name>" (or "Z-score_ < tissue_name>") of those transcripts for each respective tissue or comparison. Multiple comparisons or tissues can be accommodated by adding subsequent PPDE (or p-value) and LFC (or z-score) columns following the same annotation format (*see* Fig. 1).
4. *EXPANSION API (Optional, for Automated Usage):* A computer or server with Internet connectivity to access and run the

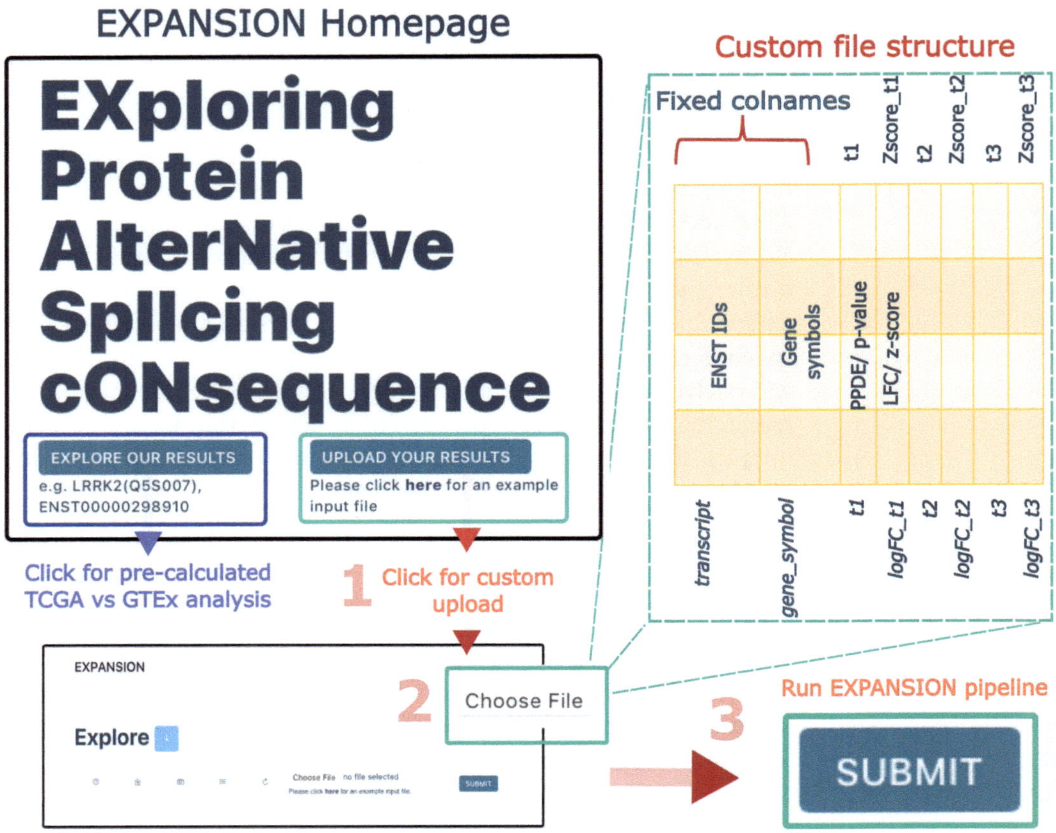

Fig. 1 The flowchart illustrates the steps for utilizing the EXPANSION web server to analyze a custom uploaded dataset. Starting from the homepage, users can follow steps 1, 2, and 3 to execute the analysis pipeline on their uploaded ".csv" file. The file must conform to the "custom file structure" format, where t1, t2, etc., represent generic names of tissues or conditions. These names should be prefixed with either "Zscore_" or "logFC_" in the even-numbered (fourth onward) columns, depending on the type of analysis being performed. Columns 1 and 2 have fixed names, that is, "*transcript*" and "*gene_symbol,*" respectively

EXPANSION API for querying information on protein-coding alternative splice variants. This requires basic knowledge of programming in Python/R to interact with the API endpoints, parse the retrieved data, and perform custom analyses based on the results obtained from the EXPANSION API. Following are the required materials for the API usage:

(a) Python/R Programming Environment: Installation of Python/R programming language on the computer or server to interact with the EXPANSION API and execute API queries.

(b) *requests* and *json* Libraries (Python): Installation of the "requests" library and "json" library in the Python environment to make Hypertext Transfer Protocol (HTTP) requests and parse JSON response from EXPANSION data.

(c) *httr* and *jsonlite* libraries (R): Alternatively, for R users, the "httr" and "jsonlite" libraries can be installed to handle HTTP requests and JSON parsing.

(d) API Documentation: Access to the EXPANSION API documentation website (https://expansion.bioinfolab.sns.it/api/) for information on API usage, supported methods, and query parameters.

(e) Query Data: Input data in the form of gene symbols, ENST IDs, or protein identifiers for querying specific information on alternative splice variants and their functional consequences.

5. *Bioinformatics Resources (Optional, for Additional Analyses):* Access to bioinformatics databases, literature, and tools for further analysis/inference of obtained results. Functional protein-coding splice variants can be linked to recently predicted structural databases of protein alternative isoforms [13]. Moreover, the information about functional alternative transcripts can be used to guide survival analysis whenever clinical metadata are available for transcriptomics dataset, such as the standard Cox regression as utilized for gene-expression data [14] or more recent spliceform-specific survival analyses [15], for identification of splice forms associated with patient survival.

3 Methods

From a web browser, navigate to the EXPANSION web server at https://expansion.bioinfolab.sns.it/.

3.1 EXPANSION Homepage

1. Explore Results: To discover the "main" results with a specific gene or transcript ID, simply click on the "Explore our results" button. This action will enable access to precalculated (The Cancer Genome Atlas [TCGA] versus Genotype-Tissue Expression [GTEx], originally retrieved from Xena [16] and subsequently processed *in-house* [11]) datasets for exploration.
2. Upload Custom Dataset (Optional): If you have a differential expression (DE) transcript dataset of interest (or z-scores from within condition comparison), you can upload it by clicking on the "Upload your results" button. This will redirect you to an upload window where you can submit your dataset for analysis. An "example" input file is also provided to structure the custom file in the required format (*see* Subheading 2, **item 3** and Fig. 1).

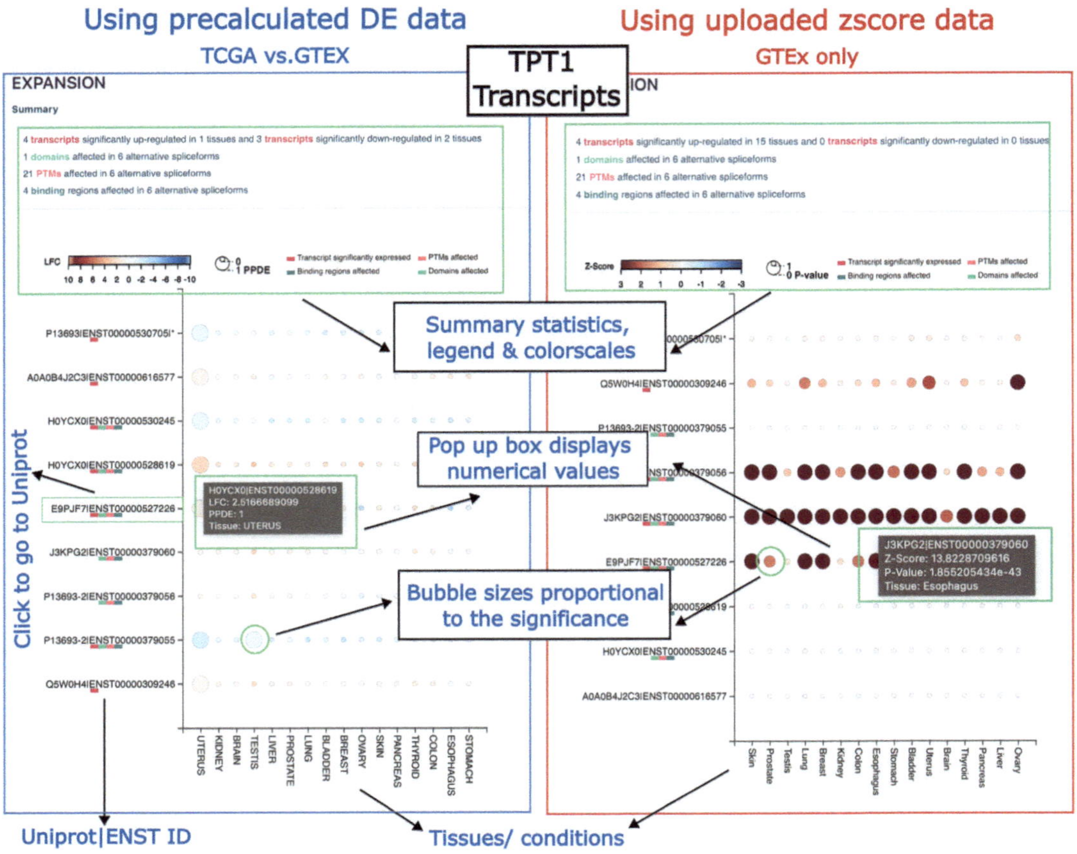

Fig. 2 The figure displays the central bubble plot output for a query submitted via the search box (left panel), specifically for the TPT1 gene. Alternatively, it shows the results for a custom dataset upload, such as z-scores for within-condition comparisons of TPT1 isoforms across various tissues (right panel). The figure includes utilities like summary statistics at the top, pop-up information when hovering over a bubble and hyperlinks to UniProt results. Additionally, there are options to download the results as high-resolution images

3.2 Querying EXPANSION

Utilize the search bar to explore the results for a query transcript of choice. User can choose to enter a gene symbol or a specific ENST ID (*see* **Notes 1–3**).

3.3 Web Server Output

1. Differential Expression Visualization: Visualize the differential expression of transcripts (y-axis) across various cancer tissues (x-axis) with the help of the central bubble plot. Each circle represents a transcript, with color indicating the log-fold change (TCGA over GTEx) (or z-score) and size proportional to significance (*see* **Notes 4** and **5** and Fig. 2). Asterisk (*) marks the canonical isoform.
2. *Splicing Variation and Posttranslational Modification (PTM) Information:* Gain insights into splicing variation and PTMs affecting domain architecture. Cartoon panels on the right

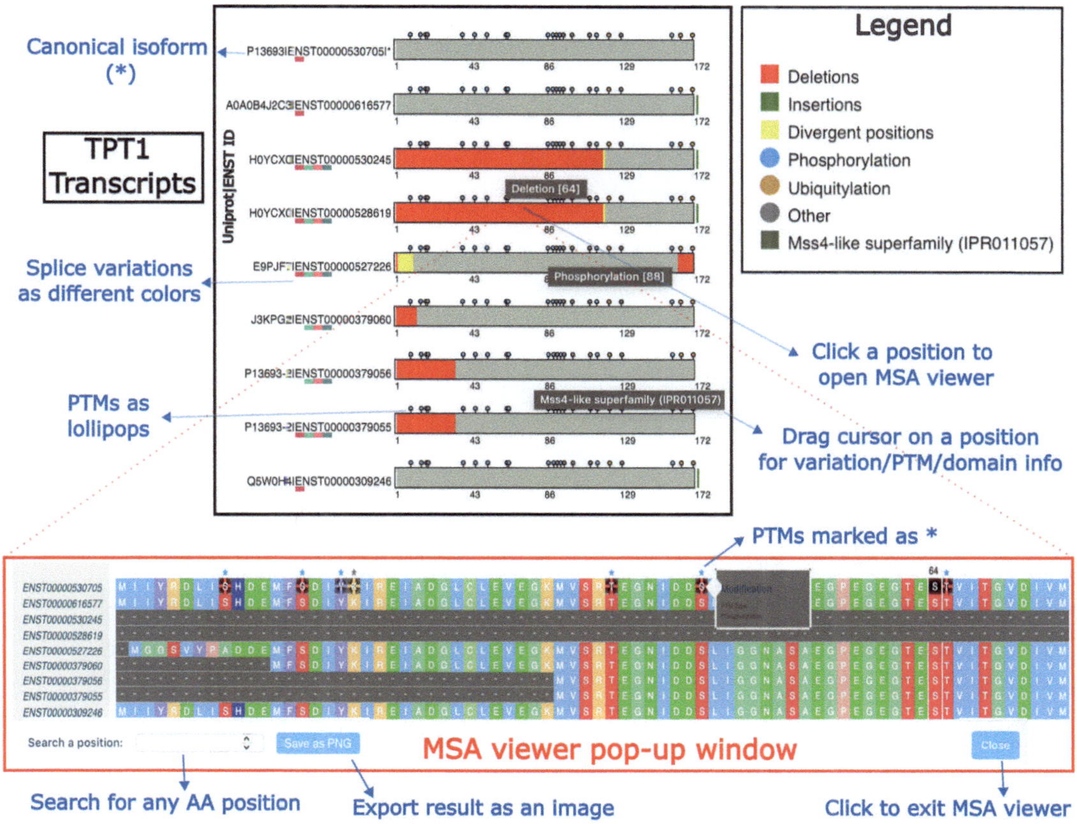

Fig. 3 The cartoon panels on the right of bubble plot offer graphical representations of splicing-related variations, affected domains, and PTMs on the canonical protein sequence, as shown by using TPT1 isoforms. A legend adjacent to the displayed panel explains the color correspondences for each feature. Detailed information about each domain or PTM is accessible by dragging the cursor over their respective regions, revealing information about the type of variation and the AA position range. Additionally, users can explore alignments by clicking on domains, posttranslational modifications (PTMs), or sequence divergences, which opens a scrollable pop-up window displaying the MSA viewer. PTMs in the MSA viewer are annotated with asterisks. The bottom search bar allows investigation of specific amino acid (AA) positions. The MSA view can be exported as a PNG image

provide a graphical representation of splicing-related divergences (colored regions), affected domains (colored regions) and PTMs (lollipops) on the canonical protein sequence. The legend adjacent to the displayed panel has information on color correspondences to each of these features (*see* **Note 6** and Fig. 3).

3. *Exporting Results:* The export buttons, at the left of the search box, allow downloading of visualization diagrams in Scalable Vector Graphic (SVG) and Portable Network Graphic (PNG) formats (see Fig. 3).

4. *Multiple Sequence Alignment (MSA) Viewer*: Explore multiple sequence alignments by clicking on domains, PTMs, or

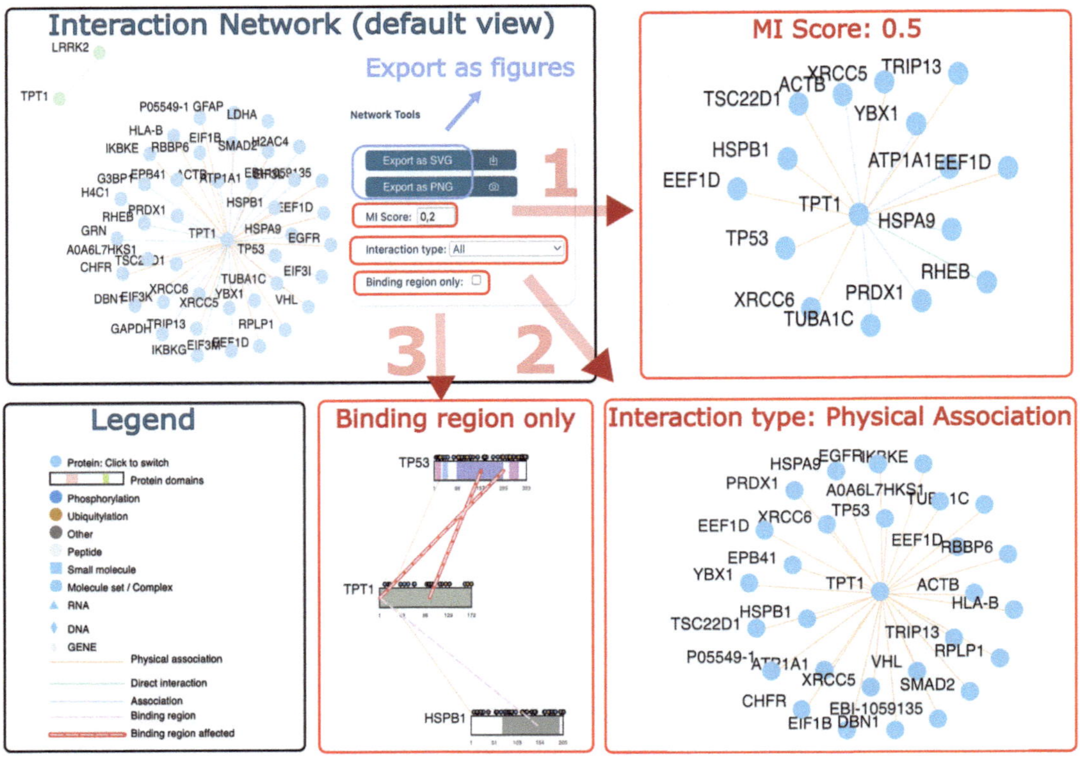

Fig. 4 The figure shows the default output "interaction network" of the EXPANSION web server, showcasing isoform-specific interactors for TPT1 splice forms. Users can adjust the network by modifying the MI score toggle, for example, to 0.5 in step 1 (default: 0.2), to filter the strength of interactions. Additionally, users can set the interaction type, for example, to "Physical Association" in step 2, to focus on direct physical interactions, with different types of interactions color-coded separately (shown in legend). By checking the "binding region only" box as in step 3, the network will highlight region-specific interaction edges based on IntAct binding region information, providing a detailed view of protein interactions. Edges affected by splicing events are highlighted in red, indicating the functional consequences of these events on protein interactions

divergences. When clicked, a scrollable pop-up window displays the MSA viewer. Clicking on transcript names will redirect you to their UniProt page for additional details. PTMs are annotated with asterisks. The bottom search bar allows one to investigate a specific amino acid (AA) position. This MSA view can also be exported as a PNG image, if required (see Fig. 3).

5. *Isoform Interaction Networks*: Access information about interaction networks mediated by the protein isoform of interest. Adjust the number of interactors via molecular interaction (MI) score and visualize domain architectures of interacting proteins. Binding region information provided by IntAct [17] is displayed (*see* **Notes 7, 8** and **9** and Fig. 4).

6. *Overrepresentation Analysis (ORA)*: Perform overrepresentation analysis (ORA) of functional categories associated with genes in the interaction network. This analysis, computed via

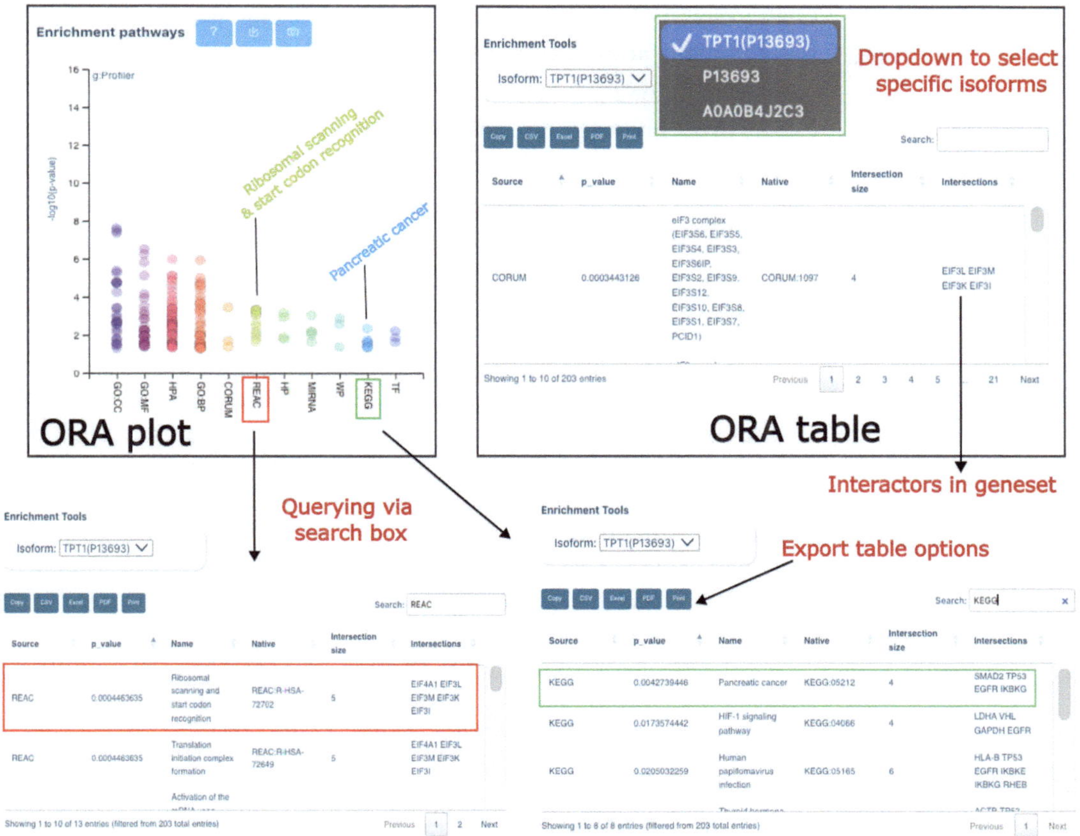

Fig. 5 The figure demonstrates the overrepresentation analysis (ORA) panel for functional categories associated with TPT1 interactors in the interaction network. The ORA plot (left) provides enrichment results for various GO terms and denoted by different colors. The supplementary ORA table on the right also provides detailed information on associated GO term, p-value, intersection size, and intersection geneset. Users can also choose a specific isoform via a drop-down menu, on the top of the ORA table. The search box enables querying the ORA table for specific pathways, such as Kyoto Encyclopedia of Genes and Genomes (KEGG) and Reactome (REAC) enrichments, to quickly identify relevant functional categories. The results can be exported in different formats via download buttons

the g:profiler python library [18], enables the comparison of distinct biological processes mediated by isoform-specific interactors when present in the network (*see* **Notes 10** and **11** and Fig. 5).

3.4 Expansion API

The EXPANSION API, accessed at https://api.expansion.bioinfolab.sns.it, allows researchers to programmatically query and retrieve data to explore the functional implications of protein-coding alternative splice variants. The API uses POST requests and requires a variable object to define the query parameters. The primary endpoints available are as follows:

- /de: Provides differentially expressed (DE) protein-coding transcripts from precalculated TCGA versus GTEx analysis.

- /da: Offers information about splicing variation affecting domain architecture as well as PTMs (insertion, deletion, or divergence).
- /ppi: Provides protein–protein interaction (PPI) networks (IntAct) mediated by the protein isoform.
- /ora: Conducts overrepresentation analysis (ORA) of isoform-specific interactors categories calculated via the g:Profiler.
- /all: Supplies a JSON of all analyses available in our web server for a gene.

The API playground (https://api.expansion.bioinfolab.sns.it/docs) can be accessed using the Swagger API documentation tool. The Swagger user interface (UI) allows users to test queries and explore the API. For instance, in Fig. 6a, the Swagger UI retrieves differentially expressed (DE) protein-coding transcripts for the gene *GNAS*. Figure 6b displays the server's response in JSON format, along with the response code and headers. The user will receive results for all related transcripts, organized by canonical isoforms such as Q5JWF2-1 and O95467-1.

Additionally, to interact with the EXPANSION API in Python, the script should be set as follows (refer to Fig. 6c):

- Base uniform resource locator (URL): Refers to the API's digital address (URL).
- Endpoint: Refers to the specific analysis to access (e.g., /ppi).
- Variables: A dictionary containing the request parameters. For instance, specify gene_symbol as "LRRK2" and interaction_type as "all." Adjust as per requirements.
- POST Request: Use the requests.post() method to dispatch the POST request to the API using the provided parameters.
- Handling the Response: In case of a successful request (status_-code == 200), the script must parse and display the JSON response. If the request has an issue, it should display an error message featuring the status code and response text to aid debugging.
- Get the response by utilizing the json.loads() method to transform the API response from JSON into a Python dictionary using the provided response.

Furthermore, users can interact with the EXPANSION API through R using specific libraries mentioned before and provide the base URL and variables as per requirements.

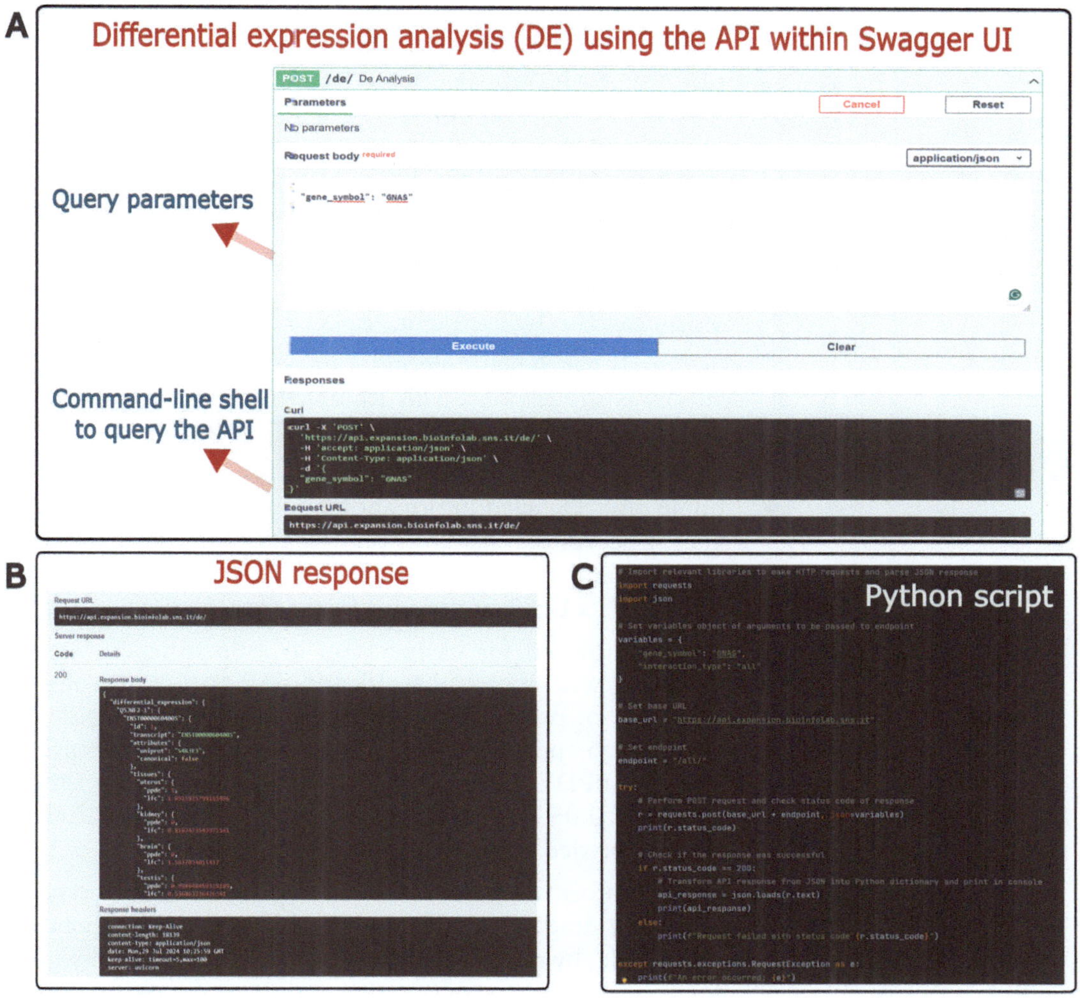

Fig. 6 (**a**) Perform a differentially expressed analysis (DE) using the application programming interface (API) within Swagger user interface (UI). Access the API documentation for the "/de" endpoint through Swagger UI, which offers a user-friendly webpage for exploring the API interactively. Users can input various query parameters and execute the query to retrieve results. Additionally, the webpage generates a client URL (cURL) command that can be executed in a command-line shell to query the API and obtain results. (**b**). Results are from a search using Swagger UI. These results were obtained by querying the "/de" API endpoint. The response body section displays the JSON response received from the API server, and the response header section shows the HTTP headers from the received request. (**c**). Python script for performing a POST request to an EXPANSION API

4 Notes

1. To detect alternative splicing accurately, we used Cluster Database at High Identity with Tolerance (CD-HIT) for sequence clustering [19]. This groups proteins/isoforms with similar sequence and functions, which help discriminating transcripts coded from complex loci, for example, *GNAS*, which share

neither sequence nor functional similarity. Therefore, the search box may display multiple UniProt IDs for a gene symbol to identify such distinct protein clusters, for example, GNAS (O95467-1) and GNAS(Q5JWF2-1). Clustering prevents aligning proteins with different roles, ensuring precise detection of alternative splicing. Each cluster is named based on the gene symbol and the UniProt accession of its longest canonical sequence in the parentheses.

2. When querying a transcript, the Ensembl ID of the canonical isoform can also be used (e.g., ENST00000313949).
3. The search bar, when entering a query, displays similar gene symbols as suggestions. This feature simplifies the process of selecting the appropriate (canonical) isoform for further analysis.
4. Bubble plot: Detailed information about each transcript's DE values/z-scores is available upon dragging the cursor over the bubble(s), and clicking on a transcript ID (y-axis labels) redirects you to its UniProt page for further exploration.
5. EBSeq, renowned for its precision in estimating differential expression, was employed to identify DE transcript isoforms in the case of precalculated datasets, with significance determined by a posterior probability of being differentially expressed (PPDE) value [12]. PPDE exceeding 0.95 is similar to p-value<0.05. A practical approach is to consider PPDE as 1-p-value or vice versa.
6. Detailed information about each domain or PTM can be accessed by dragging the cursor over their respective regions. For example, by placing the cursor over a RGB colored region informs about the type of variation and the amino acid position range.
7. In the interaction network panel, user can adjust the number of interactors shown in the network by using the toggles for MI score (with a default of 0.2) and interaction type (direct, physical, or association) colored separately. The maximum number of isoform-specific interactors displayed is set to 50.
8. Clicking on a node in the interaction network expands the nodes to represent domain architecture. If IntAct provides binding region information, the network will draw region-specific interaction edges, giving the user a detailed view of protein interactions.
9. When a splicing event affects a binding region, the network highlights the corresponding edge in red. This feature helps the user to easily interpret the functional consequences of splicing events on protein interactions.

10. When isoform-specific interactors are present in the network, the ORA panel below provides enrichment results for each group of isoform-specific interactors. The choice of the isoform can be made via a drop-down menu. This allows for the comparison of distinct biological processes mediated by these interactors, providing insights into their unique functional roles.
11. Dragging the cursor over the circles in the ORA plot displays the associated Gene Ontology (GO) term, p-value, and intersection size.

References

1. Nilsen TW, Graveley BR (2010) Expansion of the eukaryotic proteome by alternative splicing. Nature 463:457–463. https://doi.org/10.1038/NATURE08909
2. Black DL (2003) Mechanisms of alternative pre-messenger RNA splicing. Annu Rev Biochem 72:291–336. https://doi.org/10.1146/ANNUREV.BIOCHEM.72.121801.161720
3. Kahles A, Van LK, Toussaint NC et al (2018) Comprehensive analysis of alternative splicing across Tumors from 8,705 patients. Cancer Cell 34:211–224. https://doi.org/10.1016/J.CCELL.2018.07.001
4. Brown RL, Reinke LM, Damerow MS et al (2011) CD44 splice isoform switching in human and mouse epithelium is essential for epithelial-mesenchymal transition and breast cancer progression. J Clin Invest 121:1064–1074. https://doi.org/10.1172/JCI44540
5. Jiang H, Zhao W, Shao W (2014) Prognostic value of CD44 and CD44v6 expression in patients with non-small cell lung cancer: meta-analysis. Tumour Biol 35:7383–7389. https://doi.org/10.1007/S13277-014-2150-3
6. Merkin J, Russell C, Chen P, Burge CB (2012) Evolutionary dynamics of gene and isoform regulation in mammalian tissues. Science 338:1593–1599. https://doi.org/10.1126/SCIENCE.1228186
7. Betts MJ, Lu Q, Jiang Y et al (2015) Mechismo: predicting the mechanistic impact of mutations and modifications on molecular interactions. Nucleic Acids Res 43:e10. https://doi.org/10.1093/NAR/GKU1094
8. Raimondi F, Inoue A, Kadji FMN et al (2019) Rare, functional, somatic variants in gene families linked to cancer genes: GPCR signaling as a paradigm. Oncogene 38:6491–6506. https://doi.org/10.1038/S41388-019-0895-2
9. Louadi Z, Yuan K, Gress A et al (2021) DIGGER: exploring the functional role of alternative splicing in protein interactions. Nucleic Acids Res 49:D309–D318. https://doi.org/10.1093/NAR/GKAA768
10. Karakulak T, Szklarczyk D, Saylan CC et al (2023) CanIsoNet: a database to study the functional impact of isoform switching events in diseases. Bioinform Adv 3. https://doi.org/10.1093/BIOADV/VBAD050
11. Arora C, De Oliveira RN, Matic M et al (2023) EXPANSION: a webserver to explore the functional consequences of protein-coding alternative splice variants in cancer genomics. Bioinform Adv 3:vbad135. https://doi.org/10.1093/bioadv/vbad135
12. Leng N, Dawson JA, Thomson JA et al (2013) EBSeq: an empirical Bayes hierarchical model for inference in RNA-seq experiments. Bioinformatics 29:1035–1043. https://doi.org/10.1093/BIOINFORMATICS/BTT087
13. Song Y, Zhang C, Omenn GS, et al (2023) Predicting the structural impact of human alternative splicing. bioRxiv 2023.12.21.572928. https://doi.org/10.1101/2023.12.21.572928
14. Arora C, Matic M, Bisceglia L et al (2024) The landscape of cancer-rewired GPCR signaling axes. Cell Genomics 4
15. Shen S, Wang Y, Wang C, et al (2016) SURVIV for survival analysis of mRNA isoform variation Nat Commun 7:1–11. https://doi.org/10.1038/ncomms11548
16. Goldman MJ, Craft B, Hastie M et al (2020) Visualizing and interpreting cancer genomics data via the Xena platform. Nat Biotechnol 38:675–678. https://doi.org/10.1038/S41587-020-0546-8
17. del Toro N, Shrivastava A, Ragueneau E et al (2022) The IntAct database: efficient access to fine-grained molecular interaction data.

Nucleic Acids Res 50:D648–D653. https://doi.org/10.1093/NAR/GKAB1006

18. Raudvere U, Kolberg L, Kuzmin I et al (2019) G:profiler: a web server for functional enrichment analysis and conversions of gene lists (2019 update). Nucleic Acids Res 47: W191–W198. https://doi.org/10.1093/NAR/GKZ369

19. Li W, Jaroszewski L, Godzik A (2001) Clustering of highly homologous sequences to reduce the size of large protein databases. Bioinformatics 17:282–283. https://doi.org/10.1093/BIOINFORMATICS/17.3.282

INDEX

A

Activation barrier 169, 170
Active site engineering 166
ACTR–NCBD complex........... 39
AdabmDCA 2.0 83–103
Adaptive landscape flattening method 166–168, 174, 178, 180–182, 184
Adenosine triphosphate (ATP)............................ 60, 165, 172, 173, 191, 192, 194–199, 201–204
Adenylation reaction 165, 166, 171–173, 183
Adenylyl imidodiphosphates (ANPs)................ 192–199, 201–204
Adjacency matrix 127, 129, 130, 133, 136, 146
Adrenomedullin (ADM)..................................... 9, 10, 14
Affinity design 169, 175, 176, 180–183
Affinity prediction 35, 37, 45, 267–275
Alchemical free energy simulations 166
Alignment generation 52, 68, 76, 77, 79, 118
All-atom TSR algorithms......... .... 190, 192, 205
Allele frequency data 103, 110, 115, 120
AlloHub.Py................................ 141, 144, 147–149, 152
Alphabet selection 95
AlphaFold 61, 107, 113–115, 140, 215, 221, 226, 231, 237–263, 273, 280, 291
AlphaFold2 (AF2)........... 114, 225, 227, 238, 240–242, 246–253, 255–258, 261, 272
AlphaFold2-multimer 215, 238, 240, 241, 246–252, 257, 258, 261
AlphaFold3 (AF3)............ 221, 239, 249, 258–260, 280
AlphaFold Foldseek models 140
Alternative splicing.. 301–313
AMBER ff99SB force field 173, 174
AMBER parameters 171
Amino acid frequencies.......... .... 23, 65, 66, 184, 286
Amino acid grouping algorithm.................................. 190
Amino acid loading 165
Amino acids 4–7, 10, 13, 17–20, 22–24, 38, 41, 49–51, 58, 61, 65, 66, 71, 84–86, 88, 89, 93, 94, 97, 105, 113, 126, 127, 133, 135, 159, 160, 165–167, 171, 173, 178, 180, 184, 185, 190–194, 199, 201–208, 225–227, 282, 283, 286, 299, 307, 308, 312
Amino acid sequence conservation 51, 61, 85, 286
Amino acid sequences........ 4, 5, 8, 10, 19, 24, 160, 199, 225–227, 283
Amino acid substitution patterns 65
Aminoacyl-adenylate 165
Aminoacyl-tRNA synthetase 165
AMP 172, 173
Amylin...4, 5, 12, 14
Antibody-antigen complexes.............243, 246, 256–259
API usage 304, 305
Artificial intelligence (AI) 238, 241, 281
Atlas of Variant Effects (AVE) Alliance guidelines110, 120
Atom contacts 127, 133
Atomistic molecular dynamics (MD) simulation 141, 142, 145, 191, 214, 287, 289, 291, 294, 296, 297

B

Backbone substructure 194
Bacterial sensor kinases 213, 214
Basic Local Alignment Search Tool (BLAST) 3–15, 51, 52, 57–59, 67, 68, 79, 161
BeAtMuSiC 34, 36, 39, 45
Betweenness centrality (BC) 145, 146, 289, 299
Bhattacharyya coefficient 66
Big data analysis 191
Binding affinity.................................34–41, 43, 45, 106, 267–275, 288
Binding affinity changes........................... 34–36, 41, 106
Binding energy 33–46, 166, 245, 251, 252, 257, 297
Binding site...76–79, 139, 189, 190, 193, 194, 197, 200, 201, 203, 204, 207, 208, 721
BindProfX.. 34
Biomolecular recognition 190, 191
BLASTP searches 3–15
BLAST sequence retrieval.. 51
BLOSUM45 scoring matrix...................................13, 14
BLOSUM62 scoring matrix.. 9
BLOSUM90 scoring matrix.......................................13, 14
Boltzmann learning..84–91
Bridge2 algorithm 146
Brome mosaic virus (BMV).........................226, 228–232

Shahid M. Khan and Florencio Pazos (eds.), *Protein Evolution: Methods and Protocols*, Methods in Molecular Biology, vol. 2979, https://doi.org/10.1007/978-1-0716-4828-5,

C

Caenorhabditis elegans 3–8, 10, 12–15
Calcitonin (CALC) 5, 8, 9, 12, 14, 15
Calcitonin-like hormone 5, 8, 12, 14
Calcium pathway 132
cAMP-dependent kinases (cAMPDKs) 194, 198–208
Cancer genomics 303
Carlson parameters 173
Cas proteins 279–299
Catalytic efficiency 168, 170, 171, 174–176, 182–184
Catalytic efficiency design 170, 175, 182, 184
Catalytic rate constant 170
Catalytic rate design 170, 175, 181
CcdA-CcdB complex 35, 39, 41
CcdB protein purification 40
CD-HIT 20, 77, 159–161, 311
Cell death protein A (CcdA) 34–37, 39–46
Cellular functions 65, 72, 301
Chloride intracellular channel 1 (CLIC1) 15, 161, 162
Clustal omega 52, 53, 68
c-Myb–CBP complex 39
Coarse-grained folding models 299
Coevolution 65–79, 85, 282–286
Coevolutionary coupling 286
ColabFold 108, 111, 114–116, 118, 119, 225–228, 241–243, 246–248, 257, 261, 262, 272, 273
Computational docking 237
Computational identification of shape-shifting proteins 157–162
Computational methods 157, 214, 279–299
Computational model 34
Computational prediction techniques 33–46, 225, 226
Computational protein design (CPD) 165, 166, 183
Computational resources 215, 241
Compute unified device architecture (CUDA) 240, 242, 296
Conda environment 243
Conformational change 126, 147, 151, 161, 189–208
Conformational changes of amino acids 192, 193, 202
Conformational changes of kinase inhibitors or activators 192, 194
Conformational changes of kinases 189–208
Conformational diversity 157, 198, 199
Conserved domain database (CDD) 67, 68, 77
Convergence criterium 87
Convergent evolution 15
Cooperativity score 26–28
Correlated variation (Co-Var) 65–79
Correspondence score (CRS) 128, 129, 133, 136
Coupling matrix 85, 87, 88, 93, 94
Covalent modification
CRISPR/cas system 279–281
Cryo-EM 159
Cα atoms 127, 152, 190, 199

D

DALI server 141
D-amino acid 165
Data collection 158, 159
Deep learning 45, 106, 225, 226, 268, 280
Deep mutational scanning 35, 37, 39, 43, 101
Deep mutational scanning data 35, 39
Deep mutational scanning predictions 36, 101
Design 21, 27, 50, 76, 112, 116, 118, 125, 165–185, 190, 199, 274, 280
Dictionary of secondary structure in proteins (DSSP) 159, 160, 227, 229
Differential expression visualization 302, 306
Dijkstra algorithm 147
Direct coupling analysis (DCA) 19, 22, 23, 66, 83–85, 87–89, 91, 92, 98, 100–102
Distance matrix 141, 148, 149, 195
Distant homology 3–15, 126
Docker 108, 109, 119, 240
Docker container 109, 240
Domain architecture 306, 308, 310, 312
Drosophila 3, 132, 133, 135
Drug and target interactions 190, 192
Drug discovery pipelines 38
Drug resistance 192
3D structure 56–58, 112, 116, 127, 190, 194, 195, 197, 199, 225, 227, 239, 240, 246, 256, 257, 269, 281, 284, 286
3D structure mapping of functional residues 116
DynaMut2 ....282, 287, 288

E

E. coli EnvZ 214, 216
E. coli NarQ 214
E. coli PhoQ 214
Edge difference score (EDS) 128, 129, 133, 136
Eigenvalue weighted cosine score (EWCS) 128, 129, 133, 136
Eigenvalues 127, 129, 133, 146
Eigenvector centrality 145, 146, 148–150
Eigenvectors 127, 129, 133, 153
Element activation DCA (eaDCA) 84, 85, 88–89, 91–95, 103
Element decimation DCA (edDCA) 84, 85, 88, 89, 91–94, 96
Energy-based scoring function 239
Energy function 19, 85, 87, 101, 166, 174

Energy landscape theory 17
Ensembl database 5, 7–11, 303
EnvZ 214–220
Enzyme 38, 104, 114, 115, 157, 165, 167–172, 174, 178, 183, 184, 191, 267, 269, 270, 273, 280
Enzyme design 107, 125, 165–185
EquiPPIS 225–233
ESCOTT 106–110, 117–220
ESMFold 107, 113, 114, 225, 241
E-value 4, 5, 7, 9, 11, 52, 77, 79
EVcouplings 282–286, 298
Evolution 8, 18, 50, 73, 93, 106, 126, 158, 181, 282
Evolutionary conservation 34, 45, 118
Evolutionary energy fields 19, 22–24, 27
Exon-foldon correspondence 21
EXPANSION webserver 302–306, 308
Eyring equation 170

F

FASTA format 22, 60, 67, 89, 90, 107, 160, 228, 239, 283
Fiedler vector analysis 135
Fiedler vectors 130, 133, 135
FLP-12 .... 7, 8, 10, 12, 14, 15
FMRF-like peptide (FLP) gene family 4, 13
Folding dynamics simulation protocol 17–29
Foldons in protein folding 20, 24
Fold recognition
Fold-switching proteins 158, 161
FoldX 34
Full sequence design 166
Function 3, 19, 34, 49, 65, 83, 106, 125, 140, 157, 166, 189, 216, 231, 238, 267, 280, 301
Functional counterparts 4, 7, 11–13
Functional site prediction in proteins 50

G

Gap penalties 7, 9
Gaussian Restricted Boltzmann Machines (RBMs) 29
GDT-score 126
Generalized Jaccard coefficient 195, 199
Generative probabilistic models 84
Genetic diseases 59
Genome editing 277–299
Gibbs free energy of binding (ΔG°_{bind}) 37, 40, 43–45
Github 63, 73, 102, 132, 135, 208, 227, 240–242, 248, 259, 260
Global Epistatic Model for predicting Mutational Effects (GEMME) 105–120
Glucagon (GCG) 9–11
Glutamate (Glu) 192, 194, 202–206, 298, 661
G protein coupled receptor (GPCR) genes 5
G protein coupled receptors (GPCRs) 213
Graphics processing unit (GPU) 84, 92, 98, 101, 102, 140, 141, 215, 222, 227, 240–242, 247, 248, 261, 290, 296, 299
Graph spectral analysis 126, 130
Graph theory 126
GraSp-PSN webserver 136
GREASY 244, 245, 248, 252, 253
GROMACS 142, 143, 147, 151–153, 282, 289, 294–297, 299
GSATools 149

H

hhsuite 108
Hierarchical clustering 194–196, 199–201, 204–207
HIF-1α–CBP complex 39
High-Performance Computing Facility (HPC) 159, 221
HMMscan 68
Homeobox protein ARX 107, 114
Homologous counterparts 9
Homologous proteins 50, 51, 58, 65, 83, 125, 126, 130–132, 190
Homologous sequences 19, 50, 52, 68, 69, 77, 283, 284
Homology detection 67, 72, 77
Horn parameters 173
Hydrogen atoms 142, 173, 195, 204

I

IDP diseases 35, 38
IDP mutational landscape 35
iGEMME 106–108, 110–120
Importance weights 90, 95
Inhibitor/activator and protein interactions 190
Inter-molecular co-evolution 66
Inter-protein co-evolution 65–79
Intra-molecular co-evolution 65
Intrinsically disordered Protein Affinity prediction (IPApred) 33–46, 113, 158, 162
Intrinsically disordered proteins (IDPs) 33–46, 113, 158, 162
Ising model for protein folding 27
Islet amyloid polypeptide (IAPP) 4, 5, 7–12, 14, 15

J

Jalview 160
JDet software 53

K

KaiB 161
Kinase 39, 189–208, 213–222
Kinase activator 191, 192, 194
Kinase inhibitor 191, 192, 194
Kirchhoff matrix 127
Kisspeptin (KISS) 4–6, 10, 12, 14

L

Ligands 5, 40, 75, 76, 116, 130, 166, 172, 173, 176, 177, 179, 180, 183, 195, 243, 250–257, 263
Literature Search and Interpretation 158, 161
LocalColabFold 242, 243, 261, 262
Low homology BLASTP searching 12, 15
L-tyrosine 172

M

Machine learning (ML) 18, 22, 34, 46, 83, 116, 280, 299
Magnesium ion 172, 173
Mapping residues in protein 3D structures 68
Markov Chain Monte Carlo (MCMC) 87
MaxDist 195, 197–199, 201–206, 208
mCSM-PPI 34, 36, 39, 45
MD analysis 139, 205, 287
MDM-TASK 282, 288, 290, 298, 299
MD simulations 141, 142, 145, 191, 214, 217, 287–289, 291, 294, 296, 297
Membrane builder 215
Membrane protein-protein interactions 274
Membrane proteins 130, 167–275
MFIratio 40
Mg^{2+} coordination 142
Michaelis-Menten kinetics 170
Minimization 142, 143, 173, 185, 295–297
Mirror-image algorithm 190, 209
MirrorTree 66
M32K25 structural alphabet 147
MMseqs2 .... 118, 161, 241
Model confidence 246, 247, 249, 253, 254, 258, 260
Model implementation 88
Modeller 142
Molecular chirality 226
Molecular dynamics (MD) 139–145, 152, 166, 191, 195, 214, 215, 217, 250, 287–289, 291, 294–297, 299
Molecular dynamics simulation 166, 191, 214, 280, 287–289, 291
Molecular mechanics model 166
Molecular mechanisms 106, 132, 190
Molecular recognition 190–194, 206
Monte Carlo estimation of gradient 86
Monte Carlo sampling 26, 166
Monte Carlo simulation methods 178
Motif discovery 190
MPA-Pred 267–275
Multimeric ion channels 213
Multiple Alignment using Fast Fourier Transform (MAFFT) 67, 68, 79, 159, 160
Multiple sequence alignment (MSA) 12, 17, 19–25, 50–54, 57–60, 65, 66, 68, 77, 83–85, 89–91, 95–100, 102, 107, 108, 110–115, 118, 119, 159, 160, 225, 239, 241, 242, 249, 258, 261, 272, 284, 285, 302, 307, 308
Multi-protein complexes 239, 280
Mutabind2 .... 39, 45
Mutational effects 34–39, 41, 42, 45, 46, 83, 105–120
Mutational tolerance score (MTSpred) 36, 40–45
Mutual information 66, 146, 148

N

NAMD 152, 215, 222
Native sequence 166, 173, 175–177, 179–181, 183, 184
Network analysis 139–153, 287, 288
Network centrality analysis 290
Network connectivity 135, 145
Network Similarity Score (NSS) 127–130, 133, 135, 136
Neuropeptides 3–5, 9–15
Neuropeptide VF precursor (NPVF) 10, 11, 14, 15
Neuropeptide Y (NPY) 3, 4, 8–10, 12–15
Neutral model for foldon assignment 21
Next generation sequencing data 37
Noncovalent interactions 190
NPT (isothermal-isobaric ensemble) 143, 294, 296
Nuclear magnetic resonance (NMR) 159, 238
Nuclease NucB 107
NVIDIA graphics card 240
NVT (canonical ensemble) 143, 295, 296

O

OmegaFold 225, 241
One structure-one function paradigm 157, 161, 162

P

Pairwise residue coupling 24, 285
Pairwise structural alignment 160
Pancreatic polypeptide (PPY) 9, 10, 14
Parallelization strategy 190
PDB format 61, 69, 108, 132, 272–274
Pearson correlation coefficient 87, 89, 92–94, 96, 98
Peptide drug design 35, 38
Peptide hormones 3, 5, 8, 12, 14, 15

Persistent contrastive divergence (PCD) 87
pETcon plasmid 38, 39
Pfam database 19, 68
Phage display 38
Phosphorelays 139
p*Kas* 166
Planar cell polarity (PCP) pathway 132
Post-translational modifications (PTMs) 142, 152, 243, 248, 249, 253, 272, 293, 302, 306–308, 310, 312
Potts model inference 20
Preprocessing 228–230, 269
Preprocessing pipeline 84, 89
PRESCOTT 106, 110, 119, 120
PROPKA analysis 173
Protein 3–5, 7, 8, 10–15, 17–29, 33–46, 49–62, 65–79, 83, 88–90, 92, 94, 95, 98, 100–102, 105–120, 125–136, 140–145, 147, 149, 151, 153, 157–162, 165–167, 171, 183, 190–194, 198, 199, 201, 202, 213, 215, 217, 225–233, 237–263, 267–275, 279–299, 301–312
Protein-coding transcripts 309, 310
Protein data bank (PDB) 25, 26, 28, 39, 44, 56–58, 61, 68, 69, 71, 73–75, 79, 132, 158, 191, 227, 239, 256, 270, 271, 281, 290
Protein design using sequence dynamics 27
Protein-DNA co-evolution 65–79, 243, 280
Protein-DNA complexes 66, 68, 69, 72, 78, 243
Protein domains 60, 160
Protein engineering 127
Protein evolution 105, 120
Protein flexibility 147
Protein folding cooperativity 27
Protein folding temperature estimation 25–27
Protein function 8, 66, 76, 105, 113, 115, 125, 126, 135, 157, 237, 243, 288, 302, 303
ProteinGym 120
Protein interaction networks 72, 88, 288
Protein language model (pLM) 66, 106, 113, 227, 229
Protein-membrane system (EnvZ2/G-IM) 215, 216, 218, 221
Protein mutational landscapes 35, 105, 110
Protein-protein complexes 34, 66, 67, 71, 213, 237–263, 268–270
Protein-protein interaction prediction 65, 225–233
Protein-protein interactions 34, 65, 71, 74, 106, 190, 225–233, 245, 254, 274, 308, 310, 312
Protein-protein interaction site prediction 225–233
Protein receptors 3
Protein regulation 190
Protein-RNA co-evolution 65–79
Protein-RNA complexes 66, 68, 69, 72, 243
Protein sequence modeling 61, 239
Protein sequence, structure, and function relationships 199
Protein structure evolution 126
Protein structure networks (PSNs) 126, 127, 131, 133, 135, 136, 230
Protein structure prediction 225, 227
Protein synthesis 302
Proteus software 166, 183
Protonation state 143, 173, 293
Pseudocount 90–92, 97, 98
PSI-BLAST 58, 59, 77
PubMed 161
pyDock 237–263
pyDockNIP 238
pyDockRST 238
PyMOL 109, 132, 159–162, 249, 282, 288, 289, 291, 292
Pyrophosphate 165
Python scripts 109, 119, 181, 249, 311

Q

Quantitative comparison techniques 126
Quantum calculations 171

R

Rational drug design 190
Recognition codes 192–194
Reduce program 173
Regulator of fatty acid synthesis H (RfaH) 158, 161
Repeat protein folding analysis 20–21
Residue depth (RD) 35–37, 39, 41–45
RESP charges 171
Response regulators (RR) 139–140, 152
Reticulons (RTNs) 226, 231
Reweighted statistics 91
RFamide 4, 6, 8, 12, 14, 15
RFamide family 5, 15
RNA sequencing (RNA-seq) 77, 79, 90, 302, 303
Root-mean-square deviation (RMSD) 126, 133, 136, 147, 151, 153, 160–162, 251, 252, 254–257, 291, 297, 298
Root-mean-square fluctuations (RMSF) 147, 297, 298
Rosetta-based approach 34
RoseTTAFold 45, 107, 225

S

S3Det 54–56, 60
Saccharomyces cerevisiae EBY100 ..... 38, 39
Scoring matrix 9, 13, 14
SCWRL 244, 251, 263
Secondary structure element assignment 160
Selection temperature estimation 23–24

Sensor kinase 213–222
Sequence 3, 18, 34, 49, 65, 83, 105, 125, 140, 157, 166, 201, 215, 225, 239, 268, 279, 302
Sequence analysis 114, 158
Sequence-based features 227, 268–270
Sequence-based protein modeling 136
Sequence-conformation exploration 174
Sequence data 9, 86, 88, 105, 106, 114, 120, 158, 281
Sequence homology 4, 12, 51
Sequence logos 56, 61, 175, 180–182, 184
Shannon entropy 147, 148, 150
Shape-shifting proteins 157–163
Sidechain conformations 166
Signal phosphorelays 139
Signal transduction 34
Signature and characteristic substructure 192
Similarity matrix 195
Single-site mutations 19, 27, 33–46, 106, 111
Size-filtering algorithm 190
Sparse models 87–89
Specificity constant 170
Specificity-determining positions (SDPs) 49–62
Splicing variations 302, 306–307, 310
Stability design 169, 174, 175, 178–180, 184
Stability mutants 166
Statistical coupling analysis (SCA) 66
Statistical learning from MSAs 20
Stereometric differences 190
Stereospecificity inversion 165, 166, 183
Structural alignment 34, 134, 160–162, 191
Structural alphabet (SA) analysis 147
Structural analysis 160–161, 284
Structural and sequence data 158
Structural information 51, 58, 106, 107, 117, 225, 226
Structural modeling 35, 171, 232, 237–240
Structure-based features 230
Structure preparation 171–173
Subdomain emergence in folding 25
Substrate competition design 168, 170–171
Sulfhydryl-reactivity data 214

T

T-Coffee 59, 68, 161
Therapeutic peptides 35
Theta 195, 197–199, 201, 202, 204–206, 208
Thioredoxin (TRX) 107, 110–112
TIP3P model 173
TM-align tool 127
TM-score 126, 162, 249
Total network dissimilarity (TND) 125–136
Toxicity 192
Training 36, 84, 85, 87, 88, 91–94, 96–98, 102, 270
Training interruption restore 95
Transition state 165–185
Transition state model 166, 171
Transmembrane communication 214
Transmembrane domain (TMD) 213–222
Transmembrane water encapsulation 213–222
Triangular spatial relationship (TSR) algorithm 189–208
tRNA 165, 173
TSR keys 195–202, 204, 205, 207, 208
Tumor suppressor protein p53 38
Tyrosinol 173
Tyrosyl-tRNA synthetase (TyrRS) 165, 166, 171–175, 178, 180, 182–184

U

UniProtKB 160
UniRef90 .... 227, 228, 240, 283

V

Variability in folding mechanisms 29
VespaG 106
Vienna-PTM webserver 142
Viral replication 225–233
Visualization of folding curves and domains 25

W

Water encapsulation 213–222
Web server 39, 53, 67–72, 106, 108, 118, 119, 130, 136, 141, 142, 231, 238, 239, 245–246, 268, 270, 274, 282–289, 302–310
Wild-type sequence 41, 45, 102, 116
Wnt proteins 132–135
Word size for seeding alignments 9
WormBase 5–8, 13, 14

X

XLC1 158
XPLOR program 173

Y

Yeast surface display 37–40

MIX
Papier aus verantwortungsvollen Quellen
Paper from responsible sources
FSC® C105338

If you have any concerns about our products,
you can contact us on
ProductSafety@springernature.com

In case Publisher is established outside the EU,
the EU authorized representative is:
Springer Nature Customer Service Center GmbH
Europaplatz 3, 69115 Heidelberg, Germany

Printed by Libri Plureos GmbH
in Hamburg, Germany